Lecture Notes in Mathematics

Edited by A. Dold and B. Eckmann

539

A. Badrikian
J. F. C. Kingman
J. Kuelbs

Ecole d'Eté de Probabilités de Saint-Flour V-1976

Edité par P.-L. Hennequin

Springer-Verlag
Berlin · Heidelberg · New York 1976

Authors
Prof. A. Badrikian
Université de Clermont
Complexe Scientifigque des Cézeaux
Département de Mathematiques
Appliquées
B.P. 45
63170 Aubiere/France

Editor
Prof. P.-L. Hennequin
Université de Clermont
Complexe Scientifique des Cézeaux
Département de Mathematiques
Appliquées
B.P. 45
63170 Aubiere/France

Prof. J. F. C. Kingman
University of Oxford
Mathematical Institute
24–29 St. Giles
Oxford, Ox 1 3LB/Great Britain

Prof. J. Kuelbs
University of Wisconsin-Madison
Mathematics-Department
Van Vleck Hall
480 Lincoln Drive
Madison, Wisconsin 53706/USA

Library of Congress Cataloging in Publication Data

Badrikian, Albert, 1933-
Ecole d'été de probabilités de Saint-Flour V-1975.

(Lecture notes in mathematics ; 539)
French or English.
Includes bibliographies.
1. Probabilities--Congresses. 2. Banach spaces--Congresses. 3. Stochastic processes--Congresses.
I. Kingman, John Frank Charles, joint author.
II. Kuelbs, J., joint author. III. École d'ete de probabilités de Saint-Flour, 5th, 1975. IV. Title.
V. Series: Lecture notes in mathematics (Berlin) ; 539.
QA3.L28 vol. 539 [QA273.43] 510'.8s [519.2]
76-27360

AMS Subject Classifications (1970): 28A40, 46B99, 46E40, 60-02, 60B05, 60F05, 60G45, 60G50

ISBN 3-540-07858-4 Springer-Verlag Berlin · Heidelberg · New York
ISBN 0-387-07858-4 Springer-Verlag New York · Heidelberg · Berlin

Printed in Germany
Printing and binding: Beltz, Offsetdruck, Hemsbach/Bergstr.

INTRODUCTION

La cinquième Ecole d'Eté de Calcul des Probabilités de Saint Flour s'est tenue du 3 au 19 Juillet 1975 et a rassemblé, outre les conférenciers, vingt quatre participants.

Les trois conférenciers, Messieurs Badrikian, Kingman et Kuelbs, ont tenu à elaborer une rédaction définitive de leurs cours qui complète sur certains points les exposés faits à l'école. Nous les en remercions vivement.

La frappe du manuscrit a été assurée par les Départements de Madison et de Clermont et nous remercions, pour leur soin, les secrétaires qui se sont chargées de ce travail.

Nous exprimons enfin notre gratitude à la Société Springer Verlag qui permet d'accroître l'audience internationale de notre Ecole en accueillant ces textes dans la collection Lecture Notes in Mathematics.

P.L. HENNEQUIN

Professeur à l'Université de Clermont
B.P. 45

63170 AUBIERE (FRANCE)

TABLE DES MATIERES

A. BADRIKIAN : "PROLEGOMENES AU CALCUL DES PROBABILITES DANS LES BANACH"

LISTE DES AUDITEURS

Mme BADRIKIAN J.	Université de Clermont-Ferrand
Mr. BETHOUX P.	Université de Lyon
Mr. BRANDOUY P.	Université de Pau
Mr. CARMONA R.	Université de Marseille
Mr. CHEMARIN P.	Université de Lyon
Mr. CLAVILLIER A.	Université de Clermont-Ferrand
Mr. DAUBEZE P.	Université de Toulouse
Mr. DERRIENNIC Y.	Université de Rennes
Mr. FLYTZANIS E.	Ecole Polytechnique Thessaloniki (Grèce)
Mr. FOURT G.	Université de Clermont-Ferrand
Mr. GALTCHOUK L.	Université de Moscou (U.R.S.S.)
Mr. GALVES A.	Ecole Polytechnique de Paris
Mr. GRAVERSEN S.E.	Université d'Aarhus (Danemark)
Mr. HEINKEL B.	Université de Strasbourg
Mr. HENNEQUIN P.L.	Université de Clermont-Ferrand
Mr. HIRIART-URRUTY J.B.	Université de Clermont-Ferrand
Mr. KONO N.	Université de Strasbourg
Mr. MAISONNEUVE B.	Université de Grenoble
Mr. MEYER P.	Université de Strasbourg
Mle OLIVARES M.	Université Paris VI
Mr. PERENNOU G.	Université de Toulouse
Mr. SATO H.	Université de Kyushu (Japon)
Mr. WEBER M.	Université de Strasbourg
Mr. YOU Hi-sé	Korea Université (Corée)

PROLEGOMENES AU CALCUL DES PROBABILITES DANS LES BANACH

PAR A. BADRIKIAN

PROLÉGOMÈNES AU CALCUL
DES PROBABILITÉS DANS LES BANACH

INTRODUCTION

"Le Tout est de tout dire ; et je manque de mots,
et je manque de temps, et je manque d'audace."
(Paul Eluard)

Le Calcul des Probabilités dans les espaces de Banach est un sujet en plein développement. Ses liens avec la géométrie des Banach ont été déconverts depuis relativement peu de temps. Aussi a-t-il semblé bon aux participants de l'Ecole d'Eté de Saint-Flour de mettre cette question au programme de leurs studieuses vacances.

Différents mathématiciens français s'étant récusés, il nous a incombé de les suppléer (c'est un honneur que nous apprécions à sa juste valeur, mais nous n'avons pas intrigué pour l'obtenir !). Comme on ne peut tout dire en dix séances, il nous a fallu choisir ; aussi nous sommes-nous bornés à donner les fondements de la théorie. Nous regrettons d'avoir eu à nous arrêter quand les choses allaient devenir intéressantes, mais ce regret est tempéré par le fait que, en 1976, le sujet sera continué par M. HOFFMANN-JØRGENSEN qui développera ses résultats sur les lois des Grands Nombres, le théorème Central Limite, etc...

Ce cours est basé sur les travaux d'HOFFMANN-JØRGENSEN, KWAPIEN, MAUREY, PISIER et SCHWARTZ. Notre seul mérite (si mérite il y a !) consiste à les avoir rassemblés et parfois présentés d'une autre façon, espérant par là les rendre plus accessibles. Toutefois, nous n'avons pas mentionné les auteurs des résultats cités.

La rédaction présente a beaucoup profité des remarques faites par les auditeurs et d'une confrontation de nos points de vue avec Mlle Simone CHEVET. Qu'ils en soient remerciés !

Je remercie également Catherine CADIER qui s'est chargée de la dactylographie du manuscrit. Son dévouement est allé jusqu'à corriger certaines fautes de syntaxe (mais est-il prudent de l'avouer ?), ce dont nous lui sommes infiniment reconnaissants.

CHAPITRE 0

RAPPEL DE RESULTATS UTILES

Dans ce chapitre, nous donnerons un ensemble de résultats qui seront couramment utilisés. Comme ce cours ne prétend pas "prendre les mathématiques à leur début" il y aura une certaine redondance. Par exemple, les définitions et résultats sur l'intégration des fonctions à valeurs complexes pourraient se déduire de ce que nous dirons par la suite de l'intégration des fonctions à valeurs banachiques...

I - Définitions et résultats d'Analyse Fonctionnelle

Tous les espaces vectoriels que l'on considérera auront pour corps de base $\mathbb{R}$ ou $\mathbb{C}$; on notera par $\mathbb{K}$ ce corps de base.

c_o désignera l'espace vectoriel des suites de scalaires tendant vers zéro. On le munit de la norme

$$x \rightsquigarrow ||x||_\infty = \sup_n |x_n| \qquad \text{si } x = (x_n)_{n \in \mathbb{N}} .$$

Pour cette norme, on obtient un Banach. Son dual est l^1 et par conséquent son bidual est l^∞.

Si e^k désigne le vecteur de $\mathbb{C}^{\mathbb{N}}$ de composantes

$$e^k_n = \delta^k_n \ \forall n \quad (\delta^k_n \text{ symbole de Kronecker})$$

les (e^k) engendrent topologiquement c_o , mais non l^∞ .

$\mathbb{K}^{\mathbb{N}}_o$ (avec $\mathbb{K} = \mathbb{R}$ ou $\mathbb{C}$) désigne l'ensemble des suites "presque nulles" de scalaires, c'est-à-dire

$$x = (x_n) \in \mathbb{K}^{\mathbb{N}}_o \Longleftrightarrow x_n = 0$$ sauf au plus pour un nombre fini de n . $\mathbb{C}^{\mathbb{N}}_o$ est dense dans c_o , mais non dans l^∞ .

Un espace vectoriel E est dit F-espace s'il est muni d'une topologie compatible avec la structure d'espace vectoriel, métrisable et s'il est complet.

Si cet espace est localement convexe, en plus, alors on dit que c'est un espace de Fréchet.

Soit E un espace vectoriel (quelconque pour le moment) on appelle F-norme sur E une application $x \rightsquigarrow ||x||$ de E dans $[0, \infty[$ ayant les propriétés suivantes :

a) $x = 0 \Longleftrightarrow ||x|| = 0$;

b) $||\lambda\ x|| \leqslant ||x|| \quad \forall x \in E \quad \forall \lambda \in \mathbb{K}$ avec $|\lambda| \leqslant 1$;

c) $||x+y|| \leqslant ||x|| + ||y|| \quad \forall (x, y) \in E \times E$;

d) $\lambda_n \longrightarrow 0$ dans $\mathbb{K} \Longrightarrow ||\lambda_n\ x|| \to 0 \quad \forall x \in E$;

e) $||x_n|| \to 0 \Longrightarrow ||\lambda\ x_n|| \to 0 \quad \forall \lambda \in \mathbb{K}$.

Une F-norme n'est pas forcément homogène.

Etant donné une F-norme, la fonction $E \times E \to \mathbb{R}$ définie par $(x, y) \rightsquigarrow d(x, y) = ||x - y||$ est une distance sur E, définissant une topologie (métrisable bien sûr !) compatible avec la structure d'espace vectoriel.

Inversement (théorème de KAKUTANI) toute topologie d'espace vectoriel métrisable peut être définie par une F-norme.

On appelle F-normé un espace muni d'une F-norme. Un F-espace est donc un espace F-normé complet.

Comme exemples d'espaces F-normés (et de F-espaces), citons

- les espaces normés (et les Banach) ;
- les espaces p-normés (et les quasi p-Banach ou p-Banach).

Rappelons que si $0 < p \leqslant 1$ on appelle p-norme sur l'espace vectoriel E une application $x \rightsquigarrow ||x||$ telle que

a) $||x+y|| \leq ||x|| + ||y||$, $\forall (x, y) \in E \times E$;

b) $||\lambda x|| = |\lambda|^p ||x||$, $\forall \lambda \in K, x \in E$;

c) $||x|| = 0 \iff x = 0$.

Parmi les exemples fondamentaux de p-normes et d'espace p-normé, citons les espaces $L^p(X, \mathcal{F}, \mu)$ $(0 < p \leqslant \infty)$:

- Si $0 < p \leqslant 1$, la p-norme est définie par $f \rightsquigarrow \int |f|^p \, d\mu$
- Si $p \geqslant 1$: $f \rightsquigarrow (\int |f|^p \, d\mu)^{1/p}$ est une norme
- Pour $p = \infty$ la norme est la norme "sup-essential".

Ce sont des F-espaces (p-Banach dans le premier cas, des Banach dans les cas (2) et (3)).

De même, si $(\Omega, \mathcal{F}, P)$ est un espace probabilisé, la fonction

$$f \rightsquigarrow \int \frac{|f|}{1 + |f|} \, dP$$

définie sur l'espace $L^0(\Omega, \mathcal{F}, P)$ des variables aléatoires est une F-norme (prenant ses valeurs dans $[0, 1]$). La topologie qu'elle définit est celle de convergence en probabilité.

Soit E un espace F-normé ; $A \subset E$ est dit bornée si elle satisfait la condition suivante :

$$\forall \varepsilon > 0, \quad \exists \delta > 0 \text{ tel que } \sup_{x \in A} ||\delta x|| \leqslant \varepsilon .$$

Si la F-norme sur E est p-homogène $(0 < p \leqslant 1)$ cela équivaut à la condition usuelle : $\exists M \in]0, \infty[$ tel que

$$\sup_{x \in A} ||x|| \leqslant M.$$

La définition que nous venons de donner n'est que la traduction de la définition d'un ensemble borné dans un e.v.t. quelconque.

Citons également un résultat que nous utiliserons :

$A \subset L^0(\Omega, \mathcal{F}, P)$ est bornée si et seulement pour tout $\varepsilon > 0$, il existe $M < \infty$ tel que

$$\sup_{f \in A} P \left\{ |f| > M \right\} \leqslant \varepsilon$$

Si X et Y sont deux e.v.t. L (X, Y) désignera l'ensemble des applications linéaires continues de X dans Y. Parfois, on le désignera également par $\mathcal{L}(X,Y)$.

Donnons deux théorèmes qui seront constamment utilisés.

- Théorème de BANACH-STEINHAUS

Soient X et Y deux F-espaces et $A \subset L(X, Y)$ (l'espace vectoriel des applications linéaires continues de X dans Y). On suppose que pour tout $x \in X$ l'ensemble $\{u(x),\ u \in A\}$ est un ensemble borné de Y. Alors A est équicontinue.

L'on en déduit que si (u_n) est une suite d'applications linéaires continues de X dans Y (X et Y F-espaces !) convergeant simplement vers un opérateur (bien évidemment linéaire !) u, alors u est continue.

- Théorème du graphe fermé

Soient X et Y deux F-espaces et u : E → F linéaire. u est continue si et seulement si elle satisfait à :

$x_n \to 0$ dans X et $u(x_n) \to y$ dans F $\Longrightarrow$ $y = 0$.

On en déduit la conséquence suivante :

Soient $(X, \mathcal{C}_1)$ un F-espace ; $(Y, \mathcal{C}_2)$ un espace vectoriel topologique séparé. Soit $\mathcal{C}_2'$ une topologie de F-espace sur Y plus fine que $\mathcal{C}_2$. Soit u : X → Y linéaire continue relativement à $\mathcal{C}_1$ et $\mathcal{C}_2$. Alors u est $(\mathcal{C}_1, \mathcal{C}_2')$-continue.

En effet, soit $x_n \to 0$ pour $\mathcal{C}_1$ et $u(x_n) \to y$ pour $\mathcal{C}_2'$; On a alors $u(x_n) \to y$ pour $\mathcal{C}_2$; donc $y = 0$; ce qui revient à dire que u est $(\mathcal{C}_1, \mathcal{C}_2')$-continue en vertu des hypothèses.

Donnons (et démontrons) un autre résultat :

Théorème

Soit $(X, \mathcal{F}, \mu)$ un espace mesuré et soit $0 < p \leqslant \infty$. La "boule unité" de $L^p(X, \mathcal{F}, \mu)$ est fermée pour la topologie de convergence en mesure sur tout ensemble de mesure finie pour $p < \infty$. Cela reste vrai si $p = \infty$ si μ est concassable (voir plus loin).

DEMONSTRATION

Supposons d'abord $p < \infty$. Soit donc $(f_n)_{n \in \mathbb{N}} \to f$ en probabilité sur tout ensemble de mesure finie et soit $\int |f_n|^p \, d\mu \leqslant 1 \quad \forall n$; il existe une suite extraite f_{n_k} telle que $f_{n_k} \to f$ presque sûrement. Il suffit alors d'appliquer le lemme de Fatou. Le cas $p = \infty$ se traite de manière analogue.

II - Résultats et définitions de la théorie de la mesure

Sauf mention expresse du contraire, tous les espaces mesurés que l'on considérera seront des espaces mesurés "abstraits" et σ-finis. (le mot "abstrait" s'opposant à "de Radon"!). On pourrait considérer les espaces mesurés "concassables" généralisant à la fois les espaces σ-finis et les espaces mesurés de Radon ; mais nous ne le ferons pas. Quand nous aurons à introduire d'autres hypothèses (espaces mesurés finis, espaces probabilisés) nous le ferons à l'endroit où c'est utile.

Si donc $(X, \mathcal{F}, \mu)$ est un espace mesuré et si $f : X \longrightarrow [0, \infty]$ est une application (pas forcément mesurable), on définit son intégrale supérieure $\mu^*(f)$ de la façon suivante :

(1) Si f est dénombrablement étagée : $f = \sum a_n 1_{A_n}$

$a_n \in [0, \infty]$, $A_n \in \mathcal{F} \ \forall n$, A_n disjoints 2 à 2, alors :

$\mu^*(f) = \sum a_n \, \mu(A_n)$ (avec la convention $0 \cdot \infty = 0$!).

(2) Dans le cas général

$\mu^*(f) = \inf \{\mu^*(g)$, g dénombrablement étagée, $g \geq f\}$.

La fonctionnelle μ^* possède les propriétés suivantes :

- "Théorème de Fatou" : $f_n \uparrow f \Longrightarrow \mu^*(f_n) \uparrow \mu^*(f)$;

- "sous-additivité dénombrable $\mu^*(\Sigma f_n) \leq \Sigma \mu^*(f_n)$.

On écrira parfois $\int^* f\, d\mu$ au lieu de $\mu^*(f)$.

Si $A \in \mathcal{P}(X)$ on écrira $\mu^*(A)$ au lieu de $\mu^*(1_A)$.

A est μ-négligeable si $\mu^*(A) = 0$; on définit alors les propriétés vraies μ-presque partout.

Si $f : X \to [0, \infty]$, on posera

- $M_p(f) = \int^* |f|^p\, d\mu$, si $0 < p < \infty$;

- $M_\infty(f) = \inf \{a \geq 0$; $f \leq a$ μ-presque partout$\}$.

Soit $(X, \mathcal{F}, \mu)$ un espace mesuré et soit $L^0(X, \mathcal{F}, \mu, \overline{\mathbb{R}})$ l'ensemble des μ-classes d'équivalence des fonctions μ-mesurables $f : X \to \overline{\mathbb{R}}$ (i.e. égales μ-presque partout à une fonction $(\mathcal{F}, \mathcal{B}_{\overline{\mathbb{R}}})$-mesurable si $\mathcal{B}_{\overline{\mathbb{R}}}$ désigne la tribu borélienne de $\overline{\mathbb{R}}$). Bien sûr ce n'est pas un espace vectoriel, mais c'est un ensemble ordonné de manière naturelle. Il possède en outre un plus grand élément et un plus petit élément (à savoir les classes de $+\infty$ et $-\infty$ respectivement).

Proposition

Dans $L^0(X, \mathcal{F}, \mu, \overline{\mathbb{R}})$ toute famille $(f_i)_{i \in I}$ possède une borne supérieure notée $\bigvee_{i \in I} f_i$ (et aussi une borne inférieure !).

Démonstration

Compte tenu du fait que, μ étant σ-finie, une fonction sur X est mesurable si et seulement si sa restriction à tout ensemble de mesure finie est mesurable, on peut supposer μ finie.

D'autre part, il existe un isomorphisme, topologique et pour la structure d'ordre de $\overline{\mathbb{R}}$ sur $\left[-\frac{\pi}{2}, +\frac{\pi}{2}\right]$, à savoir $t \rightsquigarrow \text{Arc tg } t$. Cet isomorphisme induit alors un isomorphisme pour la structure d'ordre de $L^0(X, \mathcal{F}, \mu, \overline{\mathbb{R}})$ sur le sous ensemble de $L^\infty(X, \mathcal{F}, \mu, \mathbb{R})$ formé des f telles que $M_\infty(f) \leqslant \frac{\pi}{2}$.

Comme $L^\infty(X, \mathcal{F}, \mu, \mathbb{R}) \subset L^1(X, \mathcal{F}, \mu, \mathbb{R})$, on en déduit un isomorphisme de $L^0(X, \mathcal{F}, \mu, \overline{\mathbb{R}})$ sur une partie de $L^1(X, \mathcal{F}, \mu, \overline{\mathbb{R}})$ majorée en module par la classe de la fonction constante égale à $\frac{\pi}{2}$. Finalement, par transport, pour toute famille $(f_i)_{i \in I}$ de $L^0(X, \mathcal{F}, \mu, \overline{\mathbb{R}})$ il existe une borne supérieure et une borne inférieure.

Remarquons maintenant que la famille des enveloppes supérieures des familles finies extraites de I, est filtrante croissante et a même borne supérieure que $(f_i)_{i \in I}$. Il existe alors une suite croissante $f^{(n)}$ extraite de la famille des enveloppes supérieures finies telles que

$$\operatorname{Sup}_n f^{(n)} = \bigvee_{i \in I} f_i$$

(Il y a ici abus d'écriture, on devrait écrire $\bigvee_n f^{(n)}$ au lieu de $\operatorname{Sup} f^{(n)}$ si l'on voulait se conformer aux notations du début. Mais aucune confusion n'est possible car $\mathbb{N}$ est dénombrable, d'après le "Théorème de la Pallice"). On en déduit que $\bigvee_{i \in I} f_i$ est la plus petite μ-classe de fonctions g μ-mesurables ou non telles que pour tout i, $g \geqslant f_i$. En effet c'est la plus petite μ-classe de fonctions à être supérieure à $f^{(n)}$ pour tout n.

Remarque

Si $(X, \mathcal{F}, \mu)$ est un espace mesuré de Radon et si les f_i sont semi-continues inférieurement, en désignant par $\overset{o}{f_i}$ leur μ-classe, $\bigvee \overset{o}{f_i}$ est la μ-classe de $\operatorname{Sup}_i f_i$ (sup ponctuel).

En effet si $\mathbb{R}$ est remplacé par $\left[-\frac{\pi}{2}, +\frac{\pi}{2}\right]$ et si on suppose (f_i) filtrante, on sait que $\mathring{f}_i$ converge vers $\widehat{(\mathrm{Sup}\ f_i)}^{\circ}$ dans $L^1(X, \mathcal{F}, \mu)$.

<u>Définition</u>

Soit $(X, \mathcal{F}, \mu)$ un espace mesuré et A une partie de $L^\circ(X, \mathcal{F}, \mu)$. Soit $p \in \left]0, \infty\right]$, on dit que A est <u>latticiellement bornée</u> dans $L^p(X, \mathcal{F}, \mu)$ s'il existe $g : X \to [0, \infty]$ telle que $M_p(g) < \infty$ et que $|f| \leqslant g$ μ-presque partout pour tout $f \in A$. A est dite <u>latticiellement bornée</u> dans $L^\circ(X, \mathcal{F}, \mu)$ (ou latticiellement bornée) s'il existe $g : X \to [0, \infty]$ finie presque partout telle que $|f| \leqslant g$, $\forall f \in A$.

Une partie p-latticiellement bornée $(0 \leqslant p \leqslant \infty)$ est bornée dans $L^p(X, \mathcal{F}, \mu)$. Naturellement si A est latticiellement bornée dans $L^p(X, \mathcal{F}, \mu)$, A est contenue dans $L^p(X, \mathcal{F}, \mu)$. En outre on peut, d'après ce qui précède, supposer que $g \in L^p(X, \mathcal{F}, \mu)$.

On dira aussi p-latticiellement bornée au lieu de latticiellement bornée dans L^p. Une partie ∞-latticiellement bornée n'est autre qu'une partie bornée dans $L^\infty(X, F, \mu)$.

<u>Définition</u>

Soit $(X, \mathcal{F}, \mu)$ un espace mesuré, E un espace normé et $v : E \to L^\circ(X, \mathcal{F}, \mu)$ linéaire. On dit que v est p-latticiellement bornée si l'ensemble

$$\left\{ v(e) \; ; \; e \in E \quad \|e\| \leqslant 1 \right\} \text{ est p-latticiellement borné.}$$

Si v est p-latticiellement bornée, elle admet la factorisation

$$E \xrightarrow{\;u\;} L^p(X, \mathcal{F}, \mu) \xrightarrow{\;i\;} L^\circ(X, \mathcal{F}, \mu), \qquad v = i \circ u$$

où i est l'injection canonique ; u est en outre continue, car bornée. La réciproque est fausse en général : par exemple l'application identique de $L^p([0, 1], dx)$ dans $L^\circ([0, 1], dx)$ est continue, mais la boule unité de $L^p([0, 1], dx)$ n'est pas latticiellement bornée pour $0 < p < \infty$.

Définition

Soient E et F deux espaces normés en dualité et soit $v : E \to L^\circ(X, \mathcal{F}, \mu)$ linéaire. Soit $p \in [0, \infty]$; on dit que v est (p, F)- décomposable s'il existe $\varphi : X \to F$ telle que

a) $\langle \varphi(.), e \rangle \in v(e) \quad \forall e \in E$.

En outre si $p > 0$, on supposera

b) $M_p(\|\varphi\|_F) < \infty$ (au sens de l'intégrale supérieure).

Dire que E et F sont en dualité signifie que

$$\forall e \in E, \|e\|_E = \sup_{f \in F\ \|f\| \leq 1} |\langle e, f \rangle| \quad \text{et}$$

$$\|f\|_F = \sup_{e \in E, \|e\|_E \leq 1} |\langle e, f \rangle| \quad , \forall f \in F.$$

Dans le cas où F = E' on dira simplement p-décomposable.

Les conditions (b) et (a) impliquent que v est p-latticiellement bornée ; et si p = 0, (a) seule implique que v est latticiellement bornée.

On va maintenant donner une réciproque. Nous aurons pour cela besoin d'un lemme.

Lemme (VON-NEUMANN - MAHARAM - IONESCU-TULCEA)

Soit $(X, \mathcal{F}, \mu)$ un espace mesuré σ-fini (ou bien "concassable"). L'application identique de $L^\infty(X, \mathcal{F}, \mu)$ dans lui-même est décomposable (et même ∞-décomposable).

Démonstration

En effet (IONESCU-TULCEA "Topics in the theory of Lifting, ou bien MEYER "Probabilités et Potentiel"), on a une application linéaire ρ de $L^\infty(X, \mathcal{F}, \mu)$ dans l'espace des fonctions μ-mesurables bornées de X dans $\mathbb{R}$, soit $\mathcal{B}^\infty(X, \mu, \mathbb{R})$, telle que

(1) ρ est isométrique quand $\mathcal{B}^\infty(X, \mu, \mathbb{R})$ est muni de la norme μ usuelle ;

(2) pour tout $\mathring{f} \in L^\infty(X, \mu, \mathbb{R})$, la classe de $\rho(\mathring{f})$ est identique à $\mathring{f}$;

(3) $\rho(\mathring{a}) = a$ où $\mathring{a}$ désigne la μ-classe de la fonction constante a.

Alors pour $x \in X$; l'application $\mathring{f} \rightsquigarrow \rho(\mathring{f})(x)$ de $L^\infty(X, \mu, \mathbb{R})$ dans $\mathbb{R}$ est linéaire continue, donc définit un élément noté $\varphi(x)$ de $L^\infty(X, \mu, \mathbb{R})'$. Par définition même

$$\forall f \in L^\infty(X, \mu, \mathbb{R}), \quad <\mathring{f}, \varphi(.)> \in \mathring{f}$$

En outre pour tout $x \in X$, $||\varphi(x)|| = 1$. Donc le résultat annoncé est démontré.

Naturellement il peut exister plusieurs décompositions de l'application identique de L^∞ dans lui-même ; toute décomposition du type de celle que nous avons obtenue dans la démonstration du lemme sera appelée une décomposition de MAHARAM.

Corollaire

Toute application linéaire v continue d'un espace normé E dans $L^\infty(X, \mathcal{F}, \mu)$ est ∞-décomposable.

En effet, si φ_M est une décomposition de MAHARAM sur $L^\infty(X, \mathcal{F}, \mu)$ la fonction $\varphi : x \rightsquigarrow v' \circ \varphi_M(x)$ décompose v. En outre

$||\varphi(x)|| \leqslant ||v'|| = ||v||$. D'où le résultat.

Cela étant, on a le :

Théorème

Soit E un espace normé et $v : E \to L^0(X, \mathcal{F}, \mu)$. Les propriétés suivantes sont équivalentes pour v linéaire :

(a) v est p-latticiellement bornée ;

(b) v est p-décomposable.

Démonstration

Dire que v est p-latticiellement bornée équivaut à dire qu'il existe $g \in L^p_+(X, \mathcal{F}, \mu)$ telle que v admette la factorisation suivante

$$E \xrightarrow{\;u = \frac{v}{g}\;} L^\infty(X, \mathcal{F}, \mu) \xrightarrow{\;M_g\;} L^p(X, \mathcal{F}, \mu)$$

où $u(e) = \frac{v(e)}{g}$ (avec la convention $\frac{0}{0} = 0$) est linéaire continue et M_g est l'opérateur de multiplication par g : $f \leadsto fg$.

(a) $\Longrightarrow$ (b) : soit v p-latticiellement bornée et $g \in L^p(X, \mu, \mathbb{R})$ telle que $|u(e)| \leqslant g$ si $\|e\| \leqslant 1$. Soit ψ une décomposition de u (elle existe de par le corollaire ci-dessus) :

$$<\psi(.), e> \in \frac{v(e)}{g} \quad , \quad \forall e \in E .$$

Soit $\varphi = g\psi$; alors $<\varphi(.), e> \in g \cdot \frac{v(e)}{g} = v(e)$, $\forall e$;

En outre $\|\varphi(x)\| = g(x) \quad \|\psi(x)\| \leqslant g(x) \|u\|, \forall x \in X$.

Donc $M_p(\|\varphi\|) < \infty$; et v est p-décomposable.

(b) $\Longrightarrow$ (a). Soit v p-décomposable et $\varphi : X \to E'$ une décomposition de v. Alors si $\|e\|_E \leqslant 1$

$$|<\varphi, e>| \leq \|\varphi\|_{E'} .$$

Donc $\bigvee_{||e||\leqslant 1} |<\varphi, e>| \leqslant \sup_{||e||\leqslant 1} |<\varphi, e>| \leqslant ||\varphi||_E$.

Donc v est p-latticiellement bornée.

On appellera décomposition de MAHARAM de v toute décomposition obtenue à partir d'une décomposition de MAHARAM de L^∞ comme il est indiqué dans la démonstration du théorème. On verra au chapitre II l'intérêt de la décomposition de MAHARAM.

CHAPITRE I

CONVERGENCE DES SERIES DANS LES BANACH

N° 1 : DEFINITIONS FONDAMENTALES

Nous allons commencer par donner trois définitions.

DEFINITION 1 : Soit E un Banach, $(x_n)_{n\in\mathbb{N}}$ une suite de points de E. Cette suite est dite scalairement sommable, ou scalairement dans l^1 si, pour toute forme linéaire continue x' sur E, la suite de scalaires $(<x_n, x'>)_{n\in\mathbb{N}}$ appartient à l^1. Bien sûr, la notion de scalaire sommabilité a un sens dans n'importe quel espace vectoriel topologique.

DEFINITION 2 : La suite (x_n) des points du Banach E est dite sommable, ou bien la série $\sum x_n$ est dite commutativement convergente si les conditions équivalentes suivantes sont vérifiées :

(1) Pour toute permutation σ de $\mathbb{N}$, la série $\sum x_{\sigma(n)}$ est convergente.

(2) Pour tout $\varepsilon > 0$, il existe une partie finie I_ε de $\mathbb{N}$ telle que pour toute partie finie J de $\mathbb{N}$ ne rencontrant pas I_ε on ait

$$||\sum_{i\in J} x_i|| \leq \varepsilon$$

(2') Il existe $x \in E$ tel que pour tout $\varepsilon > 0$ il existe une partie finie I_ε de $\mathbb{N}$ ayant la propriété suivante : pour toute partie finie K de $\mathbb{N}$ contenant I_ε on a :

$$||x - \sum_{i\in K} x_i|| \leq \varepsilon$$

L'équivalence de (2) et (2') résulte du fait que E est complet. L'équivalence de (1) et (2') n'est pas triviale. La démonstration se trouve "en vente dans toutes les bonnes maisons". Si (x_n) est sommable, la somme $\sum x_{\sigma(n)}$ ne dépend pas de la permutation σ choisie.

DEFINITION 3 : La suite $(x_n)_{n \in \mathbb{N}}$ de points du Banach E est dite absolument sommable si $\sum ||x_n|| < \infty$

Il est bien connu que si E est de dimension finie, ces trois définitions sont équivalentes.

REMARQUE : Les sommabilités au sens des définitions ② et ③ sont relatives à la topologie : elles ne changent pas si on remplace la norme "initiale" par une norme équivalente.

Il est clair que ③ ⟹ ② ⟹ ①. Nous allons donner des exemples montrant qu'il n'y a pas équivalence en général.

Exemple 1 : Soit $E = c_o$ et soit $(e^k)_{k \in \mathbb{N}}$ la "base canonique"

$$e^k_n = \delta^k_n \qquad \forall n$$

Alors (e^k) est scalairement dans l^1, mais pas sommable (ni a fortiori absolument sommable).

Exemple 2 : Soit $E = l^\infty$; pour tout $m \in \mathbb{N}$, soit

$$x_m = (0, 0, \ldots 0, \underbrace{\tfrac{1}{m}}_{(m+1)\text{-ième terme}}, 0, 0, \ldots) \qquad m \geq 1$$

et $$x_o = (0, 0, 0, \ldots, 0, 0, \ldots)$$

Alors (x_m) est sommable et a pour somme $(0, 1, \frac{1}{2}, \frac{1}{3}, \ldots)$

Mais puisque $||x_m||_{l^\infty} = \frac{1}{m}$, (x_m) n'est pas absolument sommable.

On peut même démontrer (Théorème de DVORETZKY-ROGERS) que dans tout Banach de dimension infinie, il existe une suite sommable qui ne soit pas absolument sommable. Autrement dit, un Banach est de dimension finie si et seulement si toute suite sommable est absolument sommable.

On désigne par $l^1(E)$ l'espace vectoriel de suites dans E absolument sommables. Il est bien connu que c'est un Banach pour la norme $\sum ||x_n||$. $l^1(E)$ n'est autre qu'un espace de fonctions intégrables à valeurs dans le Banach E.

Nous allons étudier les suites scalairement sommables et les suites sommables de manière plus précise.

N° 2 : ETUDE DES SUITES SCALAIREMENT SOMMABLES

Dans ce numéro, E désignera un Banach, de dual E', et (x_n) sera une suite de points de E.

Il est clair que, pour l'addition terme à terme et la multiplication par un scalaire, l'ensemble des suites scalairement sommables est un espace vectoriel (sur le même corps de base que E'). Il est également clair que, si (x_n) est scalairement dans l^1, elle définit une application linéaire de E' dans l^1.

Je dis que cette application est continue. En effet, cela résulte du théorème du graphe fermé : si $x'_k \to 0$ dans E', la suite $(\langle x_n, x'_k\rangle)_{n \in \mathbb{N}}$ converge vers zéro pour la topologie de convergence suivant chaque coordonnée (quand k tend vers l'infini). Comme E' et l^1 sont métrisables et complets, cela implique la continuité.

REMARQUE : On aurait pu également utiliser le théorème de Banach-Steinhaus pour démontrer la continuité. Soit en effet pour $k \in \mathbb{N}$, u_k l'application linéaire de E' dans l^1 définie par :

$$u_k(x') = (\langle x_n, x'\rangle)_{0 \leqslant n \leqslant k}$$

On identifie le vecteur $(a_n)_{0 \leqslant n \leqslant k}$ de $\mathbb{K}^k$ au vecteur $(a_0, a_1, \ldots a_k, 0, 0, \ldots)$ de $\mathbb{K}^{\mathbb{N}}$. Par hypothèse, $\lim_{k\to\infty} u_k(x')$ existe dans l^1 pour tout $x' \in E'$; comme les u_k sont continues, le théorème de Banach-Steinhaus donne la continuité cherchée.

Soit alors (x_n) scalairement dans l^1 et soit α l'application linéaire continue de E' dans l^1 correspondante. Sa transposée α' est une application linéaire continue de l^∞ dans E''. Je dis que <u>α' envoie c_0 dans E</u>. En effet, soit (e_k) la base canonique de c_0. Pour $k \in \mathbb{N}$, on a :

$$\forall x' \in E' : \langle e_k, \alpha(x') \rangle_{\langle l^1, l^\infty \rangle} = \langle x_k, x' \rangle_{\langle E, E' \rangle} = \langle \alpha'(e_k), x' \rangle_{\langle E'', E' \rangle}$$

Donc $\qquad x'_k = \alpha'(e_k)$

Mais alors α' envoie le sous-espace fermé de l^∞ engendré par les (e_k), à savoir c_0, dans le sous-espace fermé de E'' engendré par les (x_n). En définitive, α' envoie c_0 dans E. En définitive, à toute suite (x_n) scalairement sommable, on a associé un élément de $\mathcal{L}(c_0 ; E)$.

Cette application de l'ensemble des suites scalairement sommables dans $\mathcal{L}(c_0 ; E)$ est évidemment linéaire et injective. Je dis qu'elle est surjective. En effet, soit $u : c_0 \to E$ linéaire et continue et soit $x_n = u(e_n)$ $\forall n$ ((e_n) base canonique de c_0). (x_n) est scalairement dans l^1 car u', transposée de u, envoie E' dans l^1. Si donc x' est un élément de E', pour tout n :

$$\langle x_n, x' \rangle = \langle u(e_n), x' \rangle = \langle e_n, u'(x') \rangle$$

Puisque $u'(x') \in l^1$, on obtient le résultat annoncé.

Pour résumer, on peut énoncer le

<u>THEOREME 1</u> :

Soit E un Banach ; l'espace vectoriel des suites scalairement dans l^1 est isomorphe (algébriquement) à $\mathcal{L}(c_0 ; E)$.

On en déduit qu'une suite scalairement dans l^1 est bornée, mais ne converge pas forcément vers 0 (nous l'avons vu).

Remarquons que $\mathcal{L}(c_0 ; E)$ est isomorphe à l'espace des applications linéaires de E' dans l^1 qui sont continues relativement à $\sigma(E', E)$ et $\sigma(l^1, c_0)$.

Remarquons également que si (x_n) est scalairement dans l^1, pour toute permutation σ de $\mathbb{N}$, $(x_{\sigma(n)})$ est aussi scalairement dans l_1.

THEOREME 2 :

(x_n) est scalairement dans l^1 si et seulement si pour toute suite (λ_n) de scalaires tendant vers zéro, $\sum \lambda_n x_n$ converge.

Démonstration : Soit (x_n) scalairement dans l^1 à laquelle correspond $u : c_0 \to E$. Si $(\lambda_n) \in c_0$, alors $\sum \lambda_n e_n$ converge dans c_0 vers (λ_n) (en base canonique). Donc

$$u(\sum \lambda_n e_n) = \sum \lambda_n x_n \quad \text{existe.}$$

Réciproquement, si (x_n) est une suite telle que $\sum \lambda_n x_n$ converge pour toute (λ_n) dans c_0, elle permet de définir une application linéaire u de c_0 dans E par :

$$\lambda = (\lambda_n) \quad \to \quad u(\lambda) = \sum \lambda_n x_n$$

Cette application est continue, car :

$$u(\lambda) = \lim_{k\to\infty} u_k(\lambda) \qquad \text{avec} \qquad u_k(\lambda) = \sum_{n=0}^{k} \lambda_k x_k ,$$

et du fait que les u_k sont continues, Banach-Steinhaus s'applique.

REMARQUE : Dans le théorème 2, on peut remplacer la condition " $\sum \lambda_n x_n$ converge" par " $\sum \lambda_n x_n$ converge commutativement".

Maintenant, du fait que l'espace des suites scalairement dans l^1 est un sous-espace de $\mathcal{L}(E', l^1)$, on peut le munir de la topologie induite par la norme "uniforme"

$$||(x_n)|| = \sup_{\substack{x' \in E' \\ ||x'|| \leq 1}} \sum |\langle x_n, x' \rangle|$$

EXERCICE : Démontrer que l'espace des suites scalairement dans l^1 est un Banach pour la norme ci-dessus.

On peut également obtenir très facilement une caractérisation des suites scalairement dans l^1. C'est le

THEOREME 3 :

Soit (x_n) une suite d'éléments du Banach E ; les propriétés suivantes sont équivalentes :

(1) (x_n) est scalairement dans l^1

(2) $\sup\limits_{\substack{(\lambda_n)\in \mathbb{K}_o^{\mathbb{N}} \\ \sup|\lambda_n| \leq 1}} ||\sum \lambda_n x_n|| < \infty$

Démonstration :

(1) $\Rightarrow$ (2) : c'est trivial, car on a même

$$\sup_{\substack{(\lambda_n)\in c_o \\ \sup|\lambda_n| \leq 1}} ||\sum \lambda_n x_n|| < \infty$$

d'après le théorème 1 et la caractérisation des applications linéaires continues entre Banach.

(2) $\Rightarrow$ (1) : l'application $(\lambda_n) \to \sum \lambda_n x_n$ du normé $\mathbb{K}_o^{\mathbb{N}}$ dans le Banach E est continue si (2) est satisfaite ($\mathbb{K}_o^{\mathbb{N}}$ étant muni de la topologie induite par c_o). Elle admet donc un prolongement unique en une application u linéaire continue de c_o dans E. Puisque $u(e_n) = x_n$, on en déduit, par ce qui précède, que (x_n) est scalairement dans l^1, et que

$$u(\lambda) = \sum \lambda_n x_n, \quad \lambda = (\lambda_n) \in c_o.$$

N° 3 : SUITES SOMMABLES DANS UN BANACH

Nous allons donner pour les suites d'éléments d'un Banach, des conditions équivalentes à la sommabilité. Auparavant, nous allons énoncer un lemme qui s'avèrera fondamental.

LEMME 1 (Principe de contraction) :

Soit $\{x_1, x_2, \ldots, x_n\}$ une famille finie de vecteurs d'un Banach réel. On a :

(A) $\max \{ \| \sum_{i=1}^{n} \varepsilon_i x_i \| \ \varepsilon_i = \pm 1 \ \forall i \} = \sup \{ \| \sum_{i=1}^{n} \lambda_i x_i \| ; \sup |\lambda_i| \leq 1 \}$

Démonstration : Soit l^∞_n l'espace $\mathbb{R}^n$ muni de la norme l^∞ et soit la fonction définie sur l^∞_n par :

$$\lambda \rightsquigarrow f(\lambda) = \| \sum_{i=1}^{n} \lambda_i x_i \| \qquad \lambda = (\lambda_i)_{1 \leq i \leq n}$$

f est évidemment convexe et le membre de droite de l'égalité (A) est égal à la borne supérieure de f sur la boule unité de l^∞_n.

Or cette boule n'est autre que l'enveloppe convexe fermée de l'ensemble des 2^n points dont les coordonnées sont égales à ± 1. Donc le maximum de f sur la boule de l^∞_n n'est autre que son maximum sur l'ensemble ci-dessus. Et le lemme est démontré.

On peut alors démontrer le

THEOREME 4 :

Soit (x_n) une suite de points d'un Banach E ; les conditions suivantes sont équivalentes :

(1) (x_n) est sommable

(2) Pour toute suite $(x_{n_k})_{k \in \mathbb{N}}$ extraite de (x_n), $\sum_k x_{n_k}$ converge.

(3) Pour toute suite (λ_n) de l^∞, $\sum \lambda_n x_n$ converge.

Démonstration :

(1) ⟹ (3) : Supposons (x_n) sommable et soit $\varepsilon > 0$ donné. Il existe $n_o \in \mathbb{N}$ tel que pour toute partie finie J de $\mathbb{N}$ satisfaisant à $n_o < \inf J$, on ait :
$$\| \sum_{i \in J} x_i \| \leq \frac{\varepsilon}{2}$$
Soit $\lambda \in l^\infty$; on peut supposer que $\sup |\lambda_n| \leq 1$. Si m et n sont des entiers tels que $n > m > n_o$, on a par le lemme 1 :

$$\| \sum_{i=m}^{n} \lambda_i x_i \| \leq \| \sum_{i=m}^{n} \varepsilon_i x_i \| \text{ , pour un choix convenable des } \varepsilon_i = \pm 1.$$

Mais

$$\Big\| \sum_{i=m}^{n} \varepsilon_i x_i \Big\| = \Big\| \sum_{\substack{i=m \\ \varepsilon_i>0}}^{n} x_i - \sum_{\substack{i=n \\ \varepsilon_i>0}}^{n} x_i \Big\| \leqslant \Big\| \sum_{\substack{i=n \\ \varepsilon_i>0}}^{n} x_i \Big\| + \Big\| \sum_{\substack{i=m \\ \varepsilon_i<0}}^{n} x_i \Big\|$$

$$\leqslant \frac{\varepsilon}{2} + \frac{\varepsilon}{2} = \varepsilon$$

Finalement, on voit que $(\sum \lambda_i x_i)$ satisfait à la condition de Cauchy dans le cas où sup $|\lambda_i| \leqslant 1$. Le cas général se ramène à celui-là par homogénéité.

(3) $\Longrightarrow$ (2) : Supposons (3) satisfaite et soit (x_{n_k}) une suite extraite. Soit (λ_n) définie par :

$$\lambda_n = \begin{cases} 1 \text{ si } n = n_k \\ 0 \text{ si } n \neq n_k \end{cases}$$

Il est clair que $\sum \lambda_n x_n = \sum x_{n_k}$. Donc (3) implique (2).

(2) $\Longrightarrow$ (1) : Soit (x_n) satisfaisant à (2) et supposons cette suite non sommable, c'est à dire :

Il existe $\varepsilon > 0$ et une suite (J_n) de parties finies de $\mathbb{N}$ telles que :

- $\sup J_n \leqslant \operatorname{Inf} J_{n+1} \qquad \forall\, n$
- $\Big\| \sum_{i \in J_n} x_i \Big\| \geqslant \varepsilon \qquad \forall\, n$

Considérons la suite extraite de (x_n) obtenue en prenant les éléments de $\cup J_n$ dans l'ordre croissant. Il est clair que cette suite extraite ne peut converger (elle ne satisfait pas à la condition de Cauchy) - CONTRADICTION -

<u>REMARQUE 1</u> : Les conditions du théorème 4 équivalent encore à :

(4) $\sum \varepsilon_i x_i$ converge pour <u>toute</u> famille $(\varepsilon_i)_{i \in \mathbb{N}}$ telles que $\varepsilon_i = \pm 1$.

<u>REMARQUE 2</u> : Si E est seulement normé, les conditions du théorème 4 ne sont plus équivalentes comme le montrent les exemples suivants :

Exemple 1 : Soit E = $\mathbb{R}_o^{\mathbb{N}}$ muni de la norme l^∞ et soit

$$x_n = \frac{1}{n} e_n - \frac{1}{n+1} e_{n+1} \qquad n \geq 1$$

alors $\sum x_n = e_1$.

On voit facilement que (x_n) est sommable (commutativement convergente). Par contre, $\sum x_{2n}$ ne converge pas dans $\mathbb{R}_o^{\mathbb{N}}$ (mais dans c_o).

Exemple 2 : Soit E l'espace des suites de nombres réels ne prenant qu'un nombre fini de valeurs distinctes.

$$x = (\lambda_n) \qquad \lambda_n \in \mathbb{R} \ \forall n \qquad \text{Card } \{\lambda_n, n \in \mathbb{N}\} < \infty$$

e_n désignera le vecteur de E : (0, 0, 0, 1, 0, 0...).

Munissons E de la norme $\sup_n \frac{|\lambda_n|}{n+1}$

Alors (e_n) ne satisfait pas à la condition (3) du théorème 4, mais pour toute suite extraite $\sum_k e_{n_k}$ a un sens.

COROLLAIRE

Toute suite sommable d'éléments du Banach E définit une application linéaire de l^∞ dans E.

Démonstration : Le théorème 4 permet de définir une application linéaire de l^∞ dans E. La continuité résulte alors de Banach-Steinhaus.

REMARQUE : Par analogie avec les suites scalairement sommables, on serait tenté d'identifier l'espace des suites sommables avec $\mathcal{L}(l^\infty ; E)$. Il n'en est rien comme le montre l'exemple suivant :

Soit E = l^∞ et soit u l'application identique de l^∞ dans lui-même. Supposons qu'il existe $(x_n) \in (l^\infty)^{\mathbb{N}}$ telle que

$$u(\lambda) = \sum \lambda_n x_n \qquad \forall \lambda \in l^\infty$$

On en déduirait que l^∞ est séparable, ce qui est faux.

N° 4 : ETUDE PLUS PRECISE DE LA RELATION ENTRE SOMMABILITE ET SCALAIRE SOMMABILITE

Nous avons donné un exemple de Banach dans lequel il existe une suite scalairement sommable mais non sommable, à savoir l'espace c_o (en fait, cette suite ne converge même pas vers zéro). Nous verrons plus tard que cette rencontre avec c_o n'est pas fortuite. Pour le moment, nous allons étudier quand ces deux notions sont équivalentes.

Toutefois, nous allons nous placer dans une situation un peu plus générale que dans les Banach, ce qui nous permettra d'appliquer les résultats obtenus aux espaces de variables aléatoires.

Donnons au préalable quelques définitions généralisant à d'autres espaces que les Banach les notions de suites sommables et scalairement sommables.

DEFINITIONS : Soit E un espace vectoriel topologique (pas forcément localement convexe !) et (x_n) une suite de points de E.

(1) On dit que (x_n) est une C-suite si pour tout (λ_n) dans (c_o), $\sum \lambda_n x_n$ converge.

(2) On dit que (x_n) est une C'-suite si pour tout $\lambda = (\lambda_n)$ dans 1^∞, $\sum \lambda_n x_n$ converge.

(3) On dit que E est un C-espace si toute C-suite est une C'-suite.

Il est clair que dans un Banach, une C-suite (respectivement C'-suite) n'est autre qu'une suite scalairement sommable (respectivement sommable).

Il est également clair que si (x_n) est une C-suite (respectivement C'-suite) pour tout $(a_n) \in 1^\infty$, $(a_n x_n)$ est une C-suite (respectivement C'-suite).

De même, si (x_n) est une C-suite, pour tout $(a_n) \in c_o$, $(a_n x_n)$ est une C'-suite. D'où résulte que E est un C-espace si et seulement si pour toute C-suite (x_n), $\sum x_n$ converge.

THEOREME 5 :

Soit E un espace vectoriel complet, (x_n) une suite de points de E. (x_n) est une C-suite si et seulement si l'ensemble

$$A = \{ x \in E \quad x = \sum \lambda_n x_n \quad \lambda_n \in \mathbb{R}_o^{\mathbb{N}} \quad \sup |\lambda_n| \leqslant 1 \}$$

est borné.

Démonstration :

- Suffisance : Supposons A borné et soit V un voisinage équilibré de zéro dans E. Soit $(\lambda_n) \in c_o$. A étant borné et (λ_n) convergeant vers zéro, il existe n_o tel que pour tout $n \geqslant n_o$ on ait $\lambda_n\ x \in V \quad \forall x \in A$. Si alors n et m sont deux entiers tels que $n_o < n < m$,

$$\sum_{i=n+1}^{m} \lambda_i x_i = \lambda \sum_{i=n+1}^{m} \frac{\lambda_i}{\lambda} x_i = \lambda x$$

si $\lambda = \sup\limits_{n<i\leqslant m} |\lambda_i|$ et $x = \sum\limits_{i=n+1}^{m} \frac{\lambda_i}{\lambda} x_i$ (on pose $\frac{0}{0} = 0$!)

Mais $x \in A$ et $\lambda x \in V$ (car V est équilibré) : donc la condition de Cauchy est vérifiée et $\sum \lambda_n x_n$ converge.

- Nécessité : Supposons A non borné ; il existe un voisinage de zéro équilibré dans E, une suite (λ_n) de réels tendant vers zéro, une suite (a_n) de réels telle que $|a_n| \leqslant 1$ pour tout n, et deux suites d'entiers (n_k) et (n'_k) de façon que :

$$- \ n_k < n'_k < n_{k+1} \qquad \forall k$$

$$- \ \lambda_k \sum_{n=n_k}^{n'_k} a_n x_n \notin V \qquad \forall k$$

Posons alors

$$c_n = \begin{cases} \lambda_k a_n & \text{si } n_k < n \leqslant n'_k \\ 0 & \text{sinon} \end{cases}$$

alors $(c_n) \in c_o$ et la série $\sum c_n x_n$ n'est pas convergente. Donc (x_n) n'est pas une C-suite (ici, la complétude n'a pas été utilisée).

Il est clair que l'image d'une C-suite par une application linéaire continue est une C-suite, et que tout sous-espace fermé d'un C-espace est également un C-espace.

THEOREME 6

Soit E un espace vectoriel topologique complet, E est un C-espace si et seulement si toute C-suite converge vers zéro.

Démonstration :

La nécessité est évidente.

Réciproquement, supposons que E ne soit pas un C-espace et que toute C-suite converge vers zéro.

Il existe une C-suite (x_n) telle que $\sum x_n$ ne converge pas, donc ne satisfait pas la condition de CAUCHY car E est complet.

Il existe un voisinage V de zéro et une suite croissante (n_k) d'entiers tels que pour tout k

$$y_k = \sum_{n_k}^{n_{k+1}} x_n \notin V$$

Mais (y_k) est une C-suite ne convergeant pas vers zéro. CONTRADICTION.

Nous allons maintenant donner des exemples de C-espaces.

THEOREME 7 (L. SCHWARTZ)

Soit (X, F, μ) un espace mesuré fini. L'espace $L^0 (X, F, \mu)$ est un C-espace.

Démonstration Il nous faut démontrer que toute C-suite (f_n) d'éléments de $L^0 (X, F, \mu)$ converge vers zéro en μ-mesure. En fait, on démontrera même plus : f_n converge vers zéro μ-presque partout. Cela résultera du "vieux lemme" suivant de KOLMOGOROFF-KHINCHINE.

LEMME

Soit (f_n) une suite d'éléments de L^0 $(X, \mathfrak{F}, \mu)$ (μ mesure finie) telle que $\sum \lambda_n f_n$ converge en μ-mesure pour toute $(\lambda_n) \in c_0$. Alors $\sum f_n^2 (x) < \infty$ μ-presque partout (ou encore $(f_n(x))_n$ est μ-presque partout dans l^2)

Démonstration du lemme (KWAPIEN)

Soit (R_n) le système de RADEMACHER : les R_n sont des variables aléatoires sur l'espace probabilité unité $([0, 1], dt)$, indépendantes, et de loi $\frac{1}{2} \delta_{(1)} + \frac{1}{2} \delta_{(-1)}$.

En fait, $R_n(t) = \text{sign} \left(\sin(2\pi \times 2^n t)\right) \quad \forall t \in [0, 1] \quad \forall n \in \mathbb{N}$

Soient $n \in \mathbb{N}$ et $t \in [0, 1]$ donnés. Posons :

$$F_{n,t} (.) = \sum_{i=0}^{n} R_i (t) f_i (.)$$

Les (f_n) formant une C-suite, l'ensemble $\left(F_{n,t} (.)\right)_{(n, t)}$ est une partie bornée de L^0 $(X, \mathfrak{F}, \mu)$ d'après le théorème 5. Il en résulte que, pour tout $\varepsilon > 0$, il existe une constante C positive finie telle que

(1) $$\mu \{x ; |F_{n,t} (x)| > C\} \leqslant \varepsilon \qquad \forall (n, t).$$

n étant donné, soit $B_n = \{(t, x) \ |F_{n,t} (x)| > C\}$; puis x étant fixé, soit B_n^x la coupe de B_n en x ($B_n^x \subset [0, 1]$). Posons également $A_n = B_n^c$ et $A_n^x = (B_n^x)^c$. Si λ désigne la mesure de Lebesgue sur $[0, 1]$, on déduit de (1) par Fubini que :

$\mu \otimes \lambda (B_n) \leqslant \varepsilon$, et, par conséquent d'après Fubini, à nouveau :

$$\int_X \lambda (B_n^x) \ \mu (dx) \leqslant \varepsilon$$

Il en résulte que

$$\mu \{x ; \lambda (B_n^x) \geqslant \sqrt{\varepsilon}\} \times \sqrt{\varepsilon} \leqslant \varepsilon$$

ou encore

$$\mu \{x ; \lambda (B_n^x) < \sqrt{\varepsilon}\} \geqslant \mu (X) - \sqrt{\varepsilon}$$

Puisque

$\lambda (B_n^x) = 1 - \lambda (A_n^x)$, ceci équivant encore à :

$$\mu \{x ; \lambda (A_n^x) > 1 - \sqrt{\varepsilon}\} \geqslant \mu (X) - \sqrt{\varepsilon}$$

Puisque pour tout $t \in A_n^x$ on a $|F_{n,t}(x)| \leqslant C$, on a :

$$\int_{A_n^x} \Big(\sum_{i=0}^{n} R_i(t)\, f_i(x)\Big)^2 dt \leqslant C^2 \qquad \forall x$$

ou encore :

$$(2) \qquad \int_{A_n^x} \sum_{i=0}^{n} |f_i(x)|^2 dt + 2 \int_{A_n^x} \sum_{i<j\leqslant n} f_i(x)\, f_j(x)\, R_i(t)\, R_j(t)\, dt \leqslant C^2$$

pour tout x (car $R_n^2(t) \equiv 1$ pour tout n).

Choisissons ε tel que $\sqrt{\varepsilon} < \mu(X)$; il existe des x tels que $\lambda(A_n^x) > 1 - \sqrt{\varepsilon}$ et pour ces x, on a alors :

$$(1 - \sqrt{\varepsilon}) \sum_{i=0}^{n} |f_i(x)|^2 + 2 \sum_{i<j\leqslant n} f_i(x)\, f_j(x) \int_{A_n^x} R_i(t)\, R_j(t)\, dt \leqslant C^2$$

Il résulte donc de l'inégalité de SCHWARZ que :

$$(3) \qquad (1 - \sqrt{\varepsilon}) \sum_{i=0}^{n} |f_i(x)|^2 - 2 \left\{\sum_{i<j\leqslant n} |f_i(x)\, f_j(x)|^2\right\}^{\frac{1}{2}} \times \left[\sum_{i<j\leqslant n} \Big(\int_{A_n^x} R_i(t)\, R_j(t)\, dt\Big)^2\right]^{\frac{1}{2}} \leqslant C^2$$

Il nous reste maintenant à estimer $\sum_{i<j\leq n} \big(\int_{A_n^x} R_i(t)\, R_j(t)\, dt\big)^2$.

Nous utiliserons pour cela le fait que les $(R_n(.))_{n \in \mathbb{N}}$ forment un système orthonormé (pas total) dans $L^2([0, 1], dt)$, ainsi que les $(R_m . R_n)_{m \neq n}$.
On en déduit que :

Si $i \neq j$ $\qquad \int_{A_n^x} R_i(t)\, R_j(t)\, dt = - \int_{B_n^x} R_i(t)\, R_j(t)\, dt$

car $\qquad B_n^x = (A_n^x)^c$

et que :

$$(4) \qquad \sum_{i<j\leqslant n} \Big(\int_{A_n^x} R_i(t)\, R_j(t)\, dt\Big)^2 = \sum_{i<j\leqslant n} \Big(\int_{B_n^x} R_i(t)\, R_j(t)\, dt\Big)^2$$

Puisque les RiRj $(i < j)$ forment un système orthonormé sur $L^2([0, 1], dt)$ et puisque les coefficients de Fourier de $1_{B_n^x}$ par rapport à ce système sont égaux à :

$$\int_{B_n^*} R_i(t)\, R_j(t)\, dt,$$

l'inégalité de Parseval et l'égalité (4) donnent :

$$(5) \qquad \sum_{i<j\leqslant n} \Big(\int_{A_n^x} R_i(t)\, R_j(t)\, dt\Big)^2 \leqslant \lambda(B_n^x) \leqslant \sqrt{\varepsilon}$$

Remarquons maintenant en dernier ressort que :

$$(6) \qquad \Big(\sum_{i<j\leqslant n} |f_i(x)\, f_j(x)|^2\Big)^{\frac{1}{2}} \leqslant \Big(\sum_{i\leqslant n}\ \sum_{j\leqslant n} |f_i(x)\, f_j(x)|^2\Big)^{\frac{1}{2}}$$

$$= \Big[\Big(\sum_{i=0}^{n} |f_i(x)|^2\Big)^2\Big]^{\frac{1}{2}} \qquad = \sum_{i=0}^{n} |f_i(x)|^2$$

Les inégalités (3), (4), (5) et (6) impliquent que :

$$(7) \qquad \sum_{i=0}^{n} |f_i(x)|^2 \leqslant \frac{C^2}{1 - \sqrt{\varepsilon} - 2\,\varepsilon^{\frac{1}{4}}}$$

pour tout x tel que $\lambda\,(A_n^x) > 1 - \sqrt{\varepsilon}$.

Finalement, on a démontré que, pour chaque n, on a avec une mesure $\geqslant \mu\,(X) - \sqrt{\varepsilon}$ l'inégalité (7).

Il est maintenant facile de terminer la démonstration, car si l'on pose :

$$K_n = \{x \;;\; \sum_{i=0}^{n} |f_i(x)|^2 \leqslant \frac{C^2}{1 - \sqrt{\varepsilon} - 2\,\varepsilon^{\frac{1}{4}}}\}$$

alors $\mu\,(K_n) \geqslant \mu\,(X) - \sqrt{\varepsilon}$ $\forall$ n et $K_n \downarrow$.

Donc $K = \bigcap K_n$ est tel que $\mu\,(K) \geqslant \mu\,(X) - \sqrt{\varepsilon}$; en outre, pour tout $x \in K$, on a :

$$\sum_{i=0}^{n} |f_i(x)|^2 \leqslant \frac{C^2}{1 - \sqrt{\varepsilon} - 2\,\varepsilon^{\frac{1}{4}}} \qquad \text{c.q.f.d.}$$

REMARQUE : On a en fait démontré ce qui suit : Soit $(X, \mathfrak{F}, \mu)$ un espace mesuré et soit $L^0\,(X, \mathfrak{F}, \mu)$ muni de la topologie de convergence en mesure sur tout ensemble de mesure finie. Alors pour toute C-suite (f_n) de $L^0\,(X, \mathfrak{F}, \mu)$

on a $\sum f_n^2(x) < \infty$ μ-presque partout sur tout ensemble de mesure finie (donc μ-presque partout si μ est σ-finie).

THEOREME 8 :

Soit $(X, \mathcal{F}, \mu)$ un espace mesuré ; pour tout $p \in [0, \infty[$ $L^p(X, F, \mu)$ est un C-espace (ici, $(X, \mathcal{F}, \mu)$ n'a pas besoin d'être supposé σ-fini).

Démonstration : D'après ce qui précède, il suffit de démontrer que toute C-suite (f_n) converge vers zéro dans $L^p(X, \mathcal{F}, \mu)$. Bien sûr, le cas $p = 0$ a été résolu. On supposera donc $p > 0$. Soit alors (f_n) une C-suite dans $L^p(X, \mathcal{F}, \mu)$: c'est une C-suite dans $L^0(X, \mathcal{F}, \mu)$; et par le lemme de KHINCHINE-KOLMOGOROFF, elle converge vers zéro presque partout sur tout ensemble de mesure finie (donc presque partout si $(X, \mathcal{F}, \mu)$ est σ-fini).

En utilisant le théorème d'EGOROFF et compte tenu du fait que les (f_n) sont nulles en dehors d'un ensemble σ-fini, on peut écrire :

$$X = \bigcup_{i=0}^{\infty} X_i \cup N$$

avec :

- pour tout n, f_n est nulle presque partout sur N
- $\forall i,\ X_i \in F, X_i \subset X_{i+1}$, $\mu(X_i) < \infty$
- les f_n tendent vers zéro uniformément sur chaque X_i.

(On dira pour abréger que les X_i sont des ensembles d'uniforme convergence).

Supposons que (f_n) ne tende pas vers zéro dans $L^p(X, \mathcal{F}, \mu)$. Il existe $\delta' > 0$ tel que $\int_X |f_n(x)|^p \, d\mu > \delta'$ pour une infinité de n.

On peut supposer que l'inégalité ci-dessus est vraie pour tout n. Donc, que (f_n) est une C-suite de $L^p(X, \mathcal{F}, \mu)$ telle que $\int |f_n|^p \, d\mu > \delta'$ $\forall$ n.

Soit δ tel que $0 < \delta < \delta'$. On va construire par récurrence une suite (A_k) d'ensembles de F deux à deux disjoints et une suite croissante d'entiers (n_k) telles que :

- chaque A_k soit contenu dans un ensemble d'uniforme convergence.

- Pour tout k, on a : $\int_{A_k} |f_{n_k}(x)|^p \, d\mu(x) > \delta$

- Pour tous entiers j et k tels que $j \neq k$, on a :

$$\int_{A_j} |f_{n_k}(x)|^p \, d\mu(x) < \frac{1}{2^{j+k}}$$

En effet, prenons $n_o = 0$; puisque $\int |f_o|^p \, d\mu > \delta$, il existe m_o tel que :

$$\int_{X_{m_o}} |f_{n_o}|^p \, d\mu > \delta$$

On pose $A_o = X_{m_o}$.

Supposons que $A_o, A_1, \ldots, A_k$ et $n_o, n_1, \ldots, n_k$ aient été déterminés de façon à satisfaire aux conditions ci-dessus ; soit X_{m_k} un ensemble d'uniforme convergence contenant $A_o, A_1, \ldots, A_k$. Il existe alors un ensemble d'uniforme convergence $X_{m'_k}$ avec $m'_k > m_k$ tel que :

$$\int_{X^c_{m'_k}} |f_{n_j}|^p \, d\mu \leqslant \frac{1}{2^{(k+1)+j}} \qquad \forall\, j = 0, 1, \ldots, k.$$

Ensuite, par le fait que les (f_n) convergent uniformément vers zéro sur chaque X_i, il existe un entier $n_{k+1} > n_k$ tel que :

$$\int_{X_{m'_k}} |f_{n_{k+1}}|^p \, d\mu \leqslant \frac{1}{2^{2(k+1)}} \quad \text{et} \quad \int_{X^c_{m'_k}} |f_{n_{k+1}}|^p \, d\mu > \delta .$$

On en déduit qu'il existe un entier $m_{k+1} > m'_k$ tel que

$$\int_{X_{m_{k+1}} \setminus X_{m'_k}} |f_{n_{k+1}}|^p \, d\mu > \delta$$

Posons $A_{k+1} = X_{m_{k+1}} \setminus X_{m'_k}$.

Je dis que la famille $(n_o, n_1, \ldots, n_{k+1})$ $(A_o, A_1, \ldots, A_{k+1})$ satisfait les conditions cherchées. En effet,

- $A_o, A_1, \ldots, A_{k+1}$ sont contenus dans des ensembles d'uniforme convergence.
- $\int_{A_j} |f_{n_j}|^p \, d\mu > \delta \qquad \forall\, j = 0, 1, 2, \ldots, (k+1).$

$$- \int_{A_{k+1}} |f_{n_j}|^p \, d\mu \leqslant \int_{X^c_{m',k}} |f_{n_j}|^p \, d\mu \leqslant \frac{1}{2^{(k+1)+j}} \qquad \forall j = 0, 1, \ldots, k,$$

d'une part, et d'autre part :

$$\int_{A_j} |f_{n_{k+1}}|^p \, d\mu \leqslant \int_{X_{m',k}} |f_{n_{k+1}}|^p \, d\mu \leqslant \frac{1}{2^{k+1+(k+1)}} < \frac{1}{2^{(k+1)+j}}$$

$\forall j = 0, 1, \ldots, k$. Donc la récurrence s'enclanche.

Il nous reste maintenant à terminer la démonstration. Pour cela, soit $(\lambda_n) \in c_o$ tel que $\lambda_n \geqslant 0$ $\forall n$ et $\sum \lambda_n^p = +\infty$. On va démontrer que $\sum \lambda_k f_{n_k}$ ne peut converger dans $L^p(X, \mathcal{F}, \mu)$, ce qui entrainera une contradiction. On va examiner deux cas :

① - $0 < p \leqslant 1$: soit $n \in \mathbb{N}$, on a alors :

$$\int_X |\sum_{k=0}^n \lambda_k 1_{nk}|^p \, d\mu \geqslant \sum_{l=0}^n \int_{A_l} |\sum_{k=0}^n \lambda_k 1_{n_k}|^p \, d\mu$$

$$\geqslant \sum_{k=0}^n \left[\int_{A_l} (\lambda_l^p \; |1_{n_l}|^p - \sum_{\substack{k=0 \\ k\neq l}}^n \lambda_k^p \, |1_{n_k}|^p) \, d\mu\right]$$

$$\geqslant \delta \sum_{l=0}^n \lambda_l^p - C \sum_{k,l=0}^n \frac{1}{2^{k+l}} \geqslant \delta \sum_{l=0}^n \lambda_l^p - 4\,C \;;$$

si l'on pose $C = \sup_n \lambda_n^p$. Donc la série $\sum \lambda_k f_{n_k}$ ne peut converger dans L^p, et f_{n_k} n'est pas une C-suite. Comme (f_{n_k}) est extraite d'une C-suite, on obtient une contradiction.

- ② - $1 \leqslant p < \infty$: Même méthode que dans le cas $p < 1$, en utilisant cette fois l'inégalité de MINKOWSKI.

<u>REMARQUE</u> : En général, L^∞ n'est pas un C-espace. Par exemple, l^∞, qui contient c_o, n'est pas un C-espace.

N° 5 : <u>UNE PROPRIETE CARACTERISTIQUE DES F-ESPACES NE CONTENANT PAS</u> c_o

<u>DEFINITION</u> : Soit E un espace vectoriel topologique ; on dit que E contient c_o isomorphiquement (ou en abrégé E contient c_o) s'il existe un sous-espace

de E isomorphe topologiquement, par la topologie induite par celle de E, à c_o.

Bien sûr, si E est un Banach contenant c_o, cela n'implique pas qu'il existe dans E un sous-espace isométrique à c_o. Remarquons de prime abord que si E contient c_o, il ne peut être un C-espace (car c_o n'en est pas un !). Nous allons maintenant indiquer une réciproque de ce résultat. Nous ne détaillerons pas la démonstration, ce qui nous entrainerait trop loin ; nous nous contenterons d'en donner un plan.

Indiquons le premier résultat dans cette voie.

THEOREME 10 (BESSAGA, PELCZYNSKI, ROLEWICZ)

Soit E un F-espace ayant une base. Si E n'est pas un C-espace, il contient c_o.

Rappelons que l'on dit qu'un F-espace E possède une base (e_n) si tout élément x de E possède une représentation unique, sous la forme d'une série (convergente !)

$$x = \sum \lambda_n e_n = \sum \lambda_n (x) e_n$$

On démontre que les formes linéaires $x \rightsquigarrow \lambda_n(x)$ sont continues, et même équicontinues.

Naturellement, la définition d'une base a un sens même si E n'est pas un F-espace. Toutefois, on ne peut affirmer en général que les "coefficients" (λ_n) sont des formes linéaires continues.

La seconde étape réside dans le

THEOREME 11 :

Soit F un sous-espace fermé d'un F-espace E à base. On suppose que E possède un voisinage borné. Alors si F n'est pas un C-espace, il contient c_o.

Le théorème 11 n'est pas une simple paraphrase du théorème 10, même si F est complet, car on sait depuis P. ENFLO (1972) que, même dans le cas d'un Banach, un sous-espace d'un espace à base ne possède pas forcément de base. En effet, ENFLO a donné un exemple de Banach séparable ne possédant pas de base. Or tout Banach séparable est un sous-espace de $\mathcal{C}([0, 1])$ qui lui, possède une base.

Avant d'énoncer le résultat fondamental, faisons la remarque suivante : Soit E un F-espace non-séparable. Il contient donc une C-suite (x_n) qui n'est pas une C'-suite. Soit F le sous-espace fermé engendré par les (x_n) : F est un F-espace séparable et n'est pas un C-espace. En d'autres termes, un F-espace E n'est pas un C-espace si et seulement si il contient un sous-espace séparable qui n'est pas un C-espace. A partir de là, on déduit le

THEOREME 12

Soit E un Banach. E n'est pas un C-espace si et seulement si E contient c_o.

Démonstration : On sait déjà (et c'est très général) que si E contient c_o, il ne peut être un C-espace.

Supposons alors que E n'est pas un C-espace. Il contient donc un sous-espace fermé séparable F, donc un Banach séparable, qui n'est pas un C-espace.

Mais F, nous l'avons déjà remarqué, est isomorphe à un sous-espace de l'espace de Banach à base $\mathcal{C}([0, 1])$.

Alors le théorème 10 permet d'affirmer que F, donc aussi E, contient c_o.

En particulier, les $L^p(X, F, \mu)$ $(1 \le p < \infty)$ ne peuvent contenir c_o .

N° 6 : COMPLEMENTS SUR LES C-SUITES ET LES C'-SUITES DANS DES E.V.T. METRISABLES COMPLETS

Nous avons déjà remarqué que l'on peut définir les notions de C-suite et C'-suite dans n'importe quel espace vectoriel topologique. En fait, au chapitre IV nous utiliserons ces notions dans des espaces $L^p(\Omega, \mathcal{F}, P, E)$ où E est un Banach $(0 \leq p < \infty)$.

Nous allons voir que, dans les cas qui nous intéresseront, les C'-suites ne sont autres que les suites sommables (ce qui était déjà connu dans le cas Banach). Nous donnerons également une condition pour qu'une suite soit une C-suite.

Rappelons que l'on a le résultat suivant :

THEOREME 13 :

Soit E un F-espace, (x_n) une suite de points de E ; les conditions suivantes sont équivalentes :

(1) Toute "sous-série" $\sum_{k \in \mathbb{N}} x_{n_k}$ est convergente ;

(1') Pour toute suite (α_n) de nombres réels telle que $\alpha_n = 0$ ou 1 quel que soit n, la série $\sum \alpha_n x_n$ converge.

(2) Pour toute suite $(\varepsilon_n) \in \{-1, +1\}^{\mathbb{N}}$, la série $\sum \varepsilon_n x_n$ converge.

(3) Pour toute permutation σ de $\mathbb{N}$, la série $\sum_n x_{\sigma(n)}$ converge.

(4) La suite (x_n) est sommable.

Il est trivial de voir que (1) $\Longleftrightarrow$ (1') $\Longleftrightarrow$ (2) ; (2) $\Longleftrightarrow$ (3) est un résultat bien connu d'ORLICZ (voir par exemple ROLEWICZ "Metric Linear Spaces, page 88). (3) $\Longleftrightarrow$ (4) : voir par exemple BOURBAKI (Topologie générale, chapitre III).

Nous aurons besoin maintenant de la définition suivante :

DEFINITION : Soit E un e.v.t. et $(x_n) \in E^{\mathbb{N}}$; (x_n) est dite commutativement bornée si l'ensemble

$$\{ \sum_{n\in\sigma} x_n, \ \sigma \in \Phi(\mathbb{N}) \}$$

est une partie bornée de E ($\Phi(\mathbb{N})$ désigne l'ensemble des parties finies de $\mathbb{N}$). On dira également que $\sum x_n$ est commutativement bornée.

Nous allons voir que, dans certains cas, (x_n) est commutativement bornée si et seulement si c'est une C-suite. Pour ce faire, nous aurons besoin d'un principe de contraction analogue au principe de contraction dans les espaces normés.

LEMME :

Soit $\mathcal{E}$ un espace p-normé et $\sigma \in \Phi(\mathbb{N})$; soit $(x_n)_{n\in\sigma}$ des vecteurs de $\mathcal{E}$; $(a_n)_{n\in\sigma}$ des nombres réels. On a la double inégalité :

$$\|\sum_{n\in\sigma} a_n x_n\| \le \frac{1}{2^p-1} (\sup_{n\in\sigma} |a_n|^p) \sup \{\|\sum_{n\in\sigma'} \varepsilon_n x_n\| \ ; \ \sigma' \subset \sigma ; \ \varepsilon_n = \pm 1 \quad \forall n\}$$

$$\le \frac{2}{2^p-1} (\sup_{n\in\sigma} |a_n|^p) \sup_{\sigma'\subset\sigma} \|\sum_{n\in\sigma'} x_n\| .$$

DEMONSTRATION : Par raison de p-homogénéité, on peut supposer que $\sup_{n\in\sigma} |a_n| = 1$. Alors chaque a_n admet un développement dyadique :

$$a_n = \sum_{k=0}^{\infty} \frac{\varepsilon_{nk}}{2^k} \qquad \text{avec } |\varepsilon_{n,k}| = 0 \text{ ou } 1 \ ;$$

En outre, n étant donné, tous les $\varepsilon_{n,k}$ non nuls ont même signe. Cela étant :

$$\|\sum_{n\in\sigma} a_n x_n\| = \|\sum_{n\in\sigma} \sum_{k\in\mathbb{N}} \frac{\varepsilon_{nk}}{2^k} x_n\|$$

$$= \sum_{k\in\mathbb{N}} \frac{1}{2^k} \sum_{n\in\sigma} \varepsilon_{n_k} x_n\| \le \sum_k \frac{1}{(2^k)^p} \|\sum_{n\in\sigma} \varepsilon_{nk} x_n\| .$$

Maintenant, si l'on remarque que, pour k fixé, les ε_{nk} non nuls ne peuvent prendre que les valeurs 1 ou (-1), on a

$$\forall k : \left\|\sum_{n\in\sigma} \varepsilon_{n_k} x_n\right\| \leq \sup_{\substack{\sigma'\subset\sigma \\ \varepsilon_n=\pm 1}} \left\|\sum_{n\in\sigma'} \varepsilon_n x_n\right\| ;$$

La première inégalité est donc démontrée. La seconde est alors évidente compte tenu du fait qu'une p-norme est sous-additive.

REMARQUE 1 : Sous les hypothèses du lemme ci-dessus, on peut remplacer le membre de gauche de la double inégalité par

$$\sup_{\sigma'\subset\sigma} \left\|\sum_{n\in\sigma} a_n x_n\right\| .$$

REMARQUE 2 : Le lemme ci-dessus admet évidemment une version infinie : si M est une partie infinie de $\mathbb{N}$, on a la double inégalité :

$$\sup_{\sigma\in\Phi(M)} \left\|\sum_{n\in\sigma} a_n x_n\right\| \leq \frac{1}{2^p-1} (\sup_{n\in M} |a_n|^p) \sup_{\substack{\sigma\in\Phi(M) \\ \varepsilon_n=\pm 1}} \left\|\sum_{n\in\sigma} \varepsilon_n x_n\right\|$$

$$\leq \frac{2}{2^p-1} \sup_{n\in M} |a_n|^p \sup_{\sigma\in\Phi(M)} \left\|\sum_{n\in\sigma} \varepsilon_n x_n\right\|$$

valable pour toutes les familles $(x_n) \in \mathcal{E}^M$ et $(a_n) \in \mathbb{R}^M$.

Comme conséquence, on obtient très facilement la

PROPOSITION :

(1) Soit $\mathcal{E}$ un espace p-normé et $(x_n) \in \mathcal{E}^{\mathbb{N}}$. Si (x_n) est commutativement bornée, pour toute $(\lambda_n) \in l^\infty$, $(\lambda_n x_n)$ est commutativement bornée.

(2) Si $\mathcal{E}$ est en outre supposé être un p-Banach et $(x_n) \in \mathcal{E}^{\mathbb{N}}$, alors

(a) Si (x_n) est commutativement bornée, pour toute $(\lambda_n) \in c_0$, $\sum(\lambda_n x_n)$ converge (donc aussi $(\lambda_n x_n)$ est sommable).

(b) Si (x_n) est sommable, pour toute $(\lambda_n) \in l^\infty$, $(\lambda_n x_n)$ est sommable.

Schéma de la démonstration :

(1) se déduit de la remarque 2 ci-dessus.

(2) est dû au fait que $\mathcal{E}$ étant complet, la sommabilité se vérifie au moyen de conditions de Cauchy, et le lemme ci-dessus donne le résultat.

REMARQUE 3 : Si E est un Banach et si $\mathcal{E}$ désigne l'espace $L^0(\Omega, \mathcal{F}, P, E)$ (où $(\Omega, \mathcal{F}, P)$ désigne un espace probabilisé) les conclusions de la proposition ci-dessus restent valables (voir par exemple MAUREY et PISIER - CRAS, tome 277, 1973 pp.39-42 pour le cas $E = \mathbb{R}$, ou RYLL-NARDZEWSKI et WOYCZINSKI, Proceedings of A.M.S., Vol. 53 n° 1, 1975, pp. 96-98).

En définitive, on peut énoncer sous une autre forme la proposition et la remarque 3 qui la suit :

Si $\mathcal{E}$ est un p-normé ou un espace $L^0(\Omega, \mathcal{F}, P, E)$, une C-suite n'est autre qu'une série commutativement bornée et une C'-suite n'est autre qu'une série sommable ou commutativement convergente.

En effet, on sait qu'une série commutativement bornée (resp. convergente) est une C-suite (resp. une C'-suite). Réciproquement, si (x_n) est une C-suite, en vertu du théorème 5, c'est une série commutativement bornée ; de même, de par le théorème 13, si (x_n) est une C'-suite, elle est sommable.

Dans les cas que nous aurons à utiliser par la suite, $\mathcal{E}$ sera un espace $L^p(\Omega, \mathcal{F}, P, E)$ $(0 \leq p \leq \infty)$.

REMARQUE 4 : En général, même si $\mathcal{E}$ est métrisable et complet, il n'y a pas identité entre C'-suite et suite sommable (voir par exemple ROLEWICZ, Linear Metric Spaces, P. 93).

On désigne par $\mathcal{B}(\mathcal{E})$ l'espace vectoriel des séries commutatives bornées d'éléments de $\mathcal{E}$, et par $\mathcal{C}(\mathcal{E})$ l'espace vectoriel des suites sommables d'éléments de $\mathcal{E}$. Naturellement, $\mathcal{C}(\mathcal{E}) \subset \mathcal{B}(\mathcal{E})$. On va maintenant mettre une topologie d'espace vectoriel sur $\mathcal{B}(\mathcal{E})$ et $\mathcal{C}(\mathcal{E})$. Un système fondamental de voisinage de zéro dans $\mathcal{B}(\mathcal{E})$ sera constitué des ensembles

$$\mathcal{V}(V) = \{x = (x_n) \in \mathcal{B}(\mathcal{E}) \quad \sum_{n \in \sigma} x_n \in V \quad \forall\, \sigma \in \Phi(\mathbb{N})\}$$

quand V décrit un système fondamental de voisinages de zéro dans ε.

Il est facile de voir que les $\mathcal{V}(V)$ forment un système fondamental de voisinages de zéro pour une topologie d'espace vectoriel.

$\mathcal{C}(\mathcal{E})$ sera muni de la topologie induite par celle de $\mathcal{B}(\mathcal{E})$. Le cas le plus important, et que seul nous rencontrerons par la suite, est celui où $\mathcal{E}$ est un F-espace. On a alors la

PROPOSITION :

Soit $\mathcal{E}$ un F-espace ; $\mathcal{B}(\mathcal{E})$ est aussi un F-espace, $\mathcal{C}(\mathcal{E})$ est un sous-espace fermé de $\mathcal{B}(\mathcal{E})$ (donc aussi un F-espace).

DEMONSTRATION : Il est facile de voir que, si pour $x = (x_n) \in \mathcal{B}(\mathcal{E})$ l'on pose :

$$||x||_{\mathcal{B}(\mathcal{E})} = \sup_{\sigma\in\Phi(\mathbb{N})} ||\sum_{n\in\sigma} x_n||_{\mathcal{E}}$$

(où $||.||_{\mathcal{E}}$ désigne une F-norme définissant la topologie de $\mathcal{E}$), l'application $x \rightsquigarrow ||x||_{\mathcal{B}(\mathcal{E})}$ est une F-norme définissant la topologie de $\mathcal{B}(\mathcal{E})$. Il reste à démontrer la complétude de $\mathcal{B}(\mathcal{E})$. Soit donc $(x^j)_j$ une suite de Cauchy dans $\mathcal{B}(\ )$: $x^j = (x_n^j)_{n\in\mathbb{N}}$. La topologie de $\mathcal{B}(\mathcal{E})$ étant plus fine que la topologie induite par celle de $\mathcal{E}^{\mathbb{N}}$, pour tout n, la suite $(x_n^j)_{j\in\mathbb{N}}$ converge vers, disons y_n ($\mathcal{E}$ étant complet).

Il reste à montrer que $y = (y_n)_n \in \mathcal{B}(\mathcal{E})$ et que $x^j \to y$ dans $\mathcal{B}(\mathcal{E})$.

La suite (x^j) étant de Cauchy dans $\mathcal{B}(\mathcal{E})$, on en déduit qu'elle satisfait à :

"Pour tout $\varepsilon > 0$, il existe $\delta > 0$ tel que $\sup_j \sup_{\sigma\in\Phi(\mathbb{N})} ||\delta \sum_{n\in\sigma} x_n^j|| \leq \varepsilon$".

Et par conséquent, pour $\sigma \in \Phi(\mathbb{N})$ on a $||\delta \sum_{n\in\sigma} y_n|| \leq \varepsilon$.

Donc $(y_n) \in \mathcal{B}(\mathcal{E})$; on démontre de même que $x^j \to y$ dans $\mathcal{B}(\mathcal{E})$.

Démontrons enfin que $\mathcal{C}(\mathcal{E})$ est fermé dans $\mathcal{B}(\mathcal{E})$. Tout d'abord, les suites "presque nulles" (x_n) sont denses dans $\mathcal{C}(\mathcal{E})$; et inversement toute limite, pour la topologie de $\mathcal{B}(\mathcal{E})$ de suites presque nulles appartient à $\mathcal{C}(\mathcal{E})$. Donc $\mathcal{C}(\mathcal{E})$ est fermé et c'est l'adhérence dans $\mathcal{B}(\mathcal{E})$ de l'ensemble des suites presque nulles.

La proposition est donc complétement démontrée.

<u>REMARQUE</u> : Soit $x \in \mathcal{C}(\mathcal{E})$ et posons $Tx = \sum_n x_n$. On a évidemment

$$||Tx||_{\mathcal{E}} \leq ||x||_{\mathcal{C}(\mathcal{E})} .$$

T est donc une application (évidemment) linéaire et continue de $\mathcal{C}(\mathcal{E})$ dans $\mathcal{E}$.

CHAPITRE II

MESURABILITÉ DES FONCTIONS BANACHIQUES

Dans tout ce chapitre, $(X, \mathcal{F}, \mu)$ désignera un espace mesuré (σ-fini donc, d'après nos conventions). Souvent on le notera (X, μ). E désignera un Banach et $\mathcal{B}_E$ sa tribu borélienne.

Si E est séparable, la tribu borélienne du produit $E \times E$ est identique à la tribu produit $\mathcal{B}_E \otimes \mathcal{B}_E$. Ici, le caractère Banachique n'a rien à voir, on utilise seulement le fait que E est un espace topologique à base dénombrable.

Par contre, si E n'est pas séparable, la tribu borélienne de $E \times E$ est plus grande que $\mathcal{B}_E \otimes \mathcal{B}_E$: par exemple, la diagonale Δ de $E \times E$ est un ensemble fermé, donc borélien, qui n'appartient pas à $\mathcal{B}_E \otimes \mathcal{B}_E$ (en fait, la puissance de E étant strictement supérieure à celle du continu, Δ n'appartient même pas à $\mathcal{P}(E) \otimes \mathcal{P}(E)$.)

De cela, on déduit facilement qu'en général (i.e. si E n'est pas séparable) l'ensemble des applications $(\mathcal{F}, \mathcal{B}_E)$-mesurables de X dans E n'est pas un espace vectoriel.

Toutefois, cet ensemble est stable par passage à la limite des suites, que E soit séparable ou non, comme le montre le

LEMME 1 :

Soit $(X, \mathcal{F})$ un espace mesurable, F un espace métrique et $\mathcal{B}_F$ sa tribu borélienne. Soit (f_n) une suite de fonctions de X dans F qui sont $(\mathcal{F}, \mathcal{B}_F)$-mesurables. On suppose que f_n converge simplement vers f. Alors f est $(\mathcal{F}, \mathcal{B}_F)$-mesurable.

<u>Démonstration</u> : Il suffit de démontrer que, pour tout ouvert U de F, $f^{-1}(U) \in \mathcal{F}$. Remarquons tout d'abord que si U est un ouvert de F, on a à cause de la convergence de (f_n) vers f :

$$f^{-1}(U) \subset \limsup_n f_n^{-1}(U)$$

(On rappelle que si (A_n) est une suite de parties de F, on pose

$$\limsup_n A_n = \bigcap_{n \in N} \bigcup_{k \geq n} A_k \text{).}$$

De même, si G est un fermé de F : $\limsup_n f_n^{-1}(G) \subset f^{-1}(G)$.

Soit alors U un ouvert de F. Pour tout $n \geq 1$, posons :

$$G_n = \{y \in F,\ d(y, U^c) \geq \frac{1}{n}\}$$

$$U_n = \{y \in F,\ d(y, U^c) > \frac{1}{n}\}$$

(où d désigne la distance sur F). Il est clair que G_n est fermé, U_n ouvert et que :

$$U = \bigcup_{n=1}^{\infty} U_n = \bigcup_{n=1}^{\infty} G_n .$$

Maintenant,

$$f^{-1}(U) = \bigcup_n f^{-1}(G_n) \supset \bigcup_n \limsup_k f_k^{-1}(G_n)$$

$$\supset \bigcup_n \limsup_k f_k^{-1}(U_n) \quad \text{d'une part ;}$$

et d'autre part :

$$f^{-1}(U) = \bigcup_n f^{-1}(U_n) \subset \bigcup_n \limsup_k f_k^{-1}(U_n) .$$

En définitive :

$$f^{-1}(U) = \bigcup_n \limsup_k f_k^{-1}(U_n) \text{ ; donc } f^{-1}(U) \in \mathcal{F} .$$

§ 1 : FONCTIONS FORTEMENT MESURABLES

N° 1 : DEFINITIONS ET RESULTATS FONDAMENTAUX

<u>DEFINITION 1</u> : Soit $(X, \mathcal{F}, \mu)$ un espace mesuré et E un Banach et soit $f : X \to E$. On dit que f est fortement μ-mesurable (ou

μ-mesurable) s'il existe une suite de fonctions μ-étagée (f_n) convergeant μ-presque partout vers f.

Rappelons que $q : X \to E$ est μ-étagée si $q = \sum_{i\in I} 1_{A_i} \cdot e_i$ (card. $I < \infty$, $A_i \in \mathcal{F}$ et $\mu(A_i) < \infty$ $\forall i$; $e_i \in E$. On peut supposer les A_i deux à deux disjoints).

Il est clair, d'après la définition, que les fonctions μ-fortement mesurables forment un espace vectoriel ; elles sont $(\mathcal{F}_\mu, \mathcal{B}_E)$-mesurables si $\mathcal{F}_\mu$ désigne la tribu complétée de $\mathcal{F}$ relativement à μ.

Dans la suite, nous dirons simplement "μ-mesurable" au lieu de "μ-fortement mesurable".

PROPOSITION 1 :

Soit $f : X \to E$; les conditions suivantes sont équivalentes :

(1) f est μ-mesurable ;

(2) Il existe $N \in \mathcal{F}$ avec $\mu(N) = 0$ telle que la restriction de f à N^c soit $(\mathcal{F}_{N^c}, \mathcal{B}_E)$-mesurable et que $f(N^c)$ contienne un sous-ensemble dénombrable partout dense.

Démonstration : Avant tout, rappelons que la proposition 1 est vraie pour autant que μ est σ-finie ; autrement dit, il faudrait ajouter à la condition (2) "f est nulle en dehors d'un ensemble σ-fini".

(1) ⟹ (2) : Soit $N \in \mathcal{F}$ tel que $\mu(N) = 0$ et que $\{f_n(x)\}_n$ converge vers $f(x)$ pour tout $x \in N^c$ (ou les f_n sont μ-étagées).

$$f_n = \sum_{j=1}^{k_n} 1_{A_j^n} \, a_n^j \quad .$$

Alors $f(N^c)$ est dans l'adhérence de l'ensemble dénombrable

$$D = \{a_j^n \quad n \in \mathbb{N} \quad 1 \le j \le k_n\} \quad .$$

Comme $\bar{D}$ est métrisable séparable, il en est de même de $f(N^c)$. D'autre part, les restrictions des (f_n) à N^c sont $(\mathcal{F}_{N^c}, \mathcal{B}_E)$-mesurables ; donc la restriction de f à N^c l'est, de par le lemme du début. En définitive, on a démontré que (1) implique (2).

(2) $\Rightarrow$ (1) : Dû au fait que $f : X \to E$ est $(\mathcal{F}, \mathcal{B}_E)$-mesurable si et seulement si sa restriction à chaque A de $\mathcal{F}$ de mesure finie est $(\mathcal{F}_A, \mathcal{B}_E)$-mesurable, et que si $q : X \to E$ est μ-étagée, sa restriction à chaque A de $\mathcal{F}$ est μ_A-étagée (où μ_A désigne la mesure induite par μ sur A), <u>on peut supposer $\mu(X) < \infty$</u> .

Par ailleurs, en considérant le sous-espace fermé engendré par $f(N^c)$ <u>on peut supposer E séparable</u> .

Soit alors (e_n) une suite de points, dense dans E. Si $n \in \mathbb{N}$, désignons par φ_n la fonction $E \to \{e_0, e_1, \ldots, e_n\}$ définie de la façon suivante :

$\varphi_n(e)$ est le premier e_i $(0 \leq i \leq n)$ pour lequel le minimum $\min_{0 \leq j \leq n} ||e - e_j||$ est atteint. En d'autres termes :

$\varphi_n(e) = e_i \quad (0 \leq i \leq n)$ sur l'ensemble

$$\{e \;;\; ||e-e_i|| < ||e-e_j|| \;\; \forall j \in [0, i[\;;\; ||e-e_i|| \leq ||e-e_j|| \; \forall j \in]i, n]\} .$$

Les φ_n sont boréliennes de E dans E et $\varphi_n(e) \to e$ pour tout $e \in E$, à cause de l'hypothèse de densité.

Si maintenant $f : X \to E$ satisfait aux hypothèses de (2), les fonctions $f_n = \varphi_n . f$ sont μ-étagées à valeurs dans E et $f_n \to f$ en tout point de X. Donc (1) est vérifiée.

<u>REMARQUE 1</u> : On peut améliorer la proposition 1 en supposant que :

$$||f_n(x)|| \leq 2||f(x)||, \quad \forall x \in X .$$

En effet, si la suite (e_n) qui intervient dans la démonstration de (2) $\Rightarrow$ (1) est telle que $e_0 = 0$, alors, pour tout e de E,

$||f_n(e)|| \leq 2||e||$; en outre, si $f_n(e) = e_j \quad (0 \leq j \leq n)$,

$||e - e_j|| \leq ||e||$;

Donc $\quad ||\varphi_n \cdot f(x)|| \leq 2||f(x)|| \;, \quad \forall x \in X .$

REMARQUE 2 : Si E est séparable, dire que $f : X \to E$ est μ-mesurable équivaut à dire qu'elle est $(\mathcal{F}_\mu, \mathcal{B}_E)$-mesurable.

Il est clair que l'ensemble des fonctions μ-mesurables est stable par passage à la limite des suites.

N° 2 : LES ESPACES $L^p(X, \mu, E)$ $(0 \leq p \leq \infty)$

DEFINITION 2 : $\mathcal{L}^0(X, \mathcal{F}, \mu, E)$ ou $\mathcal{L}^0(X, \mu, E)$ ou $\mathcal{L}^0(\mu, E)$ désignera l'espace vectoriel des fonctions μ-mesurables de X dans E.

$L^0(X, \mathcal{F}, \mu, E)$ ou $L^0(X, \mu, E)$ ou $L^0(\mu, E)$ désignera son quotient par la relation d'équivalence "f = g μ-presque partout". On notera souvent par la même lettre un élément de $\mathcal{L}^0(\mu, E)$ et son représentant dans $L^0(\mu, E)$.

$L^0(\mu, E)$ sera muni de la topologie de convergence en mesure sur tout ensemble de mesure finie. Un système fondamental de voisinages de zéro pour cette topologie est constitué des

$$V(\varepsilon, \delta, A) \qquad \varepsilon > 0, \delta > 0, \mu(A) < \infty .$$

$$V(\varepsilon, \delta, A) = \{f \in L^0(X,\mu, E) \;;\; \mu\{x : ||f(x)|| > \delta, \; x \in A\} \leq \varepsilon\} .$$

Cette topologie, non localement convexe si μ n'est pas discrète, est définie par la famille de jauges

$$J_{\varepsilon,A}(f) = \text{Inf}\,\{\delta > 0 \;\; \mu\{x \;;\; ||f(x)|| > \delta, \;\; x \in A\} \leq \varepsilon\}$$

$(A \in \mathcal{F} \;;\; \mu(A) < \infty \;;\; \varepsilon > 0)$.

Dû au fait que μ est σ-finie, $L^0(X, \mathcal{F}, \mu)$ est un espace complet métrisable.

Exercice : Démontrer que $M \subset L^0(\mu, E)$ est une partie bornée pour la topologie ci-dessus si et seulement si elle remplit la condition :

Quels que soient $\varepsilon > 0$ et $A \in \mathcal{F}$ avec $\mu(A) < \infty$, il existe un réel $C > 0$ (dépendant de ε et A en général) tel que, pour tout $f \in M$, on ait :

$$\mu\{x \,;\, x \in A \,;\, ||f(x)|| > C\} \leq \varepsilon .$$

Souvent, dans la suite, $(X, \mathcal{F}, \mu)$ sera un espace probabilisé, que l'on notera alors $(\Omega, \mathcal{F}, P)$. Et les fonctions P-mesurables (ou plutôt les P-classes de fonctions mesurables) à valeurs dans E seront appelées variables aléatoires à valeurs dans E. On les notera alors par X, (X_n), etc.

La topologie de $L^0(\Omega, P, E)$ peut être alors définie par la F-norme non homogène :

$$X \rightsquigarrow ||X||_{L^0} = E\left\{ \frac{||X||}{1 + ||X||} \right\}$$

Exercice : Démontrer ce résultat. Démontrer que $M \subset L^0(\Omega, P, E)$ est bornée si et seulement si pour tout $\varepsilon > 0$, il existe $\delta > 0$ tel que

$$||\delta X||_{L^0} \leq \varepsilon \qquad \forall X \in M.$$

On va maintenant définir les espaces $L^p(\mu, E)$ $(0 < p \leq \infty)$.

DEFINITION 3 : Soit $(X, \mathcal{F}, \mu)$ un espace mesuré, $p \in]0, \infty[$ et $f : X \to E$; on dit que f est de puissance p-ième intégrable si

(1) f est μ-mesurable

(2) $\int_X ||f(x)||^p \, d\mu(x) < \infty$.

On désignera par $\mathcal{L}^p(X, \mu, E)$ ou $\mathcal{L}^p(\mu, E)$ l'espace vectoriel des fonctions de puissance p-ième intégrable (encore faut-il avoir démontré que c'est un espace vectoriel, mais c'est classique).

$L^p(X, \mu, E)$ ou $L^p(\mu, E)$ est l'espace vectoriel des μ-classes d'équivalence de fonctions de puissance p-ième intégrable.

Pour $0 < p \leq 1$, $L^p(\mu, E)$ est muni de la p-norme :

$$||f||_{L^p} = \int_X ||f(x)||^p \, d\mu(x) \quad ;$$

sur $p \in [1, \infty[$, il est muni de la norme

$$||f||_{L^p} = \left(\int_X ||f(x)||^p \, d\mu(x)\right)^{\frac{1}{p}}$$

Les classes de fonctions μ-étagées sont denses dans $L^p(X, \mu, E)$ $(0 < p < \infty)$. En effet, si $f \in L^p(X, \mu, E)$, il existe une suite f_n de fonctions μ-étagées convergeant μ-presque partout vers f telles que $||f_n|| \leq 2||f||$.

Alors $||f_n - f|| \to 0$ et $||f_n - f|| \leq 3||f|| \quad \forall n$.
Il reste à appliquer le théorème de Lebesgue dans $L^p(\mu, \mathbb{R})$.

Dans $L^p(\mu, E)$ on a un théorème de convergence dominée :

Si (f_n) est une suite de fonctions de $L^p(\mu, E)$ convergeant μ-presque partout vers f, et si $||f_n|| \leq g$ pour tout n, où g est un élément de $\mathcal{L}^p(\mu, \mathbb{R})$, alors $f \in \mathcal{L}^p(\mu, E)$ et f_n converge vers f pour la topologie de $L^p(\mu, E)$.

On en déduit que $L^p(\mu, E)$ est complet ; donc un Banach si $p \geq 1$.

Enfin, $L^\infty(\mu, E)$ est l'espace des μ-classes de fonctions $X \to E$ μ-mesurables et "essentiellement bornées". Il est muni de la norme $||f||_{L^\infty} = \sup \mathrm{ess}_X ||f(x)||$. C'est un Banach.

REMARQUE : Si $0 \leq p < \infty$, l'espace vectoriel des classes de fonctions μ-étagées est dense dans $L^p(X, \mu, E)$; on l'a vu pour $p > 0$. Pour $p = 0$, cela résulte de la définition d'une fonction μ-mesurable et du fait que la convergence μ-presque partout d'une suite implique la convergence en mesure sur tout ensemble de mesure finie.

L'espace des classes de fonctions μ-étagées n'est pas dense dans $L^\infty(\mu, E)$. Toutefois ce résultat est vrai si $\mu(X) < \infty$ et E est de dimension finie.

Nous allons maintenant définir l'intégrale d'une fonction de $\mathcal{L}^1(\mu, E)$ comme suit :

a) Tout d'abord, si f est μ-étagée, $f = \sum_{i\in I} 1_{A_i} \vec{a}_i$, on pose

$$\int_X f\, d\mu = \sum \mu(A_i)\, \vec{a}_i .$$

Il est clair que $\int_X f\, d\mu$ ne dépend pas de la représentation de f sous la forme $\sum 1_{A_i} \vec{a}_i$ et que si deux fonctions μ-étagées sont égales presque partout elles ont même intégrale. Donc l'intégrale est définie sur les classes de fonctions μ-étagées.

Elle est évidemment linéaire et contractante quand l'espace des classes de fonctions μ-étagées est muni de la norme induite par celle de $L^1(X, \mu, E)$.

b) Si $f \in L^1(X, \mu, E)$, on définit $\int_X f\, d\mu$ par prolongement à $L^1(\mu, E)$ de l'intégrale définie dans a). En effet, l'espace des classes de fonctions μ-étagées est partout dense dans $L^1(X, \mu, E)$ et l'intégrale étant linéaire et continue sur les fonctions μ-étagées, pour la norme N_1, elle admet un prolongement continu à $L^1(X, \mu, E)$.

On désigne encore par $\int_X f\, d\mu$ ou $\int_X f\, d\mu$ l'intégrale de $f \in L^1(X,\mu, E)$.

On voit facilement que :

- $\langle \int_X f\, d\mu, e' \rangle = \int_X \langle f, e' \rangle\, d\mu, \quad \forall e' \in E'$;

- Plus généralement, si f est un Banach et $u : E \to F$ est linéaire et continue, alors :

$$u\left(\int_X f\, d\mu\right) = \int_X u(f)\, d\mu \qquad \forall f \in L^1(X, \mu, E).$$

Indiquons sans démonstration un dernier résultat :

PROPOSITION 2 :

Le dual de $L^1(X, \mu, E)$ est isomorphe (algébriquement) à l'espace $B\left(L^1(\mu), E\right)$ des applications bilinéaires continues sur $L^1(\mu) \times E$.

La démonstration de ce résultat fait intervenir des notions du type "produit tensoriel topologique". La donner introduirait une trop importante digression.

Indiquons toutefois comment se réalise l'isomorphisme ci-dessus :

Si $u \in L^1(X, \mu, E)'$, la forme bilinéaire associée est donnée par :

$$B_u(f, e) = u(f . e) \quad f \in L^1(\mu) \; ; \; e \in E.$$

REMARQUE : L'on a également :

$$L^1(\mu, E)' = L\big(E, L^\infty(\mu)\big) = L\big(L^1(\mu), E'\big) ,$$

En effet $\quad B\big(L^1(\mu), E\big) = L\big(E, L^\infty(\mu)\big) = L\big(L^1(\mu), E'\big)$.

Indiquons brièvement comment se réalisent ces isomorphismes :

Soit $u \in L^1(\mu, E)'$; si $e \in E$, l'application $f \rightsquigarrow u(f_e)$ est une forme linéaire continue sur $L^1(\mu)$; elle est donc définie par un élément de $L^\infty(\mu)$ (μ étant σ-finie !) soit $L_u(e)$; et l'application $e \rightsquigarrow L_u(e)$ est linéaire continue de E dans $L^\infty(\mu)$.

De même si $f \in L^1(\mu)$, $e \rightsquigarrow u(f.e)$ est une forme linéaire continue sur E donc un élément de E'.

N° 3 : ESPERANCES CONDITIONNELLES ET MARTINGALES BANACHIQUES

Soit dans ce numéro $(\Omega, \mathcal{F}, P)$ un espace probabilisé, E un Banach et $X \in L^1(\Omega, P, E)$. Soit $\mathcal{G}$ une sous-tribu de $\mathcal{F}$. On définit l'espérance mathématique de X quand $\mathcal{G}$ de la façon suivante :

a) Si X est étagée : $X = \sum_{i \in I} 1_{A_i} \vec{e}_i$, on pose

$$E^{\mathcal{G}}(X) = \sum_{i \in I} P^{\mathcal{G}}(A_i) . \vec{e}_i$$

où $P^{\mathcal{G}}(A_i)$ désigne la probabilité conditionnelle quand $\mathcal{G}$ de l'évènement A_i.

$E^{\mathcal{G}}(X)$ est évidemment une variable aléatoire $\mathcal{G}$-mesurable (mais pas forcément $\mathcal{G}$-étagée).

Il est facile de voir que $E^{\mathcal{G}}(X)$ ne dépend pas de la représentation de X choisie et que l'application $X \to E^{\mathcal{G}}(X)$ est linéaire. En outre elle est contractante pour la topologie induite par celle de $L^1(\Omega, P, E)$.

b) D'où il résulte que l'application $X \to E^{\mathcal{G}}(X)$ définie dans a) admet un prolongement unique à $L^1(\Omega, P, E)$.

Résumant ce qui précède, on arrive facilement au

LEMME :

Pour tout $X \in L^1(\Omega, \mathcal{F}, P, E)$ il existe un élément unique $E^{\mathcal{G}}(X) \in L^1(\Omega, \mathcal{G}, P, E)$ ayant les propriétés suivantes :

(1) $\int_A E^{\mathcal{G}}(X)\, dP = \int_A X\, dP \quad , \quad \forall A \in \mathcal{G}$;

et plus généralement pour tout $h \in L^\infty(\Omega, \mathcal{G}, P)$:

$$\int_\Omega h E^{\mathcal{G}}(X)\, dP = \int_\Omega h X\, dP \ ;$$

(2) $X \to E^{\mathcal{G}}(X)$ est linéaire ;

(3) $\|E^{\mathcal{G}}(X)\|_{L^1} \leq \|X\|_{L^1}$ et $\|E^{\mathcal{G}}(X)\|_E \leq \|X\|_E$;

(4) Si h est une fonction de $L^\infty(\Omega, \mathcal{G}, P)$, on a $E^{\mathcal{G}}(hX) = hE^{\mathcal{G}}(X)$.

REMARQUE : Si $X \in L^\infty(\Omega, \mathcal{F}, P)$, il est facile de voir que $E^{\mathcal{G}}(X)$ appartient à $L^\infty(\Omega, \mathcal{G}, F)$ et que

$$\|E^{\mathcal{G}}(X)\|_{L^\infty} \leq \|X\|_{L^\infty} .$$

Donc la restriction de $E^{\mathcal{G}}$ à $L^\infty(\Omega, P, E)$ est une application linéaire contractante de $L^\infty(\Omega, \mathcal{F}, P, E)$ dans $L^\infty(\Omega, \mathcal{G}, P, E)$. Plus généralement, on peut voir facilement que $E^{\mathcal{G}}$ induit une application contractante de $L^p(\Omega, \mathcal{F}, P, E)$ dans $L^p(\Omega, \mathcal{G}, P, E)$ si $p \geq 1$. En effet,

$$\|E^{\mathcal{G}}(X)\|^p_{L^p} = \int_\Omega \|E^{\mathcal{G}}(X)\|^p_E\, dP \leq \int_\Omega [E^{\mathcal{G}}(\|X\|)]^p\, dP$$
$$\leq \int_\Omega \|X\|^p\, dP = \|X\|^p_{L^p} .$$

On peut maintenant passer à la définition d'une martingale.

Soit $(\Omega, \mathcal{F}, P)$ un espace probabilisé et $(\mathcal{G}_n)$ une suite croissante de sous-tribus de $\mathcal{F}$.

On note par $\mathcal{G}_\infty = \bigvee_{n \in N} \mathcal{G}_n$ la tribu engendrée par les $\mathcal{G}_n$. Soit E un Banach et (X_n) une suite de variables aléatoires sur Ω à valeurs dans E.

<u>DEFINITION</u> : Avec les hypothèses ci-dessus, l'on dit que la suite (X_n) est une martingale relativement aux $(\mathcal{G}_n)$ si

(1) $\forall n,\ X_n \in L^1(\Omega, \mathcal{G}_n, P, E)$;

(2) $\forall n,\ X_n = E^{\mathcal{G}_n}(X_{n+1})$.

Nous allons donner trois exemples de martingales que nous retrouverons par la suite.

<u>EXEMPLE 1</u> : <u>Martingale de Paley-Walsh associée à un arbre</u>

Nous allons auparavant définir la notion d'<u>arbre</u> dans un espace de Banach (en fait, on pourrait définir cette notion dans un espace vectoriel quelconque, mais cela ne présente pas d'intérêt !)

Soit $\mathbb{Z}_2$ le groupe multiplicatif $\{-1, +1\}$ et soit l'ensemble suivant :

$$\mathcal{Z} = \mathbb{Z}_2 \cup \mathbb{Z}_2^{(2)} \cup \mathbb{Z}_2^{(3)} \cup \ldots$$

$\mathcal{Z}$ est la somme directe des puissances finies de $\mathbb{Z}_2$.

Un élément de $\mathcal{Z}$ s'écrit sous la forme $(\varepsilon_1\ \varepsilon_2\ \ldots\ \varepsilon_k)$ pour un certain $k \geq 1$ avec $\varepsilon_j = \pm 1$ pour tout $j = 1, 2, \ldots, k$.

Maintenant, soit E un Banach et x_o un point de E ; on appelle <u>arbre</u> d'origine x_o la donnée de ce point x_o et d'une application de $\mathcal{Z}$ dans E telle que :

- $x_o = \dfrac{x_1 + x_{-1}}{2}$;

- Pour tout $k \geq 1$ entier,

$$x_{\varepsilon_1 \ldots \varepsilon_k} = \frac{1}{2}\left[x_{\varepsilon_1 \ldots \varepsilon_k, 1} + x_{\varepsilon_1 \ldots \varepsilon_k, -1}\right]$$

L'arbre est dit <u>fini de longueur n</u> si, pour tout $m > n$, on a

$$x_{\varepsilon_1 \ldots \varepsilon_n\, \varepsilon_{n+1} \ldots \varepsilon_m} = x_{\varepsilon_1 \ldots \varepsilon_n} .$$

Etant donné un arbre dans E, il est facile de lui associer une martingale à valeurs dans E comme suit :

Soit $\Omega = \mathbb{Z}_2^{\mathbb{N}_+}$ (c'est le produit dénombrable d'une famille dénombrable de groupes égaux à $\mathbb{Z}_2$). Ω est un groupe compact dont la mesure de Haar normalisée est :

$$P = \bigotimes_{i=1}^{\infty} \left(\frac{1}{2}\,\delta_{(1)} + \frac{1}{2}\,\delta_{(-1)}\right) .$$

Soit $\mathcal{G}_0 = \{\emptyset, \Omega\}$ et soit $\mathcal{G}_n$ $(n \geq 1)$ la tribu sur Ω engendrée par les n premières applications coordonnées ; $\mathcal{G}_n$ est engendrée par le partition formée des ensembles :

$$A_{\varepsilon_1^o \ldots \varepsilon_n^o} = \{ \omega = (\varepsilon_i)_{1 \leq i \leq \infty} \in \Omega \; ; \; \varepsilon_i = \varepsilon_i^o \quad i = 1, 2, \ldots, n\}$$

où $(\varepsilon_i^o)_{1 \leq i \leq n}$ est un élément de $\mathbb{Z}_2^{(n)}$.

Supposons donc donné un arbre d'origine x_o ; définissons une suite $(X_n)_{n \geq 0}$ de variables aléatoires sur Ω comme suit :

$$- \; X_o(\omega) = x_o \qquad \forall \; \omega \in \Omega$$

$$- \; X_n(\omega) = x_{\varepsilon_1^o \ldots \varepsilon_n^o} \qquad \text{si } n \geq 1 \text{ et si } \omega \in A_{\varepsilon_1^o \ldots \varepsilon_n^o}$$

Pour tout n, (X_n) est évidemment $\mathcal{G}_n$-mesurable (et même $\mathcal{G}_n$-étagée). D'autre part, en vertu de

$$A_{\varepsilon_1^o \ldots \varepsilon_n^o} = A_{\varepsilon_1^o \ldots \varepsilon_n^o, 1} \cup A_{\varepsilon_1^o \ldots \varepsilon_n^o, -1} \; ,$$

et de

$$P(A_{\varepsilon_1^o \ldots \varepsilon_n^o, 1}) = P(A_{\varepsilon_1^o \ldots \varepsilon_n^o, -1}),$$

on déduit

$$\int_{A_{\varepsilon_1^o \ldots \varepsilon_n^o}} X_{n+1} \, dP = \frac{1}{2} P(A_{\varepsilon_1^o \ldots \varepsilon_n^o}) \left(x_{\varepsilon_1^o \ldots \varepsilon_n^o, 1} + x_{\varepsilon_1^o \ldots \varepsilon_n^o, -1}\right)$$

$$= \int_{A_{\varepsilon_1^o \ldots \varepsilon_n^o}} X_n \, dP \; ;$$

Cela signifie que (X_n) est une martingale relativement aux $\mathcal{G}_n$.

<u>EXEMPLE 2</u> (Généralisation du précédent) :

Soit $\mathcal{N}$ l'ensemble somme direct de $N^1 \cup N^2 \cup N^3 \cup \ldots$

Un élément de $\mathcal{N}$ est de la forme $n_1 \, n_2 \ldots n_k$ pour un certain k (avec $n_i \in N$, $1 \leq i \leq k$).

Soit E un Banach et $x_o \in E$; on considère une application de $\mathcal{N}$ dans une partie bornée de E : $n_1 \ldots n_k \rightsquigarrow x_{n_1 \ldots n_k}$ et une application de $\mathcal{N}$ dans [0, 1] : $n_1 \ldots n_k \rightsquigarrow \alpha_{n_1 \ldots n_k}$ telles que :

- Pour tout $k \geq 1$, et tous $n_1, n_2, \ldots, n_{k-1} \in \mathcal{K}$,

$$\sum_{n_k \in \mathbb{N}} \alpha_{n_1 \ldots n_{k-1}\, n_k} = 1 ;$$

$$- x_o = \sum_{n_k \in \mathbb{N}} \alpha_{n_1} x_{n_1}$$

- Pour tout $k \geq 1$, $x_{n_1 \ldots n_k} = \sum_{n_{k+1} \in \mathbb{N}} \alpha_{n_1 \ldots n_k\, n_{k+1}} x_{n_1 \ldots n_k\, n_{k+1}}$

Soit $\Omega = \mathbb{N}^\infty$ et $\mathcal{F}$ la tribu produit des $\mathcal{P}(\mathbb{N})$ sur Ω. Soit pour $k \geq 1$ $\mathcal{F}_k$ la tribu engendrée par les k premières applications coordonnées ; et soit $\mathcal{F}_o = \{\emptyset, \Omega\}$.

$\mathcal{F}_k$ est engendrée par la partition dénombrable formée des ensembles :

$$A_{n_1^o \ldots n_k^o} = \{ \omega = (n_i)_{1 \leq i \leq \infty} \quad n_i = n_i^o \quad i = 1, 2, \ldots, k\}$$

où $\quad (n_i^o)_{1 \leq i \leq k} \in \mathbb{N}^k$

Définissons la suite de variables aléatoires sur $(\Omega, \mathcal{F})$ de la façon suivante :

- $X_o(\omega) = x_o \qquad \forall\, \omega \in \Omega$

- $X_k(\omega) = x_{n_1^o \ldots n_k^o}$ si $k \geq 1$ et si $\omega \in A_{n_1^o \ldots n_k^o}$

(X_k) est évidemment $\mathcal{G}_k$-mesurable pour tout $k \geq 0$. Il reste maintenant à définir une probabilité P sur $(\Omega, \mathcal{F})$ par rapport à laquelle les $(X_k, \mathcal{F}_k)$ forment une martingale.

Puisque $(\Omega, \mathcal{F})$ est limite projective des espaces $(\mathbb{N}^k, \mathcal{F}^k)$ $k \geq 1$, il suffit de définir sur chaque $\mathbb{N}^k$ une probabilité P_k, avec les conditions de cohérence, pour définir une probabilité sur $(\Omega, \mathcal{F})$. Définissons par récurrence :

$$- P_1 = \sum_{n_1} \alpha_{n_1} \delta_{(n_1)}$$

$$- P_{k+1}(n_1, n_2, \ldots n_k, n_{k+1}) = P_k(n_1, n_2, \ldots n_k)\, \alpha_{n_1 \ldots n_k\, n_{k+1}}$$

$(k \geq 1)$.

Il est clair que les (P_k) forment un système cohérent de probabilités sur les $\mathbb{N}^k$.

Si Π_k désigne l'application canonique de Ω sur $\mathbb{N}^k$, il est clair que

$$A_{n_1^o \ldots n_k^o} = \Pi_k^{-1}\{(n_1^o \ldots n_k^o)\} .$$

Donc $$P\{A_{n_1\ldots n_k n_{k+1}}\} = P\{A_{n_1\ldots n_k}\} \times \alpha_{n_1\ldots n_k n_{k+1}}$$

Dû au fait que les $\{x_{n_1\ldots n_k}\ ;\ k \geq 1\}$ sont dans une partie bornée, les (X_k) sont P-intégrables.

Il reste à vérifier que les $(X_k, \mathcal{F}_k)$ forment une martingale relativement à P. Mais c'est immédiat, car si $k \geq 1$:

$$\int_{A_{n_1\ldots n_k}} X_{k+1}\, dP = \sum_{n_{k+1}} \int_{A_{n_1\ldots n_k n_{k+1}}} X_{k+1}\, dP$$

$$= \sum_{n_{k+1}} x_{n_1\ldots n_k n_{k+1}}\, P(A_{n_1\ldots n_k n_{k+1}})$$

$$= \sum_{n_{k+1}} P(A_{n_1\ldots n_k})\, \alpha_{n_1\ldots n_k n_{k+1}}\, x_{n_1\ldots n_k n_{k+1}}$$

$$= P(A_{n_1\ldots n_k}) \sum_{n_{k+1}} \alpha_{n_1\ldots n_k n_{k+1}}\, x_{n_1\ldots n_k n_{k+1}}$$

$$= P(A_{n_1\ldots n_k})\, x_{n_1\ldots n_k} = \int_{A_{n_1\ldots n_k}} X_k\, dP .$$

EXEMPLE 3 : Martingale construite à partir d'un "système de Haar"

Soit $(\Omega, \mathcal{F}, P)$ un espace probabilisé et (X_n) une suite de v.a. réelles sur cet espace, ayant une espérance.

On suppose qu'il existe une suite croissante $(\mathcal{G}_n)$ de sous-tribus de $\mathcal{F}$ telles que :

- X_m est $\mathcal{G}_n$-mesurable si $m \leq n$
- $E^{\mathcal{G}_n}(X_{n+1}) = 0 \quad \forall n.$

Le système de Haar sur $([0, 1], dx)$ a ces propriétés.

Soit (x_n) une suite d'éléments du Banach E.

Posons $$Y_n = \sum_{m=0}^{n} X_m x_m \qquad (n \in \mathbb{N})$$

Il est clair que $(X_n, \mathcal{G}_n)$ est une martingale, car

$$E^{\mathcal{G}_n}(Y_{n+1} - Y_n) = E^{\mathcal{G}_n}(X_{n+1} x_{n+1}) = x_{n+1} E^{\mathcal{G}_n}(X_{n+1}) = 0.$$

§ 2 : FONCTIONS SCALAIREMENT MESURABLES

Il y a une définition naturelle de la notion de fonction scalairement mesurable. Mais elle est trop générale et assez peu restrictive. Les définitions essentielles se trouveront donc aux numéros 2 et 3 de ce paragraphe. Le n° 1 sera donc surtout un "faire-valoir" des suivants, et c'est à ce titre qu'il peut avoir une utilité.

N° 1 : GENERALITES SUR LES FONCTIONS SCALAIREMENT MESURABLES

Soit (E, F) un couple d'espaces normés en dualité. Les principaux exemples que l'on rencontrera dans la suite sont :

- E est un Banach, F = E' est son dual fort ;
- $E = F'_1$ est le dual d'un Banach fort ; $F = F_1$.

DEFINITION 4 : Soient $(X, \mathcal{F}, \mu)$ un espace mesuré, (E, F) un couple d'espaces normés en dualité, et $\varphi : X \to E$. On dit que φ est F-scalairement mesurable si pour tout $f \in F$ la fonction scalaire $\langle\varphi(\cdot), f\rangle$ est μ-mesurable.

De même on dit provisoirement que φ est F-scalairement dans $\mathcal{L}^p(\mu)$ (ou F-scalairement de puissance p-ième intégrable si $0 < p < \infty$) si pour tout $f \in F$, $\langle\varphi(\cdot), f\rangle$ appartient à $\mathcal{L}^p(\mu)$.

Dans le cas où $(X, \mathcal{F}, \mu) = \left(\mathbb{N}, \mathcal{P}(\mathbb{N}), \sum_n \delta_n\right)$ et où F = E', une fonction F-scalairement dans $L^1(\mu)$ est une suite scalairement sommable.

Naturellement, la notion de F-scalaire mesurabilité a un sens même si E et F ne sont pas normés.

Si F = E' (E Banach) on dit simplement scalairement mesurable au lieu de E'-scalairement mesurable.

Si E est un Banach, une fonction $(\mathcal{F}, \mathcal{B}_E)$-mesurable, ou bien $(\mathcal{F}_\mu, \mathcal{B}_E)$-mesurable est scalairement mesurable. Une réciproque partielle de ce résultat

PROPOSITION 3 :

Soit E un Banach séparable , $(X, \mathcal{F}, \mu)$ un espace mesuré et $\varphi : X \to E$.

L'on a les équivalences :

- (1) φ est μ-Bochner mesurable ;
- (2) φ est mesurable relativement à $\mathcal{F}_\mu$ et $\mathcal{B}_E$;
- (3) φ est scalairement μ-mesurable .

DEMONSTRATION :

(1) $\Leftrightarrow$ (2) résulte immédiatement de la proposition 1 compte tenu du fait que E est séparable.

Il est clair par ailleurs que (2) $\Rightarrow$ (3). Il reste donc à démontrer que (3) $\Rightarrow$ (2). Soit donc φ satisfaisant à (3). Compte tenu du fait que, E étant séparable et métrisable, la tribu borélienne est engendrée par les boules, il nous suffit de démontrer que l'image réciproque d'une boule de E appartient à $\mathcal{F}_\mu$.

Or il existe dans la boule unité de E' un sous-ensemble dénombrable faiblement partout dense, soit D'. Soit $a > 0$ et $e \in E$, alors :

$$\{x \in X \; ; \; ||\varphi(x) - e|| \leq a\} = \{x \in X \; |<\varphi(x) - e, f>| \leq a \quad \forall f \in D'\}$$
$$= \bigcap_{f \in D'} \{x \; ; \; |<\varphi(x) - e, f>| \leq a\} \in \mathcal{F}_\mu .$$

Et la proposition est démontrée.

REMARQUE : Le résultat ci-dessus est faux en général comme le montre l'exemple suivant : X = [0, 1] muni de la mesure de Lebesgue ; E est un Hilbert (évidemment non-séparable) ayant une base orthonormée $(e_x)_{x \in X}$ équipotente à X. Soit φ l'application $x \rightsquigarrow e_x$, alors :

- φ est scalairement mesurable, car pour tout $y \in H$, $(\varphi(\cdot)/y)_H$ est nulle en dehors d'un ensemble au plus dénombrable de points ;

- φ n'est pas fortement mesurable, car elle ne prend pas p.p ses valeurs dans un sous-espace séparable.

<u>DEFINITION 5</u> : Soit $(X, \mathcal{F}, \mu)$ un espace mesuré et soit (E, F) un couple d'espaces normés en dualité. Soient φ_1 et φ_2 deux fonctions de X dans E ; elles sont dites F-scalairement égales μ-presque partout si pour tout $f \in F$ les fonctions scalaires $<\varphi_i(\cdot), f>$ $(i = 1, 2)$ sont égales μ-presque partout.

La définition ci-dessus ne suppose aucune mesurabilité (même scalaire pour les φ_i). Toutefois, si l'une d'elles est scalairement μ-mesurable, il en est de même pour l'autre.

Si φ_1 et φ_2 sont égales μ-presque partout, elles le sont F-scalairement. Toutefois la réciproque est fausse comme le montre l'exemple déjà donné ci-dessus. En effet, φ est nulle scalairement presque partout, par contre φ n'est jamais nulle.

Soient φ_1 et φ_2 F-scalairement égales presque partout et F-scalairement mesurables ; on a l'égalité :

$$\bigvee_{||f||\leq 1} < \overset{\circ}{\widehat{\varphi_1, f}} > = \bigvee_{||f||\leq 1} < \overset{\circ}{\widehat{\varphi_2, f}} >$$

(voir ces notations au chapitre zéro). Ici $< \overset{\circ}{\widehat{\varphi_i, f}}>$ représente la classe de μ-équivalence de la fonction $< \varphi_i, f >$ $(i = 1, 2)$.

Par contre, nous l'avons vu, on n'a pas forcément

$$\sup_{||f||\leq 1} |< \varphi_1, f >| = \sup_{||f||\leq 1} |< \varphi_2, f >| \qquad \mu\text{-presque partout.}$$

Remarquons enfin que si $\varphi : X \to E$ est F-scalairement dans $L^1(\mu)$, on peut définir une intégrale "à la PETTIS" par la formule suivante :

$$\forall f \in F, < \int \varphi(x)\, d\mu(x), f > = \int < \varphi(x), f > d\mu(x).$$

Par définition même, $\int \varphi(x)\, d\mu(x)$ est un élément du dual algébrique F^* de F et on ne peut affirmer que $\int \varphi(x)\, d\mu(x) \in E$.

Bien sûr, si $\varphi \in L^1(X, \mu, E)$, son intégrale de PETTIS est égale à son intégrale de BOCHNER et appartient à E.

Enfin si φ_1 et φ_2 scalairement intégrables sont égales scalairement presque partout, elles ont même intégrale de PETTIS.

Comme on le voit, tout ce qui précède est très facile, mais trop général pour être vraiment utile. On aura besoin de restreindre la classe des fonctions scalairement de puissance p-ième intégrable. Faisons au préalable une remarque : une fonction F-scalairement dans $\mathcal{L}^p(\mu)$, soit φ, définit une application linéaire, soit u_φ de F dans $L^p(\mu)$ par la formule :

$$f \rightsquigarrow u_\varphi(f) = \widehat{\langle \varphi, f \rangle}^{\circ}$$

Réciproquement, étant donnée une application linéaire de F dans $L^p(\mu)$, en général elle n'est pas associée à une fonction scalairement dans $\mathcal{L}^p$ (voir le chapitre zéro).

Cela étant, on peut poser la :

NOTATION ET DEFINITION : Soient $(X, \mathcal{F}, \mu)$, (E, F) comme dans la définition 4 et soit $p \in [0, \infty]$. On désigne par $\mathcal{L}^p_{s,F}(X, \mu, E)$ l'espace vectoriel des $\varphi : X \to E$ F-scalairement dans $\mathcal{L}^p$ et telles que l'application linéaire correspondante u_φ soit p-latticiellement bornée (voir chapitre zéro).

Il est clair que si $\varphi_1 \in \mathcal{L}^p_{s,F}(X, \mu, E)$ et si $\varphi_2 = \varphi_1$ F-scalairement presque partout, $\varphi_2 \in \mathcal{L}^p_{s,F}(X, \mu, E)$.

Si $\varphi \in \mathcal{L}^p_{s,F}(X, \mu, E)$, on dira simplement que φ est scalairement dans $\mathcal{L}^p$ (ce qui est un abus de langage par rapport à la définition 4).

On désignera par $L^p_{s,F}(X, \mu, E)$ ou $L^p_{s,F}(\mu, E)$ l'espace vectoriel des classes d'équivalence d'éléments de $\mathcal{L}^p_{s,F}(\mu, E)$ par la relation "$\varphi_1 = \varphi_2$ F-scalairement presque partout".

Et bien sûr, tant qu'aucune confusion ne sera possible, on identifiera un élément de $\mathcal{L}^p_{s,F}(\mu, E)$ avec sa classe d'équivalence ; les principaux cas que nous rencontrerons dans la suite sont :

- F = E' (avec E Banach) : un élément de $\mathcal{L}^p_{s,E'}(\mu, E)$ est dit scalairement dans $\mathcal{L}^p$ et on notera simplement $\mathcal{L}^p_s(\mu, E)$ et $L^p_s(\mu, E)$ au lieu de $\mathcal{L}^p_{s,E'}(\mu, E)$ et $L^p_{s,E'}(\mu, E)$ respectivement.

- $E = F'$ (avec F Banach) ; donc E est un dual. Un élément de $\mathcal{L}^p_{s,F}(\mu, F')$ sera dit $*$-scalairement dans $\mathcal{L}^p$ et on écrira $\mathcal{L}^p_*(\mu, F')$ et $L^p_*(\mu, F')$ les espaces correspondants aux espaces définis plus haut.

REMARQUE 1 : Si $p = 0$, les notions de fonctions F-scalairement mesurables au sens des définitions 4 et 5 coïncident, car il est clair que :

$$\Phi = \bigvee_{||f|| \leq 1} |\widehat{< \varphi, f >}| \qquad \text{est p.s finie.}$$

(en fait, $\sup_{||f|| \leq 1} |< \varphi(\cdot), f >| = ||\varphi(\cdot)||$ majore presque sûrement Φ).

REMARQUE 2 : Quand $p = \infty$, dire que φ est scalairement dans L^∞, revient à dire que l'application correspondante u_φ envoie continuement F dans $L^\infty(X, \mu)$.

Si $\varphi \in L^p_{s,F}(\mu, E)$, on notera par Φ la variable aléatoire

$$\Phi = \bigvee_{||f|| \leq 1} |\widehat{< \varphi, f >}| \quad .$$

On mettra sur les $L^p_{s,F}(\mu, E)$ les topologies définies par les F-normes suivantes :

(1) $\quad \varphi \rightsquigarrow (\int |\Phi|^p d\mu)^{\frac{1}{p}}$ $\qquad$ si $1 \leq p < \infty$

(2) $\quad \varphi \rightsquigarrow \int \Phi^p \, d\mu$ $\qquad$ si $0 < p \leq 1$

(3) $\quad \varphi \rightsquigarrow \sup \text{ess } \Phi = ||\Phi||_{L^\infty}$ $\qquad$ si $p = \infty$

(4) $\quad$ Si $p = 0$, la topologie de $L^0_{s,F}(\mu, E)$ sera définie par la famille de jauges :

$$\varphi \rightsquigarrow J_{\alpha,A}(\varphi) = \inf \{M \; ; \; \mu\{x \in A \; ; \; \Phi > M\} \leq \alpha\}$$

avec $0 \leq \alpha \leq \infty$, $A \in \mathcal{F}$ et $\mu(A) < \infty$.

Le résultat suivant, bien que simple, sera fondamental par la suite :

PROPOSITION 5 :

Soient E et F deux espaces normés en dualité ; on suppose en outre F Banach. Soit $\varphi \in \mathcal{L}^0_{s,F}(E)$ et $\psi \in \mathcal{L}^0(X, \mu, F)$. Alors la fonction $x \rightsquigarrow < \varphi(x), \psi(x) >$ est μ-mesurable.

$\widehat{< \varphi, \psi >}$ ne dépend que des classes de φ dans $\mathcal{L}^0_{s,F}(E)$ et de ψ dans $\mathcal{L}^0(F)$ respectivement.

En outre si $\varphi \in \mathcal{L}^p_{s,F}(E)$ et $\psi \in \mathcal{L}^{p'}_{s,F}(\mu, F)$ $(\frac{1}{p} + \frac{1}{p'} = 1, p \geq 1)$, l'inégalité d'Hölder est vérifiée.

DEMONSTRATION : La première partie de la proposition est triviale si ψ est μ-étagée. Le cas général s'en déduit immédiatement en approchant ψ par des ψ_n μ-étagées.

Maintenant, si ψ est nulle presque partout, $< \varphi, \psi >$ est nulle presque partout. Donc $\widehat{< \overset{\circ}{\varphi}, \psi >}$ ne dépend que de la classe de ψ dans $L^0(\mu, F)$. Mais si φ est nulle scalairement presque partout et ψ est μ-étagée, il est trivial que $< \varphi, \psi >$ est nulle presque partout. Le cas général s'en déduit immédiatement. Finalement, $\widehat{< \overset{\circ}{\varphi}, \psi >}$ ne dépend que des classes de φ et ψ.

Il reste à démontrer l'inégalité de type Hölder. Soit Φ associée à φ comme il a été indiqué plus haut. Dû au fait que ψ prend presque sûrement ses valeurs dans un sous-espace séparable G de F, on a :

$$|< \varphi, \psi >| \leq \sup_{\substack{||f|| \leq 1 \\ f \in G}} |< \varphi, f >| \ . \ ||\psi|| \quad \text{p.s.}$$

Mais dans la formule ci-dessus on peut prendre f dans un sous-ensemble dénombrable D, partout dense dans la boule unité de G. Donc, avec un abus d'écriture (que nous permettrons souvent) :

$$\sup_{\substack{||f|| \leq 1 \\ f \in G}} |< \varphi, f >| = \sup_{\substack{||f|| \leq 1 \\ f \in D}} |< \varphi, f >| = \bigvee_{f \in D} |< \varphi, f >|$$

$$\leq \bigvee_{\substack{f \in F \\ ||f|| \leq 1}} |< \varphi, f >| = \Phi \ .$$

Donc $|< \varphi, \psi >| \leq \Phi \ . \ ||\psi||$ et la relation d'Hölder usuelle donne le résultat.

N° 2 : FONCTIONS SCALAIREMENT DANS L^p ET A VALEURS DANS UN DUAL DE BANACH

Soit E un Banach et soit $\varphi \in L^p_*(X, \mu, E')$; on a vu (et c'est très général), qu'il lui correspond une fonction linéaire $u_\varphi : E \to L^p(\mu)$. (Jusqu'à présent, on aurait pu remplacer E' par un espace normé en dualité avec E).

La réciproque est vraie. Plus précisément :

THEOREME :

Soit $v : E \to L^p(X, \mu)$ p-latticiellement bornée ; il existe $\varphi \in L^p_*(X, \mu, E')$ (unique à une scalaire μ-équivalence près) telle que :

$$v(e) = \overset{\circ}{\widehat{<\varphi(\cdot), e>}} \quad , \quad \forall e \in E.$$

Ceci en effet n'est autre que le théorème de IONESCU-TULCEA rappelé au chapitre zéro : il suffit de prendre pour φ la décomposition donnée par la décomposition de MAHARAM.

On va maintenant se poser le problème suivant : Soit $\varphi \in \mathcal{L}^p_*(X, \mu, E')$ et u_φ correspondante. Soit φ_M la décomposition de MAHARAM de u_φ (on ne peut affirmer que $\varphi = \varphi_M$ presque partout, mais seulement que $\varphi = \varphi_M$ scalairement presque partout). Comment caractériser la décomposition de MAHARAM ?

Tout d'abord, soit $v : E \to L^0(X, \mu)$ linéaire et soit $\Phi = \bigvee_{||e|| \le 1} |v(e)|$; Supposons que v soit décomposable par $\varphi : X \to E'$. Alors il est clair que

$$||\varphi|| = \sup_{||e|| \le 1} <\varphi, e> \ge \Phi .$$

Ceci est vrai pour toute décomposition de v.

Le résultat suivant va donner une propriété de minimalité de la décomposition de MAHARAM.

PROPOSITION 6 :

Soit $\varphi \in L^p_*(X, \mu, E')$ $(0 \le p \le \infty)$ et soit φ_M une décomposition de MAHARAM de la fonction aléatoire linéaire u_φ associée à φ. Alors :

$$||\varphi_M|| = \Phi \quad \mu\text{-presque partout.} \text{ (Il y a ici abus d'écriture).}$$

En particulier, $||\varphi_M||$ est μ-mesurable.

DEMONSTRATION : L'on sait déjà que $||\varphi_M|| \ge \Phi$ μ-presque partout. D'autre part, u_φ admet la factorisation suivante :

$$E \xrightarrow{\;\frac{u_\varphi}{\Phi}\;} L^\infty(X, \mu) \xrightarrow{\;M_\Phi\;} L^0(X, \mu)$$

(avec les notations du chapitre zéro). D'autre part, la norme de l'opérateur $\frac{u_\varphi}{\Phi}$ est ≤ 1.

Soit ψ_M le relèvement de MAHARAM de $\frac{u_\varphi}{\Phi}$. Puisque

$$||\psi_M(x)||_{E'} \le ||\frac{u_\varphi}{\Phi}|| \le 1 \quad \text{et que} \quad \varphi_M = \Phi \, . \, \psi_M \text{ , alors on déduit}$$

que
$$||\varphi_M|| = \Phi \, ||\psi_M|| \le \Phi \, .$$

D'où le résultat.

REMARQUE : Pour la décomposition de MAHARAM, on a :

$$M_p \, (||\varphi_M||) = M_p \, (\Phi).$$

Cela permet de donner une autre définition de $L^p_*(X, \mu, E')$.

Supposons d'abord $p > 0$ et soit $\varphi \in L^p_*(X, \mu, E')$. On ne peut affirmer que $M_p(\varphi) < \infty$, mais on sait qu'il existe dans la classe $\mathring{\varphi}$ de scalaire-équivalence de φ une fonction φ_1 telle que $M_p(\varphi_1) < \infty$. Par conséquent :

$$||\mathring{\varphi}||_{L^p_*(X,\mu,E')} = \inf \{M_p(\varphi_1), \ \varphi_1 \in \mathcal{L}^p_*(X, \mu, E') \ \ \varphi_1 \in \mathring{\varphi}\}$$

si $0 < p \le 1$; on a des définitions analogues pour $p = 0$.

REMARQUE 2 : Supposons $\varphi \in L^0(X, \mu, E')$ (autrement dit, φ est μ-Bochner-mesurable à valeurs dans le Banach E') ; elle prend presque sûrement ses valeurs dans un sous-espace séparable de E', soit G.

Soit $D' \subset G$ dénombrable partout dense dans G. Pour tout $f \in D'$ il existe D_f dénombrable dans la boule unité de E telle que

$$||f|| = \sup_{e \in D_f} |\langle f, e \rangle| .$$

Alors $D = \bigcup_{f \in D'} D_p$ est dénombrable. En outre, pour toute $f \in D'$, on a

$$||f|| = \sup_{e \in D} |\langle f, e \rangle| ;$$

on a donc la même relation pour toute f dans G. En conséquence :

$$||\varphi|| = \sup_{e \in D} |\langle \varphi, e \rangle| = \bigvee_{e \in D} |\langle \varphi, e \rangle| \leq \bigvee_{\substack{e \in E \\ |e| \leq 1}} |\langle \varphi, e \rangle| = \Phi$$

(toujours avec un abus d'écriture).

Donc $||\varphi|| = ||\Phi||$ presque sûrement.

En définitive, si dans la classe d'un élément de $L_*^o(X, \mu, E')$ il existe une fonction μ-mesurable, elle réalise la norme minimum et deux fonctions μ-mesurables de la même classe de scalaire équivalence sont presque partout égales. On peut donc écrire :

$$L^p(X, \mu, E') \subset L_*^p(X, \mu, E').$$

L'introduction des espaces $L_*^p(X, \mu, E')$ (réalisée par L. SCHWARTZ, à notre connaissance) permettra de simplifier l'exposition de la propriété de RADON-NIKODYM. Nous le verrons plus tard.

Mais avant, nous allons voir que $L_*^q(X, \mu, E')$ est le dual de $L^p(X, \mu, E)$ $(\frac{1}{p} + \frac{1}{q} = 1,\ p \geq 1)$, ce qui sera une raison supplémentaire d'avoir introduit les L_*^p.

§ 3 : DUALITE ENTRE ESPACES DE FONCTIONS MESURABLES ET ESPACES DE FONCTIONS SCALAIREMENT MESURABLES

Soient E un Banach, E' son dual, $(X, \mathfrak{F}, \mu)$ un espace mesuré (donc σ-fini d'après nos conventions).

Soient $p \in [1, \infty[$ et q sa quantité conjuguée. Nous allons démontrer que le dual de $L^p(E)$ est identique à $L^q_*(E')$. Faisons les remarques préalables suivantes :

(1) Soit $\psi \in \mathcal{L}^p(X, \mu, E)$ et $\varphi \in \mathcal{L}^q_*(X, \mu, E')$

Nous avons vu que la fonction à valeurs scalaires $x \rightsquigarrow \langle \varphi(x), \psi(x) \rangle$ est μ-intégrable, que sa classe ne dépend que des classes de ψ et φ et que :

$$\left|\int \langle \varphi, \psi \rangle \, d\mu\right| \leq ||\psi||_{L^p(E)} \quad ||\varphi||_{L_*(E')}$$

Cela signifie qu'il existe une application (évidemment linéaire) de $L^q_*(E')$ dans $L^p(E)'$. Cette application est évidemment contractante (et on verra plus loin que c'est une isométrie surjective).

(2) Soit $u \in L^p(E)'$; elle définit une application linéaire L_u de E dans $L^q(\mu)$.

En effet, si $\vec{e} \in E$ et $f \in L^p(\mu)$, $f\vec{e} \in \mathcal{L}^p(X, \mu, E)$ et $u(f.\vec{e})$ a alors un sens. En outre, l'application $f \rightsquigarrow f\vec{e}$ est continue (et linéaire !).

$f \rightsquigarrow u(f\vec{e})$ est alors une forme linéaire sur L^p (t) donc définissable au moyen d'un élément de $L^q(\mu)$, que l'on désignera par $L_u(\vec{e})$.

$$\langle L_u(\vec{e}), f \rangle = u(f\,\vec{e}) \qquad \forall\ f \in L^p(\mu).$$

L'application $L_u : E \to L^q(\mu)$ est linéaire et continue et sa norme satisfait à :

$$||L_u||_{\mathcal{L}(E,\ L^q(\mu))} \leq ||u||_{L^p(E)'}$$

Donc on a une application linéaire contractante de $L^p(E)'$ dans $\mathcal{L}(E, L^q(\mu))$.

Si $\varphi \in L^q(E')$, on désignera u_φ l'élément de $L^p(E)'$ correspondant et L_{u_φ} l'élément de $\mathcal{L}(E, L^q(\mu))$ correspondant à φ.

(3) Rappelons enfin que l'injection naturelle de $L^q(E')$ dans $L^q_*(E')$ est une isométrie.

Démontrons comme première étape le

LEMME :

Soit E un Banach de dimension finie et $p \in]1, \infty[$. Si q désigne la quantité conjuguée de p, on a :

$$L^p(E)' = L^q_*(E') = L^q(E') \quad \text{algébriquement et isométriquement.}$$

DEMONSTRATION : Dû au fait que E est de dimension finie, on a $L^q_*(E') = L^q(E')$ algébriquement ; et aussi isométriquement par la remarque (3) précédente.

On va maintenant démontrer que $\varphi \rightsquigarrow u$ est une isométrie de $L^q(E')$ dans $L^p(E)'$. Soit donc $\varphi \in L^q(E')$.

- Supposons d'abord φ μ-étagée. Il existe $h \geq 0$ constante sur les étages de φ telle que :

$$||h||_{L^p(\mu)} = 1 \quad \text{et} \quad \int ||\varphi||h \, d\mu = ||\varphi||_{L^q(E')}$$

Ensuite il existe $\psi \in \mathcal{L}^p(E)$ μ-étagée et constante sur les étages de φ telle que, ε étant donné

$$||\psi|| = h \quad \text{et} \quad <\varphi, \psi> \geq ||\varphi|| \, h \, (1-\varepsilon)$$

alors $\quad ||\psi||_{L^p(E)} = 1$, et par conséquent

$$||u\varphi||_{L^p(E)'} \geq |\int <\varphi, \psi> d\mu| \geq (1-\varepsilon) \int ||\varphi|| \, h \, d\mu = (1-\varepsilon)||\varphi||_{L^q(E')}$$

d'où l'on déduit, ε étant arbitraire, que

$$||u_\varphi||_{L^p(E)'} \geq ||\varphi||_{L^q(E')}$$

L'inégalité inverse étant vraie, d'après les remarques de ce numéro, on a démontré l'égalité.

- Dans le cas général, il existe des φ_n μ-étagées telles que

$$||\varphi - \varphi_n||_{L^q(E')} \xrightarrow[n \to \infty]{} 0 \quad ;$$

donc aussi $||u_\varphi - u_{\varphi_n}||_{L^p(E)'}$.

On en déduit que $||u \quad ||_{L^p(E)'} = ||\varphi||_{L^q(E')}$.

On a démontré l'isométrie. Reste à démontrer la surjectivité. Soit $u \in L^p(E)'$; il nous faut trouver $\varphi \in L^q(E')$ telle que $u = u_\varphi$.

Soit $L_u : E \to L^q(\mu)$ correspondant à u comme il a été dit plus haut. Une application linéaire d'un Banach de dimension finie dans un $L^q(\mu)$ étant q-latticiellement bornée (car la boule unité de ce Banach est contenue dans l'enveloppe convexe d'un nombre fini de points), on en déduit que L_u est q-décomposable par, disons φ, $\varphi \in L^q_*(X, \mu, E')$. Alors $u = u_\varphi$. Le lemme est donc démontré.

REMARQUE : Le résultat du lemme (1) reste vrai si p = 1 (voir la démonstration du théorème suivant).

THEOREME :

Soit $p \in [1, \infty[$; E un Banach ; E' son dual. L'on a algébriquement et isométriquement :

$$L^p(X, \mu, E)' = L^q_*(X, \mu, E')$$

quel soit l'espace mesuré σ-fini $(X, \mathcal{F}, \mu)$.

DEMONSTRATION : Le cas p = 1 résulte de la proposition 2 et de la remarque qui le suit, car toute application linéaire continue de E dans un $L^\infty(\mu)$ est latticiellement bornée.

On peut donc supposer p > 1.

L'on sait déjà qu'il existe une application linéaire contractante $L^q_*(E')$ dans $L^p(E)'$. Il reste à démontrer qu'elle est surjective et isométrique.

Soit $u \in L^p(E)'$ et $L_u : E \to L^q(\mu)$ l'application qui lui est associée comme il a été dit plus haut. Montrons qu'elle est q-latticiellement bornée.

Soit F un sous-espace de E de dimension finie et soit i_F l'injection naturelle de $L^p(F)$ dans $L^p(E)$

Soit $u_F = u \circ i_F$; alors $u_F \in L^p(F)'$ et

$$||u_F||_{L^p(F)'} \leq ||u||_{L^p(E)'}$$

u_F définit une application q-latticiellement bornée L_{u_F} de F dans $L^q(\mu)$

d'après le lemme 1 et

$$\Big\| \bigvee_{\substack{e\in F \\ \|e\|\leq 1}} L_u(e) \Big\|_{L^q(\mu)} = \|u_F\|_{L^p(F)'} \leq \|u\|_{L^p(E)'}$$

Cette inégalité étant vraie pour tout F, on en déduit :

$$\Big\| \bigvee_{\substack{e\in E \\ \|e\|\leq 1}} L_u(e) \Big\|_{L^q(\mu)} \leq \|u\|_{L^p(E)'}$$

Donc L_u est q-latticiellement bornée ; donc décomposable par φ et $u = u_\varphi$. La surjectivité est démontrée.

Maintenant :

$$\|u_\varphi\|_{L^p(E)'} \leq \|\varphi\|_{L^q_*(E')} = \Big\| \bigvee_{\substack{e\in E \\ \|e\|\leq 1}} L_u(e) \Big\|_{L^q(\mu)} .$$

Comme on a déjà constaté l'inégalité inverse, on a l'égalité.

CHAPITRE III

SÉRIES DE VARIABLES ALÉATOIRES BANACHIQUES

N° 1 : DEFINITIONS FONDAMENTALES

Dans ce chapitre, on se donnera un espace probabilisé $(\Omega, \mathcal{F}, P)$ et un espace mesurable $(E, \mathcal{B})$ où E est un espace vectoriel (pas forcément un Banach pour le moment) et $\mathcal{B}$ une tribu sur E. Les v.a. considérées sont basées sur cet $(\Omega, \mathcal{F}, P)$. On supposera que $\mathcal{F}$ est P-complète, ce qui permettra, dans les cas que nous considérerons, d'affirmer qu'une limite p.s d'une suite de variables aléatoires est une variable aléatoire. Toutefois, nous ferons la restriction suivante : toutes les variables aléatoires basées sur $(\Omega, \mathcal{F}, P)$ et à valeurs dans $(E, \mathcal{B})$ que nous considérerons, seront supposées appartenir à un même espace vectoriel. Ce sera la cas si

- les opérations $(x, y) \rightsquigarrow x + y$ et $(\lambda, x) \rightsquigarrow \lambda x$ sont mesurables relativement à $\mathcal{B} \otimes \mathcal{B}$ et $\mathcal{B}_{\mathbb{R}} \otimes \mathcal{B}$: dans ce cas l'ensemble des v.a. à valeurs dans $(E, \mathcal{B})$ est un espace vectoriel.

- E est un Banach et l'on ne considérera que les variables Bochner-mesurables, ou P-mesurables, c'est à dire égales P-presque sûrement à une v.a. fortement mesurable.

Si (X_n) est une suite de v.a. dans $(E, \mathcal{B})$, elle définit une v.a. à valeurs dans $(E^{\mathbb{N}}, \underset{n\in\mathbb{N}}{\otimes} \mathcal{B})$.

Enfin, nous supposons connues la définition de loi d'une v.a. prenant ses valeurs dans $(E, \mathcal{B})$ et de l'indépendance des variables aléatoires.

Si E est normé, on supposera que si X est une v.a., $||X||$ est une v.a. réelle.

DEFINITION 1 : Une v.a. X à valeurs dans E est dite symétrique si $P_X = P_{(-X)}$ (où P_X désigne la loi de la v.a. X).

Soit (X_n) une suite de v.a. à valeurs dans E, l'on dit que cette suite est symétrique si pour tout (ε_n) de nombres égaux à +1 ou -1, la variable aléatoire $(\varepsilon_n X_n)$ à valeurs dans E^N a même loi que la variable aléatoire (X_n).

Une famille finie de v.a. $(X_0, X_1, \ldots, X_n)$ est dite symétrique si la suite $(X_0, X_1, \ldots, X_n, 0, 0, 0, \ldots)$ est symétrique.

REMARQUE 1 : Si la suite (X_n) est symétrique, la v.a. à valeurs dans E^N qu'elle définit est évidemment symétrique. Toutefois ces deux conditions ne sont pas équivalentes : la première est beaucoup plus forte que la seconde.

Par exemple, si X est une v.a. symétrique réelle non p.s égale à zéro, la v.a. (X, X) à valeurs dans $\mathbb{R}^2$ est symétrique, mais (X, X) et $(-X, X)$ n'ont pas même loi.

REMARQUE 2 : Si la suite (X_n) est symétrique, ou même si la v.a. (X_n) est symétrique, pour chaque n, la v.a. X_n est symétrique.

Bien évidemment, la réciproque est fausse.

Remarquons que la suite (X_n) est symétrique si et seulement si toute sous-famille finie $(X_{n_0}, X_{n_1}, \ldots, X_{n_k})$ est symétrique.

Remarquons enfin, et ce sera très important pour la suite, que si les (X_n) sont indépendantes, on a l'équivalence de :

(1) chaque X_n est symétrique ;

(2) la suite (X_n) est une suite symétrique ;

(3) la variable aléatoire dans E^N définie par la suite (X_n) est symétrique.

DEFINITION 2 : Soit X une v.a. sur un $(\Omega, \mathcal{F}, P)$ à valeurs dans E ; on appelle symétrisée de X une variable aléatoire X^s à valeurs dans E, définie sur un $(\Omega', \mathcal{F}', P')$ telle que

$$X^s = X' - X''$$

où X' et X" (définies sur $(\Omega', \mathcal{F}', P')$) sont indépendantes et ont même loi que X.

Il est facile de voir qu'une symétrisée de X est une v.a. symétrique. En "agrandissant" l'espace probabilisé initial, on peut supposer que X et X^s sont basées sur le même espace probabilisé. C'est ce que nous ferons par la suite.

REMARQUE 3 : Soit (X_n) une suite de v.a. à valeurs dans E, définissant une v.a. à valeurs dans $E^{\mathbb{N}}$. Une symétrisée de cette variable aléatoire n'est pas forcément une suite symétrique.

C'est toutefois une suite symétrique, si la suite (X_n) est indépendante.

N° 2 : LES LEMMES FONDAMENTAUX

LEMME 1 :

Soit (X_n) une suite de variables aléatoires sur un $(\Omega, \mathcal{F}, P)$ à valeurs dans E. Soit (Y_n) la suite de variables aléatoires définies sur $(\Omega \times [0,1], P \otimes dt)$ par

$$Y_n(\omega, t) = X_n(\omega)\, R_n(t) ,$$

où (R_n) désigne la suite de Rademacher, la suite (Y_n) est alors symétrique.

Si de plus (X_n) est une suite symétrique, (Y_n) et (X_n) ont même loi considérées comme variables aléatoires à valeurs dans $E^{\mathbb{N}}$.

DEMONSTRATION : Soit $(\varepsilon_n) \in \{-1, +1\}^{\mathbb{N}}$ et soit $A \in \mathcal{B}^{\otimes \mathbb{N}}$. Alors

$$P_{Y_n}(A) = \int_\Omega P_{\{R_n(\cdot) X_n(\omega)\}}(A)\, .\, P(d\omega)$$

$$= \int_\Omega P_{\{R_n(\cdot)\ \varepsilon_n X_n(\omega)\}}(A)\, .\, P(d\omega) = P_{(\varepsilon_n Y_n)}(A).$$

(Ici P désigne la probabilité de l'espace Ω sur lequel sont basées les X_n, et pour tout ω, $P_{R_n(\cdot)|X_n(\omega)|}$ désigne la loi de la v.a. $t \mapsto (R_n(t)\, X_n(\omega))$.)

Donc (Y_n) est une suite symétrique.

Si maintenant la suite (X_n) est supposée symétrique, l'on a

$$P_{(Y_n)}(A) = \int_0^1 P_{\{R_n(t)\, X_n(\cdot)\}}(A)\, dt = \int_P P_{(X_n)}(A)\, dt = P_{(X_n)}(A).$$

LEMME 2 :

Si, dans le lemme précédent, on suppose que $E = \mathbb{R}$ et que la suite (X_n) est symétrique, et si l'on pose pour tout n

$$Z_n(t, \omega) = R_n(t)\, |X_n(\omega)| = R_n \otimes |X_n|(t, \omega),$$

la suite (Z_n) de v.a. sur $\Omega \times [0, 1]$ a même loi que (X_n).

DEMONSTRATION : Soit A un borélien de $R^{\mathbb{N}}$; alors

$$P_{(R_n \otimes |X_n|)}(A) = \int_\Omega P_{\left(R_n(\cdot)|X_n(\omega)|\right)}(A)\, P(d\omega)$$

$$= \int_\Omega P_{\left(R_n(\cdot)X_n(\omega)\right)}(A)\, P(d\omega) = P_{(X_n)}(A)$$

LEMME 3 :

Soient E un Banach, X et Y deux variables aléatoires de $L^p(E)$ $(0 \leq p \leq \infty)$. On suppose que l'une au moins des conditions suivantes est réalisée :

(a) la famille (X, Y) est symétrique ;

(b) X et Y sont indépendantes et X est symétrique ;

(c) $p \geq 1$, $E(X) = 0$ et X et Y indépendantes.

L'on a alors :

$$\frac{1}{2}\, ||2Y||_{L^p(E)} \leq ||X + Y||_{L^p(E)}$$

DEMONSTRATION :

- Démontrons le résultat dans le cas (a) ou (b) et $0 \leq p \leq 1$. On a pour tout $(x, y) \in E \times E$, $||2y|| \leq ||y-x|| + ||y+x||$. Et par conséquent, si φ est une fonction croissante borélienne et sous-additive de R_+ dans R_+, on en déduit :

$$E\{\varphi(||2Y||)\} \leq E\{\varphi(||X+Y||)\} + E\{\varphi(||X-Y||)\}.$$

Mais il résulte des hypothèses que X + Y et X - Y ont même loi, et par conséquent $E\{\varphi(||2Y||)\} \leq 2E\{\varphi(||X+Y||)\}$.

On obtient le résultat annoncé pour p = 0 en prenant pour φ la fonction $t \rightsquigarrow \frac{t}{1+t}$, et dans le cas $0 < p \leq 1$ en prenant pour φ la fonction $t \rightsquigarrow t^p$.

Si maintenant (toujours dans les cas (a) et (b)) l'on suppose $1 < p < \infty$, on déduit de $||2y|| \leq ||y-x|| + ||y+x||$:

$$E\{||2Y||^p\}^{\frac{1}{p}} \leq E\{(||Y-X|| + ||Y+X||)^p\}^{\frac{1}{p}} \leq E\{||Y-X||^p\}^{\frac{1}{p}} + E\{||Y+X||^p\}^{\frac{1}{p}}$$

et on termine comme plus haut.

- Supposons maintenant le cas (c) réalisé et $p < \infty$. Pour tout $y \in E$, on a :

$$||y||^p = ||y+E(X)||^p = ||\int_E (y+x)\, P_X(dx)||^p$$

$$\leq \left(\int_E ||y+x||\, P_X(dx)\right)^p \leq \int_E ||y+x||^p\, P_X(dx)\ ;$$

et en conséquence :

$$E\{||Y||^p\} = \int ||y||^p\, P_Y(dy) \leq \int_{E\times E} ||y+x||^p\, P_X(dx)\, P_Y(dy)$$

$$= E\{||X+Y||^p\}.$$

Le lemme est encore démontré dans ce cas.

- Il nous reste enfin à examiner le cas $p = \infty$; mais alors cela résulte facilement du fait que si $Z \in L^\infty(E)$, on a :

$$||Z||_{L^\infty(E)} = \lim_{p\to\infty} ||Z||_{L^p(E)}.$$

LEMME 4 :

Soit E un Banach et X une v.a. à valeurs dans E ; soit $a \in E$ et X^s une symétrisée de X ; alors pour tout $t \geq 0$, on a :

$$P\{||X^s|| \geq t\} \leq 2P\{||X-a|| \geq \frac{t}{2}\}.$$

DEMONSTRATION : Si $X^s = X - X'$ (X' indépendante de X et ayant même loi que X), alors $X^s = (X-a) - (X'-a)$.

Donc $$\{||X^s|| \geq t\} \subset \{||X-a|| \geq \frac{t}{2}\} \cup \{||X'-a|| \geq \frac{t}{2}\} ;$$

et $$P\{||X^s|| \geq t\} \leq 2P\{||X-a|| \geq \frac{t}{2}\} .$$

LEMME 5 :

Soient X, X^s, E comme dans le lemme 4 et soient $t, u \geq 0$. Alors :

$$P\{||X|| \leq u\}\, P\{||X|| > t+u\} \leq P\{||X^s|| > t\} .$$

DEMONSTRATION : Avec les notations ci-dessus, on a :

$$\{||X'|| \leq u \;;\; ||X|| > t+u\} \subset \{||X^s|| > t\} .$$

Donc $$P\{||X|| \leq u\}\, P\{||X|| > t+u\} \leq P\{||X^s|| > t\} . \qquad \text{CQFD.}$$

LEMME 6 :

Soit (X_n) une suite symétrique de v.a. à valeurs dans un Banach E, et soit, pour tout n :

$$S_n = X_o + X_1 + \ldots + X_n .$$

Alors :

(1) $P\{\max_{j\leq n} ||S_j|| \geq t\} \leq 2P\{||S_n|| \geq t\}$, $\forall\, t \geq 0$, $\forall\, n$;

(2) $P\{\sup_n ||S_n|| \geq t\} \leq 2 \liminf_n P\{||S_n|| \geq t\}$, $\forall\, t \geq 0$;

(3) Si $M \subset \mathbb{N}$ est infinie,

$$P\{\sup_n ||S_n|| \geq t\} \leq 2 \sup_{n\in M} P\{||S_n|| \geq t\} \quad , \forall\, t \geq 0.$$

DEMONSTRATION : (2) et (3) se déduisent très facilement de (1). Il reste à démontrer (1).

Soit $A = \{\sup_{j\leq n} ||S_j|| \geq t\}$ et $B = \{||S_n|| \geq t\}$.

Partitionnons A en les $(n+1)$ ensembles suivants :

$$A_o = \{||S_o|| \geq t\}$$

$$A_1 = \{||S_o|| < t \ ; \ ||S_1|| \geq t\}$$

$$A_2 = \{||S_o|| < t \ ; \ ||S_1|| < t \ ; \ ||S_2|| \geq t\} \ ;$$

et ainsi de suite.

Soit m fixé tel que $0 \leq m \leq n$ et soit

$$S_m^n = S_m - X_{m+1} - X_{m+2} \ldots - X_n \ ;$$

Alors $S_m^n + S_n = 2S_m$; et la suite étant symétrique, S_m^n et S_n ont même loi.

Mais alors $\{||S_m|| \geq t\} \subset \{||S_m^n|| \geq t\} \cup \{||S_n|| \geq t\}$;

et

$$A_m = [A_m \cap \{||S_m^n|| \geq t\}] \cup [A_m \cap \{||S_n|| \geq t\}] \ .$$

Mais à cause de la symétrie, on a :

$$P\ [A_m \cap \{||S_m^n|| \geq t\}] = P\ [A_m \cap \{||S_n|| \geq t\}] \quad \text{et}$$

$$P(A_m) \leq 2P\ [A_m \cap \{||S_n|| \geq t\}] \ , \ \forall\, m \in [0, n] \ .$$

En sommant sur m on obtient le résultat.

Le lemme 6 va nous permettre de passer de la convergence en probabilité à la convergence presque-sûre pour les séries associées à des suites symétriques.

Remarquons au préalable qu'une série de v.a. $(X_n)_n$ à valeurs dans un Banach est p.s. convergente si et seulement si la suite suivante converge vers zéro en probabilité :

$$Z_n = \sup_{\substack{n',n'' \\ n\leq n'\leq n''}} ||X_{n'} + X_{n'+1} + \ldots + X_{n''}|| \qquad \forall\, n \in \mathbb{N} \ .$$

On obtient facilement, à partir de ce qui précède, le

<u>THEOREME 1</u> :

Soit (X_n) une suite symétrique de variables aléatoires à valeurs dans le Banach E et soit, pour tout n, $S_n = X_o + X_1 + \ldots + X_n$. Les conditions suivantes sont équivalentes :

(a) (S_n) converge p.s. ;

(b) (S_n) converge en probabilité ;

(c) Il existe une suite extraite de (S_n) convergeant en probabilité.

DEMONSTRATION : Il est clair que (a) $\Rightarrow$ (b) $\Rightarrow$ (c).

Reste donc à démontrer (c) $\Rightarrow$ (a).

Soit $(S_{n_k})_{k\in\mathbb{N}}$ une suite extraite convergeant en probabilité. Cela signifie que, pour tout $\varepsilon > 0$, il existe un entier k_ε tel que :

$$k,\ l \geq k_\varepsilon\ ,\quad k < l \Rightarrow P\{||S_{n_l} - S_{n_k}|| \geq \varepsilon\} \leq \varepsilon$$

Considérons alors la suite (symétrique) $(Y_m)_{m\geq 1} = (X_{n_{k_\varepsilon}+m})_{m\geq 1}$ et soit $M_\varepsilon = \{n_i\ ,\ i \geq k_\varepsilon\} \subset \mathbb{N}$. Alors :

$$S_m^\varepsilon = Y_1 + Y_2 + \ldots + Y_m = X_{n_{k_\varepsilon}+1} + X_{n_{k_\varepsilon}+2} + \ldots + X_{n_{k_\varepsilon}+m}$$
$$= S_{n_{k_\varepsilon}+m} - S_{n_{k_\varepsilon}}\ ;$$

et pour tout $m \in M_\varepsilon$ on a, d'après l'hypothèse, $P\{||S_m^\varepsilon|| \geq \varepsilon\} \leq \varepsilon$.

Donc, en vertu du lemme 6.(3) : $P\{\sup_{m\in\mathbb{N}^+} ||S_m^\varepsilon|| \geq \varepsilon\} \leq 2\varepsilon$.

Mais alors, pour tout $m \geq n_{k_\varepsilon}$, $P\{\sup_{\substack{n',n'' \\ m\leq n'\leq n''}} ||S_{n''} - S_{n'}|| > \frac{\varepsilon}{2}\} \leq 2\varepsilon$,

ce qui, d'après la remarque précédant le théorème 1, signifie que (S_n) converge presque sûrement.

REMARQUE 2 : Sous les hypothèses du théorème (1), l'on peut démontrer de la même façon les équivalences :

(a) (S_n) est p.s. bornée ;

(b) (S_n) est bornée en probabilité ;

(c) L'on peut extraire de (S_n) une suite bornée en probabilité.

C'est en effet immédiat si l'on remarque que :

- (S_n) est bornée en probabilité si et seulement si, pour tout $\varepsilon > 0$, il existe $M < \infty$ tel que $\sup_n P\{||S_n|| \geq M\} \leq \varepsilon$;

- (S_n) est presque sûrement bornée si, pour tout $\varepsilon > 0$, il existe $M > 0$ tel que :

$$P\{\sup_n \|S_n\| \geq M\} \leq \varepsilon .$$

Maintenant, le lemme (6) permet de passer de la première inégalité à la seconde.

Dans le cas où les variables aléatoires ne forment pas une suite symétrique, mais si l'on suppose qu'elles sont indépendantes, on va voir que la convergence en probabilité de $(\sum X_n)$ équivaut à la convergence presque-sûre. C'est l'objet du

THEOREME 2 :

Soit (X_n) une suite indépendante de v.a. à valeurs dans le Banach E. L'on a l'équivalence de :

(1) $\sum X_n$ converge presque sûrement ;

(2) $\sum X_n$ converge en probabilité.

DEMONSTRATION : La seule chose à démontrer est l'implication (2) ⟹ (1). Supposons donc (2) vérifiée. Alors pour toute suite symétrisée (X_n^s) de (X_n), $\sum_n X_n^s$ converge en probabilité (par exemple grâce au lemme 4 appliqué à $a = 0$ et $X = \sum_{m \leq l \leq n} X_l$). Et, en vertu du lemme 1, $\sum_n X_n^s$ converge presque sûrement ; En particulier, $\sum_n \left(X_n(\omega) - X_n(\omega')\right)$ converge $P \otimes P$ presque sûrement sur $\Omega \times \Omega$; il existe donc une suite (a_n) d'éléments de $\mathbb{R}$ telle que $\sum_n \left(X_n(\omega) - a_n\right)$ converge P-presque sûrement. Mais alors $\sum_n a_n = \sum_n X_n - \left(\sum_n (X_n - a_n)\right)$ converge en probabilité ; donc $\sum a_n$ converge dans E, et a fortiori $\sum_n X_n$ converge presque sûrement. CQFD.

REMARQUE 3 : On peut démontrer que, sous l'hypothèse d'indépendance, on a l'équivalence :

(1) (S_n) est p.s. bornée ;

(2) (S_n) est bornée en probabilité.

(Démonstration tout à fait analogue à celle du théorème 2).

REMARQUE 4 : Sous les hypothèses du théorème 2, on ne peut affirmer, comme dans le théorème 1, que l'on a équivalence de

(a) $\sum X_n$ converge p.s.

(b) (S_{n_k}) converge en probabilité pour une certaine sous-suite (n_k).

Il suffit, pour s'en convaincre, de considérer l'exemple suivant :
Les (X_n) sont des éléments presque certains et pour tout $n, X_n = (-1)^n$ p.s. Alors $\sum X_n$ ne converge pas p.s. mais la suite $(S_{2n})_n$ converge p.s.

Les résultats qui précèdent montrent que, sous les hypothèses de symétrie ou d'indépendance, la convergence (resp. la bornitude) en probabilité équivaut à la convergence (resp. la bornitude) presque-sûre.

Nous allons maintenant donner des critères de convergence ou de bornitude dans $L^0(E)$, et plus particulièrement dans $L^p(E)$. Il nous faudra distinguer le cas $p < \infty$ du cas $p = \infty$.

N° 3 : CONVERGENCE ET BORNITUDE DE (S_n) DANS $L^p(E)$ $(0 \leq p < \infty)$

Dans ce numéro, on se donnera une suite (X_n) de v.a. indépendantes à valeurs dans le Banach E.

L'on posera $S_n = \sum_{j=0}^{n} X_j$; $N = \sup ||X_n||$; $M = \operatorname{Sup} ||S_n||$. N et M sont des v.a. à valeurs dans $[0, \infty]$.

D'autre part, si Y est une v.a. numérique, on désignera par F_Y la fonction $t \rightsquigarrow P\{Y > t\}$ (c'est le complément à l'unité de la fonction de répartition usuelle). F_Y sera appelée la "fonction de queue" de Y. Cela étant, on a le

LEMME 7 :

Soit $(X_n)_{n\geq 0}$ une suite <u>indépendante et symétrique</u> de variables aléatoires à valeurs dans E. Pour tout entier k, on a :

$$F_{||S_k||}(s+t+u) \leq F_N(s) + 4F_{||S_k||}(t)\, F_{||S_k||}(u) \quad (s,t,u \geq 0).$$

DEMONSTRATION : Soit T le "temps d'arrêt" :

$$T = \inf \{n \; ; \; ||S_n|| > t\} .$$

Si $||S_k|| > s+t+u$, alors $T \leq k$. Et par conséquent :

$$F_{||S_k||}(s+t+u) = \sum_{j=0}^{k} P\{||S_k|| > s+t+u \; ; \; T = j\} .$$

D'autre part : $\{T = j\}$ avec $j \leq k$ et $\{||S_k|| > s+t+u\}$ impliquent $\{||S_{j-1}|| \leq t\}$ et $||S_k - S_j|| \geq ||S_k|| - ||S_{j-1}|| - ||X_j|| > u+s-N$. (où l'on pose $S_{-1} = 0$).

Donc $\quad P\{T=j,\ ||S_k|| > t+s+u\} \leq P\{T=j \; ; \; ||S_k - S_j|| > u+s-N\}$

$$\leq P\{T=j \; ; \; N > s\} + P\{T=j \; ; \; ||S_k - S_j|| > u\}$$

$$= P\{T=j \; ; \; N > s\} + P\{T=j\}\, P\{||S_k - S_j|| > u\}$$

(à cause de l'indépendance). D'où l'on déduit :

$$F_{||S_k||}(s+t+u) \leq F_N(s) + \sum_{j=0}^{k} P\{T=j\}\, P\{||S_k - S_j|| > u\} .$$

Posons $Z_o = S_k - S_j$; $Z_1 = S_j$; Z_o et Z_1 sont symétriques et $Z_o + Z_1 = S_k$. Donc, par le lemme 6.(1) appliqué à Z_o, Z_1, $S_k = Z_o + Z_1$, on obtient :

$$P\{||S_k - S_j|| > u\} \leq 2P\{||S_k|| > u\} \; ;$$

De l'égalité $\sum_{j=0}^{k} \{T=j\} = \{\max_{j\leq k} ||S_j|| > t\}$, on déduit alors

$$F_{||S_k||}(s+t+u) \leq F_N(s) + 2F_{||S_k||}(u)\, P\{\max_{j\leq k} ||S_j|| > t\} .$$

D'où, en appliquant encore le lemme (6, 1) :

$$F_{||S_k||}(s+t+u) \leq F_N(s) + 4F_{||S_k||}(u)\, F_{||S_k||}(t) . \quad \text{CQFD.}$$

THEOREME 3 :

Soit (X_n) une suite indépendante de v.a. à valeurs dans le Banach E. Pour tout q tel que $0 < q$, on a l'équivalence (avec les notations ci-dessus) :

(1) (S_n) est bornée en probabilité et $N \in L^q(\mathbb{R})$

(2) $M \in L^q(\mathbb{R})$.

DEMONSTRATION : Supposons d'abord la suite (X_n) symétrique ; pour tous réels u et t et tout entier naturel n, on a d'après le lemme 7 :

$$P\{||S_n|| > 2t + u\} \leq P\{N > t\} + 4P^2\{||S_n|| > t\} ,$$

et par conséquent, d'après le lemme 6.(2),

$$\frac{1}{2} P\{M > 2t + u\} \leq P\{N > t\} + 4P^2\{M > t\} .$$

D'où, pour tout réel $A > 0$,

$$(1) \quad \frac{1}{2}\int_0^A P\{M > 3t\}\, d(t^q) \leq \int_0^A P\{N > t\}\, d(t^q) + \int_0^A 4P^2\{M > t\}\, d(t^q) .$$

Jusqu'à présent, nous n'avons pas fait intervenir le fait que $\{S_n, n \in \mathbb{N}\}$ est bornée en probabilité. Si l'on tient compte de ce fait, il existe $t_0 > 0$ tel que :

$$P\{M > t_0\} < \frac{1}{16 \times 3^q} .$$

Choisissons alors $A > 3t_0$; on a alors, en vertu de l'inégalité (1) :

$$\int_0^A P\{M > t\}\, d(t^q) = 3^q \times q \int_0^{A/3} t^{q-1} P\{M > 3t\}\, dt$$
$$\leq 2q \times 3^q \int_0^{A/3} t^{q-1} P\{N > t\} dt + 8q \times 3^q \int_0^{A/3} t^{q-1} P^2\{M > t\}\, dt .$$

Mais
$$\int_0^{A/3} P\{N > t\}\, d(t^q) \leq \int_0^{\infty} P\{N > t\}\, d(t^q) \leq E\{N^q\} ;$$

et d'autre part

$$\int_0^{A/3} P^2\{M > t\}\, d(t^q) = \int_0^{t_0} P^2\{M > t\}\, d(t^q) + \int_{t_0}^{A/3} P^2\{M > t\}\, d(t^q)$$
$$\leq \int_0^{t_0} d(t^q) + \int_0^{A/3} P\{M > t_0\} P\{M > t\}\, d(t^q) .$$

Donc, en définitive, pour tout $A > t_0$, on a :

$$(2) \quad \int_0^A P\{N > t\}\, d(t^q) \leq 2 \times 3^q E\{N^q\} + 8 \times 3^q t_0^q + \frac{1}{2}\int_0^A P\{M > t\}\, dt$$

(car $P\{M > t_o\} \leq \frac{1}{16 \times 3^p}$).

L'inégalité (2) étant une inégalité entre éléments de $\mathbb{R}$, on en déduit

$$\frac{1}{2}\int_0^A P\{N > t\}\, d(t^q) \leq 2 \times 3^q\, E\{N^q\} + 8 \times 3^q\, t_o^{\ q} ,$$

ceci étant vrai pour tout $A > 3t_o$, on en déduit que :

$$\int_0^\infty P\{N > t\}\, d(t^q) = E\{M^q\} < \infty .$$

Dans le cas symétrique, on a démontré que (1) $\Longrightarrow$ (2).

Dans le cas général, on procède par symétrisation : Soit $X_n^s = X_n - X'_n$ ($n \in \mathbb{N}$) une symétrisation de (X_n), soient S'_n, M', N' (resp. S_n^s, M^s, N^s) définis à partir des X'_n (resp. (X_n^s)) comme S_n, M, N ont été définis à partir de (X_n).

M et M' d'une part, N et N' d'autre part, ont même répartition. Donc $N^s \in L^q(\mathbb{R})$ et $\{S_n^s, n \in \mathbb{N}\}$ est bornée en probabilité. Il resulte encore de la première partie que $M^s \in L^q(\mathbb{R})$.

Soit μ la loi de la v.a. (X_n) (c'est une probabilité sur $E^{\mathbb{N}}$). On a alors

$$E\{(M^s)^q\} = \int_{E^{\mathbb{N}}} E\{\sup_n ||S'_n - \sum_{j=o}^n x_j||^q\}\, \mu(dx).$$

Mais (S'_n) est bornée presque sûrement (car les X'_n sont indépendantes), et il existe $x \in E^{\mathbb{N}}$ tel que

$$\alpha = \sup_n ||\sum_{j=0}^n x_j|| < \infty$$

et

$$E\{\sup_n ||S_n - \sum_{j=0}^n x_j||^q\} < \infty .$$

Alors

$$M \leq \alpha + \sup_n ||S_n - \sum_{j=0}^n x_j|| \text{ ; donc } M \in L^q(\mathbb{R}),$$

et l'implication (1) $\Longrightarrow$ (2) est démontrée.

L'implication (2) $\Longrightarrow$ (1) est évidente (car $N < 2M$) et le theorème est alors complètement démontré.

<u>COROLLAIRE 1</u> : Soit (X_n) une suite indépendante de v.a. à valeurs dans le Banach E ; soit $q \in]0, \infty[$; on a l'équivalence :

(a) (S_n) est bornée dans $L^q(E)$;

(b) $M \in L^q(\mathbb{R})$

(c) $N \in L^q(\mathbb{R})$ et (S_n) est bornée en probabilité.

DEMONSTRATION : Il est clair que (b) ⟹ (a) (même sans l'hypothèse d'indépendance). Je dis maintenant que (a) ⟹ (b). En effet, supposons les (X_n) symétriques. L'on sait que $P\{M > t\} \leq 2 \liminf_n P\{\|S_n\| > t\}$. Et, grâce au lemme de Fatou :

$$E\{M^q\} = \int P\{M > t\}\, d(t^q) \leq 2 \int \liminf_n P\{\|S_n\| > t\}\, d(t^q)$$

$$\leq 2 \liminf_n \int P\{\|S_n\| > t\}\, d(t^q) = 2 \liminf_n E\{\|S_n\|^q\}$$

Dans le cas général, on symétrise comme dans la démonstration du théorème 3.

Par ailleurs, on a vu que (b) ⟺ (c).

REMARQUE 1 : Dans le cas $p = \infty$, on a l'équivalence (et c'est trivial) de :

$\{(S_n),\ n \in \mathbb{N}\}$ est bornée dans $L^\infty(E)$ ⟺ $M \in L^\infty(\mathbb{R})$.

COROLLAIRE 2 : Soit (X_n) une suite indépendante de variables aléatoires à valeurs dans le Banach E. On suppose que $S_n \to S$ presque sûrement (ou en probabilité). Soit $p \in \,]0, \infty[$. On a les équivalences :

(1) $S_n \to S$ dans $L^p(E)$

(2) $S \in L^p(E)$

(3) $M \in L^p(\mathbb{R})$

(4) $N \in L^p(\mathbb{R})$

(5) $\{S_n,\ n \in \mathbb{N}\}$ est bornée dans $L^p(E)$.

DEMONSTRATION : Il est clair que $S_n \to S$ en probabilité implique que $\{S_n,\ n \in \mathbb{N}\}$ est bornée en probabilité.

(3) ⟺ (4) ⟺ (5) : C'est le corollaire 1.

(1) ⟹ (2) est trivial et (3) ⟹ (1) par Lebesgue. La seule chose qui

reste à démontrer est donc l'implication (2) => (3). C'est trivial si la suite (X_n) est symétrique, car en vertu du lemme 6.(2), on a :

$$P\{M \geq t\} \leq 2 \liminf_n P\{||S_n|| \geq t\} \leq 2 \limsup_n P\{||S_n|| \geq t\}.$$

et $(||S_n||)_n$ convergeant vers $||S||$ en loi

$$\limsup_n P\{||S_n|| \geq t\} \leq P\{||S|| \geq t\}.$$

Le cas non symétrique se traite par symétrisation.

REMARQUE 2 : Quand $p = \infty$, il est clair que sous les autres hypothèses du corollaire 2 on a l'équivalence :

$$S \in L^{\infty}(E) \iff M \in L^{\infty}(\mathbb{R}),$$

par le même raisonnement que celui qui a été fait pour l'implication (2) => (3) dans ce corollaire.

En résumé, l'on peut dire qu'étant donnée une série presque sûrement convergente de variables aléatoires banachiques indépendantes si la somme appartient à $L^p(E)$, $(p < \infty)$, toutes les sommes partielles appartiennent à $L^p(E)$ et on a la convergence dans $L^p(E)$ (le cas $p = 0$ est évidemment trivial et bien connu).

Nous allons maintenant examiner ce qui se passe pour $p = \infty$ (en dehors des remarques faites plus haut).

REMARQUE 3 : Jusqu'à présent, nous n'avons donné que des critères de convergence et de bornitude des (S_n). Mais les démonstrations données permettent très facilement d'obtenir des critères de convergence commutatives et de "bornitude" commutative des séries $\sum X_n$. Par exemple, le théorème 3 se traduit par :

THEOREME 3' : Avec les conditions du théorème (3), on a l'équivalence de :

(1) $\sum X_n$ est commutativement bornée en probabilité et $N \in L^q(\mathbb{R})$;

(2) $M_1 = \sup_{\sigma \in \Phi(N)} ||\sum_{n \in \sigma} X_n|| \in L^q(\mathbb{R})$.

De même, le corollaire 1 de ce théorème se transcrit sous la forme suivante :

COROLLAIRE 1' : Sous les hypothèses du corollaire 1, on a l'équivalence de

(a) $\sum X_n$ est commutativement bornée dans $L^q(E)$

(b) $M_1 \in L^q(\mathbb{R})$

(c) $N \in L^q(\mathbb{R})$ et $\sum X_n$ est commutativement bornée en probabilité.

On pourrait également transcrire le corollaire 2 suivant le même modèle.

N° 4 : CONVERGENCE DANS $L^\infty(E)$

On va chercher sous quelles conditions $S \in L^\infty(E)$ implique $S_n \to S$ dans $L^\infty(E)$. Remarquons au préalable que cette propriété n'est pas vraie "en général", comme le montre le contre-exemple suivant :

Soit $E = l^\infty$ et soit P une probabilité sur $(\Omega, \mathcal{F}) = (\mathbb{N}, \mathcal{P}(\mathbb{N}))$ chargeant tout élément de $\mathbb{N}$.

Se donner une variable aléatoire $X : \Omega \to l^\infty$ c'est se donner pour tout $n \in \mathbb{N}$ une suite $(\lambda_n(k))_{k\in\mathbb{N}}$ de scalaires telles que $\sup_k |\lambda_n(k)| < \infty \ \forall n$.
De même, se donner une famille $(X^i)_{i\in\mathbb{N}}$ de v.a. dans l^∞, c'est se donner une famille de scalaires

$$\left(\lambda_n^i(k)\right) \quad (i,n,k) \in \mathbb{N}^3 \text{ , telle que}$$

$$\sup_k |\lambda_n^i(k)| < \infty \text{ , } \forall i, \ \forall n.$$

Soit donc $(X^i)_i$ une suite indépendante de variables aléatoires à valeurs dans l^∞ et soit $S^i = \sum_{j\leq i} X^j$.

Dire que $S^i \to S$ presque sûrement revient à dire que :

$$\forall n, \quad \sup_{k\in\mathbb{N}} |S_n^i(k) - S_n(k)| \xrightarrow[i\to\infty]{} 0 \qquad \text{(A)}$$

Dire que $S^i \to S$ dans $L^\infty(l^\infty)$ équivaut à dire que :

$$\sup_n \sup_k |S_n^i(k) - S_n(k)| \xrightarrow[i\to\infty]{} 0 \qquad (B)$$

Il est alors facile de construire des contre-exemple tels que (A) soit vrai sans que (B) le soit.

Si on avait pris c_o au lieu de l^∞, on aurait pu construire un contre-exemple analogue.

Par conséquent, on a démontré que dans tout Banach E contenant c_o isomorphiquement, il existe une suite de v.a. indépendantes à valeurs dans E telle que

$\sum X_n \to S$ presque sûrement et $\sum X_n \not\to S$ dans $L^\infty(E)$.

En d'autres termes, si E n'est pas un C-espace, $S_n \to S$ p.s. n'implique pas forcément $S_n \to S$ dans $L^\infty(E)$.

Nous allons voir que la réciproque est vraie. Plus précisement

THEOREME 4 :

Soit (X_n) des v.a. indépendantes à valeurs dans le Banach E (et donc Bochner-mesurable d'après nos conventions) ; on suppose que

- $S_n \to S$ presque sûrement ;
- $S \in L^\infty(\Omega, \mathcal{F}, P, E)$.

Alors si E est un C-espace, on a $\sum X_n = S$ au sens de la convergence dans $L^\infty(E)$.

DEMONSTRATION : Soit donc $S \in L^\infty(E)$ et supposons que S_n ne converge pas vers S pour la topologie de $L^\infty(E)$. Démontrons maintenant que E n'est pas un C-espace en construisant une C-suite (b_k) ne tendant pas vers zéro dans E.

- Remarquons tout d'abord que $S \in L^1(E)$, donc par le corollaire 2 du théorème 3, $S_n \to S$ dans $L^1(E)$ et en particulier $\sum_{j=0}^{n} E(X_j) \to E(S)$.

Si l'on pose pour tout $n \in \mathbb{N}$, $X'_n = X_n - E(X_n)$ et $S'_n = X'_0 + X'_1 + \ldots + X'_n$, on a l'équivalence :

$$S'_n \to S' \text{ dans } L^\infty(E) \iff S_n \to S \text{ dans } L^\infty(E) ;$$

de sorte que l'on peut supposer les (X_n) centrées.

- (S_n) ne peut être une suite de Cauchy dans $L^\infty(E)$, sinon elle convergerait vers S. Il existe donc $\varepsilon > 0$ et une suite extraite $(S_{n_k})_{k\in\mathbb{N}}$ telle que :

$$(*) \qquad ||S_{n_{k+1}} - S_{n_k}||_{L^\infty(E)} > \varepsilon \quad \forall k.$$

On peut supposer $n_o = 0$. Posons $Y_k = S_{n_{k+1}} - S_{n_k}$. Les Y_k sont indépendants et centrés tout comme les X_n.

Alors $\sum_{j=0}^{k} Y_j = S_{n_{k+1}} - S_o = S_{n_{k+1}} - X_o$. Supposons pour simplifier $X_o = 0$. Il est clair que $\sum Y_k = S$ presque sûrement. La démonstration va maintenant se terminer en deux étapes.

1) Supposons d'abord les (Y_k) dénombrablement étagées :

$$Y_k = \sum_{n\in\mathbb{N}} 1_{A_n^k} a_n^k .$$

Remarquons qu'il existe une constante finie K telle que

$$||\sum_{k=0}^{n} Y_k||_{L^\infty(E)} \leq K \qquad \forall n \in \mathbb{N} ,$$

car $\sum Y_n = S$ et $S \in L^\infty(E)$. Maintenant, pour toute partie σ de $\{1, 2, \dots, n\}$, on a :

$$||\sum_{k\in\sigma} Y_k||_{L^\infty(E)} \leq ||\sum_{k=0}^{n} Y_k||_{L^\infty(E)} \leq K .$$

(Cela résulte de $||Y||_{L^\infty(E)} \leq ||X + Y||_{L^\infty(E)}$ si X a une moyenne nulle et X et Y indépendantes).

Par conséquent, il existe $\Omega_o \in \mathcal{F}$ avec $P(\Omega_o) = 1$ tel que

$$\forall \omega \in \Omega_o , \forall \sigma \in \Phi(\mathbb{N}), ||\sum_{k\in\sigma} Y_k(\omega)|| \leq K .$$

($\Phi(\mathbb{N})$ désigne l'ensemble des parties finies de $\mathbb{N}$).

Maintenant, nous allons faire intervenir le fait que $||Y_k||_{L^\infty(E)} > \varepsilon$ pour tout k. De ce fait, on déduit que pour tout $k \in \mathbb{N}$, il existe un entier n(k) tel que :

$$||a_{n(k)}^k||_E > \varepsilon \quad \text{et} \quad P(A_{n(k)}^k) > 0 .$$

Posons $b_k = a^k_{n(k)}$ et $B_k = A^k_{n(k)}$. Alors

$$P(B_0 \cap B_1 \cap \ldots \cap B_k) = \prod_{i=0}^{n} P(B_i) > 0 .$$

Donc, pour tout k il existe $\omega_k \in \Omega_0 \cap B_0 \cap \ldots \cap B_k$.

Il est clair que $Y_j(\omega_k) = b_j$ $(0 \leq j \leq k)$. Donc

$$\|\sum_{j\in\sigma} b_j\| \leq K , \quad \forall \sigma \in \Phi(\mathbb{N}).$$

Pour toute partie finie σ de $\mathbb{N}$, on a donc $\|\sum_{j\in\sigma} \varepsilon_j b_j\| \leq 2K$ $(\varepsilon_j = \pm 1)$; et, en vertu des principes de contraction, (b_k) est une C-suite. Mais $\|b_k\| \geq \varepsilon$, $\forall k$; donc, cette C-suite ne peut converger vers zéro. D'où contradiction avec le fait que E est un C-espace.

2) <u>Cas général</u> : Pour tout k, il existe une v.a. dénombrablement étagée Z_k telle que

$$\|Z_k - Y_k\| < \frac{\varepsilon}{2^{k+1}} \quad \text{(si } \varepsilon \text{ est le nombre défini plus haut dans (*))}.$$

On peut en outre supposer les Z_k indépendantes et $E(Z_k) = 0$ $\forall$ k. Mais

$$\sum \|Z_k - Y_k\|_{L^\infty(E)} < \varepsilon \text{ ; donc } \sum Z_k \text{ existe p.s. et}$$

$$\sum S_k = S \in L^\infty(E) \iff \sum Z_k \in L^\infty(E).$$

On est donc ramené au cas précédent, et le théorème est démontré.

Maintenant, nous allons étudier la condition "$\{S_n\}$ est bornée dans $L^\infty(E)$". Remarquons préalablement que dire que $\{S_n\}$est bornée dans $L^\infty(E)$ équivaut à :

"Il existe $\Omega_0 \in \mathcal{F}$ avec $P(\Omega_0) = 1$, et $K < \infty$ tel que

$$\forall \omega \in \Omega_0, \quad \forall n, \quad \|S_n(\omega)\|_E \leq K \text{ "}.$$

Et cela équivaut encore à

"Il existe $\Omega_0 \in \mathcal{F}$ avec $P(\Omega_0) = 1$ et $K < \infty$ tel que, pour tout $x' \in E'$, tout $\omega \in \Omega_0$ et tout n,

$$|\sum_{j=0}^{n} \langle x', X_j(\omega)\rangle| \leq K \|x'\| \text{"}.$$

Bien sûr, ces équivalences sont vraies sans condition d'indépendance. Dans le cas de l'indépendance, on a le

THEOREME 5 :

Soit (X_n) une suite indépendante de v.a. à valeurs dans le Banach E. On suppose que, pour tout entier n, $E(X_n)$ existe et $E(X_n) = 0$.

Les conditions suivantes sont équivalentes :

(1) $\{S_n\}$ est bornée dans $L^\infty(E)$;

(2) Il existe $K < \infty$ et $\Omega_0 \in \mathcal{F}$ avec $P(\Omega_0)$ tel que

$$\forall x' \in E', \ \forall \omega \in \Omega_0, \ \sum_{j=0}^{\infty} |\langle x', X_j(\omega)\rangle| \leq K \, ||x'|| \ ;$$

(3) Pour tout $x' \in E'$, il existe $K(x') < \infty$ tel que

$$P\{\omega \ ; \ \sum_{j=1}^{\infty} |\langle x', X_j(\omega)\rangle| \leq K(x')\} = 1 \ ;$$

(4) Pour tout $x' \in E'$, il existe $K(x') < \infty$ tel que

$$\forall n, \ P\{\omega \ ; \ |\sum_{j=0}^{n} \langle x', X_j(\omega)\rangle| \leq K(x')\} = 1.$$

DEMONSTRATION :

(1) ⟹ (2) : Soit K tel que $||\sum_{j=0}^{n} X_j||_{L^\infty(E)} < K$, $\forall n$. Dû au fait que les (X_j) sont indépendantes et centrées, comme plus haut on en déduit que pour toute $\sigma \in \Phi(\mathbb{N})$ on a :

$$||\sum_{j\in\sigma} X_j||_{L^\infty(E)} \leq K .$$

Il existe donc $\Omega_0 \in \mathcal{F}$ avec $P(\Omega_0) = 1$ et tel que $\forall \sigma \in \Phi(\mathbb{N})$, $\forall \omega \in \Omega_0$, $||\sum_{j\in\sigma} X_j(\omega)||_E \leq K$. Cela signifie que, pour presque tout ω, $(X_j(\omega))_{j\in\mathbb{N}}$ est une C-suite ; donc elle est scalairement dans l^1, ce qui est précisément (2).

(2) ⟹ (3) ⟹ (4) : C'est évident même si les (X_n) ne sont pas indépendantes, ni centrées.

(4) ⟹ (1) : En effet (4) implique que, pour tout n, on a une application A_n linéaire de E' dans $L^\infty(\mathbb{K})$ ($\mathbb{K} = \mathbb{R}$ ou $\mathbb{C}$), à savoir :

$$A_n x' = \langle x', S_n\rangle .$$

Je dis que cette application est continue. Cela résulte en effet du théorème du graphe fermé, car si :

$$x_i' \to x' \quad \text{et} \quad A_n x_i' \to f \ \text{ dans } L^\infty ,$$

alors $\langle x_i', S_n(\omega)\rangle \to \langle x', S_n(\omega)\rangle, \ \forall \omega \in \Omega$.

D'autre part, (4) s'écrit :

$$\sup_n \|A_n x'\|_{L^\infty} < \infty \ \forall x' \ ;$$

C'est à dire encore que les (A_n) sont bornées dans $\mathcal{L}(E', L^\infty)$ muni de la topologie de la convergence simple. Mais alors, par le théorème de Banach-Steinhaus :

$$\sup_n \|A_n\|_{\mathcal{L}(E', L^\infty)} < \infty \ .$$

Ceci signifie qu'il existe $K < \infty$ telle que

$$P\{|\langle x', S_n\rangle| \le K\} = 1 \quad \forall n \ge 1, \quad \forall \|x'\| \le 1.$$

Maintenant, dû au fait que les (X_n) sont Bochner-mesurables et qu'elles prennent donc presque sûrement leurs valeurs dans un sous-espace séparable, on peut supposer E séparable. Dans ce cas, la boule-unité de E' contient un sous-ensemble dénombrable $(x'_k)_{k\in\mathbb{N}}$ tel que

$$\forall x \in E, \ \|x\|_E = \sup_k |\langle x'_k, x\rangle| \ .$$

Par conséquent, il existe $\Omega_o \in \mathcal{F}$ avec $P(\Omega_o) = 1$ et tel que

$\forall k, \ \forall n, \ \forall \omega \in \Omega_o$, on ait $|\langle x'_k, S_n(\omega)\rangle| \le K$.

Mais ceci signifie que $\|S_n(\omega)\| \le K \quad \forall n, \ \forall \omega \quad \Omega_o$.

Donc (1) est vérifié (remarquons que l'implication (4) => (1) a été démontrée sans hypothèse de centrage et d'indépendance).

On peut, à partir de là, obtenir très facilement le

THEOREME 6 :

Soit (X_n) une suite de v.a. à valeurs dans le Banach E, intégrables, centrées et indépendantes. Supposons que $\{S_n, n \in \mathbb{N}\}$ est bornée dans $L^\infty(E)$. Si E est un C-espace, (S_n) converge dans $L^\infty(E)$.

DEMONSTRATION : Il résulte du théorème (5) que presque sûrement $(X_n(\omega))_{n\in\mathbb{N}}$ est scalairement dans L^1.

Donc, E étant un C-espace, $S = \sum X_n$ existe presque sûrement, et bien sûr $S \in L^\infty(E)$ puisque $\{S_n, n \in \mathbb{N}\}$ est bornée dans $L^\infty(E)$. Il suffit alors d'appliquer le théorème 4.

N° 5 : COMPARAISON DES CONVERGENCES DES SERIES DE VARIABLES ALEATOIRES

Nous aurons souvent à considérer le problème suivant :

Soit (X_n) une suite de variables aléatoires à valeurs dans le Banach E et définie sur un $(\Omega, \mathcal{F}, P)$. Soit (Y_n) une autre suite de variables aléatoires à valeurs dans E, définies sur un $(\Omega', \mathcal{F}', P')$ au moyen des (X_n) (par différents procédés que nous expliciterons).

Quand peut-on dire que $\sum Y_n$ converge, ou encore que $\{\sum_{k\leq n} Y_k, n \in \mathbb{N}\}$ est bornée en probabilité, ou que $\sum Y_n$ est commutativement bornée dans $L^p(\Omega', \mathcal{F}', P', E)$ $(0 \leq p \leq \infty)$?

Par exemple, si $\sum X_n$ est commutativement bornée dans $L^p(\Omega, \mathcal{F}, P, E)$ $(0 \leq p \leq \infty)$, $\sum a_n X_n$ est commutativement bornée (resp. convergente) dans $L^p(\Omega, \mathcal{F}, P, E)$ pour tout $(a_n) \in l^\infty$ (resp. C_o).

Le résultat fondamental qui nous permettra de faire des comparaisons est le

LEMME 8 :

Soit (X_n) une suite de v.a. à valeurs dans le Banach E. On suppose que l'une des conditions suivantes est réalisée :

(a) (X_n) est une suite symétrique et $(X_n) \in L^p(\Omega, \mathcal{F}, P, E)$ $\forall n$ $(o\leq p\leq\infty)$.

(b) Les (X_n) sont indépendantes ; $X_n \in L^p(\Omega, \mathcal{F}, P, E)$ $\forall n$ avec $1 \leq p \leq \infty$ et $E(X_n = 0)$.

Sous ces hypothèses, on a alors :

(1) $$\sup_{\sigma\in\Phi(\mathbb{N})} \left\|2 \sum_{k\in\sigma} X_k\right\|_{L^p(E)} \leq 2 \sup_n \left\|\sum_{i\leq n} X_i\right\|_{L^p(E)}$$

(2) Pour tout $n \in \mathbb{N}$ et toute suite $(a_k)_{k\in\mathbb{N}} \in \mathbb{R}^{\mathbb{N}}$, on a si $p \neq 0$:

$$\left\|\sum_{k\leq n} a_k X_k\right\|_{L^p(E)} \leq A_p \sup |a_k|^{\min(p,1)} \left\|\sum_{i\leq n} X_i\right\|_{L^p(E)}$$

avec $$A_p = \frac{4}{2^{\min(p,1)} - 1} \times \frac{2}{2^{\min(p,1)}} .$$

<u>DEMONSTRATION</u> : (1) résulte trivialement du lemme 3 de ce chapitre, car si $\sigma \subset \{0, 1\ldots n\}$, en posant $Y = \sum_{k\in\sigma} X_k$; $X = \sum_{k\in\sigma'} X_k$ où $\sigma' = \{0, 1\ldots n\}\setminus\sigma$, X et Y sont comme dans ce lemme.

(2) Par le chapitre I, n° 5, lemme , l'on sait que :

$$\|\sum_{k\leq n} a_k X_k\|_{L^p(E)} \leq \frac{2}{2^{\min(p,1)}-1} \sup_{k\leq n} |a_k|^{\min(p,1)} \sup_{\sigma\subset\{0,\ldots,n\}} \|\sum_{k\in\sigma} X_k\|_{L^p(E)}$$

et par le lemme 3 on sait que :

$$\sup_{\sigma\subset\{0,1,\ldots,n\}} \|\sum_{k\in\sigma} X_k\|_{L^p(E)} \leq \frac{2}{2^{\min(p,1)}} \|\sum_{i\leq n} X_i\|_{L^p(E)} ,$$

d'où le résultat.

On en déduit immédiatement la

<u>PROPOSITION 1</u> : Sous l'une des conditions du lemme 8 :

(1) Si $\{S_n, n \in N\}$ est bornée dans $L^p(E)$, $\sum X_n$ est commutativement bornée dans $L^p(E)$ $(0 \leq p \leq \infty)$.

(2) Si (S_n) converge dans $L^p(E)$, elle converge commutativement dans $L^p(E)$ $(0 \leq p \leq \infty)$.

Remarquons que pour $p \geq 1$, la constante A_p qui figure dans le lemme 8 est égale à 2.

<u>COROLLAIRE 1</u> : Soit (X_n) une suite indépendante de variables aléatoires dans le Banach E, p-intégrables avec $p \geq 1$, et centrées ($E(X_n) = 0$ $\forall n$). Soit $(a_n) \in \mathbb{R}^{\mathbb{N}}$; pour tout $n \in \mathbb{N}$, on a la double inégalité :

$$(1) \qquad \frac{1}{2} \inf_{k\leq n} |a_k| . \|\sum_{k\leq n} X_k\|_{L^p(E)} \leq \|\sum_{k\leq n} a_k X_k\|_{L^p(E)} \leq 2 \sup_{k\leq n} |a_k| . \|\sum_{k\leq n} X_k\|_{L^p(E)} .$$

Et par conséquent, si (ξ_n) est une suite de v.a. numériques telles que les (ξ_n) sont indépendantes des X_n, on a pour tout n :

$$(2) \quad \frac{1}{2} \|\inf_{k\leq n} |\xi_k|\|_{L^p(E)} \|\sum_{k\leq n} X_k\|_{L^p(E)} \leq \|\sum_{k\leq n} \xi_k X_k\|_{L^p(E)} \leq 2 \|\sup_{k\leq n} |\xi_k|\|_{L^p(E)} \|\sum_{k\leq n} X_k\|_{L^p(E)} .$$

DEMONSTRATION : L'inégalité de droite dans (1) a déjà été démontrée. L'inégalité de gauche est triviale si $\inf |a_k| = 0$. Sinon, en posant $Y_k = a_k X_k$, on obtient :

$$\|\sum_{k\leq n} X_k\|_{L^p(E)} = \|\sum_{k\leq n} \frac{1}{a_k} a_k X_k\|_{L^p(E)} \leq 2 \sup \frac{1}{|a_k|} \|\sum_{k\leq n} a_k X_k\|_{L^p(E)} ,$$

ce qui donne le résultat cherché.

(2) se déduit immédiatement de (1) en intégrant les membres de (1) par rapport à la loi de $(\xi_1, \xi_2, \ldots, \xi_n)$.

Le corollaire 1.(2) sera principalement appliqué dans le cas où les (ξ_n) ne peuvent prendre que les valeurs 1 et -1 presque sûrement.

Auquel cas $\inf |\xi_k| = \sup |\xi_k| = 1$.

Ce sera le cas si les (ξ_n) sont des Bernoulli (par exemple les Rademacher).

COROLLAIRE 2 : Soit $1 \leq p < \infty$ et soit (ε_n) une suite de Bernoulli. Supposons qu'il existe une constante $C < \infty$ telle que pour tout entier n et toute famille $(x_o, x_1, \ldots, x_n)$ d'éléments de E^n on ait :

$$(3) \qquad \|\sum_{k\leq n} \varepsilon_k x_k\|_{L^p(E)} \leq C \left(\sum_{k\leq n} \|x_k\|_E^p\right)^{\frac{1}{p}}$$

Alors pour toute suite de variables aléatoires à valeurs dans E, indépendantes, p-intégrables et centrées, on a pour tout n :

$$(4) \qquad \|\sum_{k\leq n} X_k\|_{L^p(E)} \leq 2C \left(\sum_{k\leq n} \|X_k\|_{L^p(E)}^p\right)^{\frac{1}{p}} .$$

DEMONSTRATION : La suite (X_n) étant donnée, choisissons la suite (ε_n) de Bernoulli indépendante de (X_n). En intégrant les deux membres de (3) par rapport à la loi des (ε_n), puis la loi des (X_n), on obtient :

$$\|\sum_{k\leq n} \varepsilon_k X_k\|_{L^p(E)} \leq C \left(\sum_{k\leq n} \|X_k\|_{L^p(E)}^p\right)^{\frac{1}{p}}$$

D'autre part, par le corollaire (3, 2) et la remarque ci-dessus :

$$\|\sum_{k\leq n} X_k\|_{L^p(E)} \leq 2 \|\sum_{k\leq n} \varepsilon_k X_k\|_{L^p(E)}$$. D'où le résultat.

A partir du corollaire 1, on peut très facilement déduire des résultats de comparaison. Nous laissons au lecteur éventuel le soin de la faire.

COROLLAIRE 3 : Soit $1 \leq p < \infty$ et soit (η_n) une suite de v.a. réelles indépendantes et p-intégrables. Soit E un Banach et (X_n) une suite de v.a. à valeurs dans E, indépendantes et p-intégrables. Soit (ε_n) une suite de Bernoulli. L'on suppose que :

- les (η_n) sont indépendantes de (X_n) ;
- (ε_n) est indépendante de $\left((\eta_n), (X_n)\right)$;
- $E\{\eta_n X_n\} = 0 \quad \forall n$.

On pose $N = \sup_n |\eta_n|$. Alors pour tout n on a l'inégalité

(5) $$E(\|\sum_{k\leq n} \eta_k X_k\|^p) \leq 4^p E(N^p)\, E(\|\sum_{k\leq n} \varepsilon_k X_k\|^p).$$

(Naturellement, cette inégalité n'a d'intérêt que si $E(N^p) < \infty$).

DEMONSTRATION : Puisque pour tout n, $\eta_n X_n \in L^p(E)$, que les $\eta_n X_n$ sont centrées et que (ε_n) est indépendante de $(\eta_n X_n)$, on a par le corollaire 1 :

$$E(\|\sum_{k\leq n} \eta_k X_k\|^p) \leq 2^p E(\|\sum_{k\leq n} \varepsilon_k \eta_k X_k\|^p).$$

D'autre part, puisque (η_n) est indépendante de $(\varepsilon_n X_n)$ et que la suite $(\varepsilon_n X_n)$ est indépendante centrée, on a

$$E(\|\sum_{k\leq n} \eta_k \varepsilon_k X_k\|^p) \leq 2^p E(\max_{k\leq n} |\eta_k|^p) \,.\, E(\|\sum_{k\leq n} \varepsilon_k X_k\|^p).$$

D'où le résultat.

On peut maintenant en déduire le

LEMME 10 : Soient p, (η_n), (X_n), E comme dans le corollaire 3 ci-dessus. Soit (ξ_n) une suite de v.a. réelles, indépendantes et p-intégrables. On suppose que :

- les (ξ_n) et les (X_n) sont indépendantes.
- $E(\xi_n X_n) = 0 \quad \forall n$.

et l'on pose $a = \inf_n E(|\xi_n|)$.

Pour tout n, on a alors :

(6) $$E\{\|\sum_{k\le n} \eta_k X_k\|^p\} \le \left(\frac{16}{a}\right)^p E(N^p)\, E\{\|\sum_{k\le n} \xi_k X_k\|^p\}.$$

(Cette inégalité n'a d'intérêt que si a > 0).

<u>DEMONSTRATION</u> : Soit (ε_n) une suite de Bernoulli indépendante de $((\xi_n), (\eta_n), (X_n))$. Posons $a_n = E(|\xi_n|)$. Par le corollaire 1, on a :

$$E(\|\sum_{k\le n} \varepsilon_k X_k\|^p) \le 2^p \max_{k\le n} \left(\frac{1}{a_k}\right)^p E(\|\sum_{k\le n} a_k \varepsilon_k X_k\|^p)$$

et, par le corollaire 3 :

$$E(\|\sum_{k\le n} \eta_k X_k\|^p) \le \left(\frac{8}{a}\right)^p E(N^p)\, E(\|\sum_{k\le n} a_k \varepsilon_k X_k\|^p).$$

Soit ε_n^* la suite de v.a. réelles définies par :

$$\varepsilon_n^* = \begin{cases} -\varepsilon_n & \text{si } \xi_n < 0 \\ \varepsilon_n & \text{si } \xi_n \ge 0 . \end{cases}$$

Il est facile de voir que les ε_n^* ne prennent que la valeur -1 et +1 et que $$\xi_n \cdot \varepsilon_n^* = |\xi_n| \cdot \varepsilon_n \quad \forall n.$$

Par ailleurs, on vérifie facilement que (ε_n^*) est indépendante de (ξ_n). Donc la suite (ε_n^*) est indépendante de $((\xi_n), (X_n))$.

Soit $\mathcal{F}_n = \sigma\{\varepsilon_k, X_k ; k \le n\}$ la tribu engendrée par des ε_k et les X_k $(k \le n)$. Alors :

$$\begin{aligned} \|\sum_{k\le n} a_k \varepsilon_k X_k\|_E^p &= \|E(\sum_{k\le n} |\xi_k| \varepsilon_k X_k \mid \mathcal{F}_n)\|_E^p \\ &\le E(\|\sum_{k\le n} |\xi_k| \varepsilon_k X_k\|_E^p \mid \mathcal{F}_n) \\ &= E\{\|\sum_{k\le n} \xi_k \varepsilon_k^* X_k\|_E^p \mid \mathcal{F}_n\} \end{aligned}$$

(ces inégalités ayant lieu presque sûrement). Donc, par intégration :

$$E\{\|\sum_{k\le n} \eta_k X_k\|^p\} \le \left(\frac{8}{a}\right)^p E(N^p)\, E(\|\sum_{k\le n} \varepsilon_k^* \xi_k X_k\|^p)$$

Mais, comme nous l'avons remarqué, (ε_n^*) est indépendante de $((\xi_n), (X_n))$;

et par le corollaire 1, nous obtenons

$$E(\|\sum_{k\leq n} \eta_k X_k\|^p) \leq (\frac{16}{a})^p E(N^p) E(\|\sum_{k\leq n} \xi_k X_k\|^p) .$$

Le lemme est donc démontré.

Nous pouvons maintenant obtenir sans peine le :

THEOREME :

Soit $1 \leq p < \infty$ et soit E un Banach. Soit (ξ_n), (η_n) deux suites de v.a. réelles indépendantes et p-intégrables. Soit (X_n) une suite de v.a. à valeurs dans E, indépendantes et p-intégrables. L'on suppose que

- $N = \sup_n |\eta_n| \in L^p(R)$
- $\inf_n E(|\xi_n|) > 0$
- les (ξ_n) sont indépendantes des (X_n) et les (η_n) sont indépendantes des (X_n)
- $E(\xi_n X_n) = E(\eta_n X_n) = 0 \quad \forall n.$

Alors on a les implications :

(a) $\{\sum_{k\leq n} \xi_k X_k, n \in \mathbb{N}\}$ bornée dans $L^p(E)$ $\Rightarrow$ $\{\sum_{k\leq n} \eta_k X_k ; n \in \mathbb{N}\}$ bornée ;

(b) $\sum \xi_k X_k$ converge dans $L^p(E)$ $\Rightarrow$ $\sum \eta_k X_k$ converge dans $L^p(E)$;

(c) $\{\sum_{k\leq n} \xi_k X_k, n \in N\}$ bornée dans $L^p(E)$ et $(\eta_n) \to 0$ p.s. $\Rightarrow$ $\sum \eta_k X_k$ converge dans $L^p(E)$.

C'est en effet immédiat d'après ce qui précède.

REMARQUE : On peut ajouter "commutativement" dans les termes des implications du théorème.

CHAPITRE IV

ESPACES DE SUITES ASSOCIÉS À UNE LOI SUR $\mathbb{R}$

N° 1 : DEFINITIONS FONDAMENTALES

Dans tout ce chapitre, E désignera un espace de Banach. (ξ_n), (η_n), etc..., désigneront des suites de variables aléatoires réelles indépendantes. (ε_n) désignera une suite de Bernoulli. Si μ est une probabilité sur $\mathbb{R}$, on dira que μ est dégénérée si $\mu = \delta_{(0)}$.

Enfin, une probabilité sur $\mathbb{R}$ sera appelée simplement une loi.

DEFINITION 1 : Soit E, (ξ_n) comme plus haut et soit $p \in [0, \infty]$. On posera :

- $\mathcal{C}^p_{(\xi_n)}(E) = \{(x_n) \in E^{\mathbb{N}} \; ; \sum \xi_n x_n$ converge commutativement dans $L^p(E)\}$.
- $\mathcal{B}^p_{(\xi_n)}(E) = \{(x_n) \in E^{\mathbb{N}} \; ; \sum \xi_n x_n$ est commutativement bornée dans $L^p(E)\}$.

En d'autres termes, dire que $(x_n) \in \mathcal{C}^p_{(\xi_n)}(E)$ équivaut à dire, avec les notations du chapitre I, que $(\xi_n x_n) \in \mathcal{C}(L^p(E))$. Interprétation analogue pour $\mathcal{B}^p_{(\xi_n)}(E)$.

Si toutes les (ξ_n) ont même loi μ, on écrira $\mathcal{C}^p_{\mu}(E)$ (resp. $\mathcal{B}^p_{\mu}(E)$) au lieu de $\mathcal{C}^p_{(\xi_n)}(E)$ (resp. $\mathcal{B}^p_{(\xi_n)}(E)$).

REMARQUE 1 : On pourrait de même considérer l'ensemble des $(x_n) \in E^{\mathbb{N}}$ telles que :

- $\sum \xi_n x_n$ converge dans $L^p(E)$
- ou bien $\{\sum_{k \leq n} \xi_k x_k \; ; \; n \in \mathbb{N}\}$ est bornée dans $L^p(E)$.

Mais cela ne présentera guère d'intérêt par la suite.

D'ailleurs, dans les deux cas suivants,

(a) les (η_n) sont symétriques et appartiennent à $L^p(\mathbb{R})$;

(b) les (η_n) appartiennent à $L^p(\mathbb{R})$ $(1 \leq p < \infty)$ et $E(\xi_n) = 0$ pour tout n.

Ces définitions coïncident d'aprés ce que l'on a vu au chapitre précédent.

On notera par β la répartition de Bernoulli : $\beta = \frac{1}{2}\,\delta_{(1)} + \frac{1}{2}\,\delta_{(-1)}$.
Alors $\mathcal{C}^p_{(\varepsilon_n)}(E) = \mathcal{C}^p_\beta(E)$ et $\mathcal{B}^p_{(\varepsilon_n)}(E) = \mathcal{B}^p_\beta(E)$.

Naturellement, si les (ξ_n) et les (ξ'_n) ont même répartition (sur $\mathbb{R}^{\mathbb{N}}$), on obtient les mêmes espaces de suites.

Les propriétés suivantes résultent presque immédiatement de ce que l'on a vu plus haut, au chapitre III :

(1) Si la suite (ξ_n) est symétrique,

$$\mathcal{C}^o_{(\xi_n)}(E) = \{(x_n) \in E^{\mathbb{N}} \ ; \ \sum \xi_n x_n \text{ converge p.s.}\}$$

et $$\mathcal{B}^o_{(\xi_n)}(E) = \{(x_n) \in E^{\mathbb{N}} \ ; \ \sup_n \|\sum_{k\leq n} \xi_k x_k\| < \infty \text{ p.s.}\}.$$

(2) $$\mathcal{C}^p_{(\xi_n)}(E) = \{(x_n) \in \mathcal{C}^o_{(\xi_n)}(E) \text{ et } \sup_n \|\xi_n x_n\| \in L^p(\mathbb{R})\} \ ;$$

$$\mathcal{B}^p_{(\xi_n)}(E) = \{(x_n) \in \mathcal{B}^o_{(\xi_n)}(E) \text{ et } \sup_n \|\xi_n x_n\| \in L^p(\mathbb{R})\} ,$$

comme cela résulte du théorème 3' et de ses corollaires du chapitre III.

(3) Si μ est non-dégénérée, $\mathcal{C}^o_\mu(E) \subset C_o(E)$ et $\mathcal{B}^o_\mu(E) \subset l^\infty(E)$ ($C_o(E)$ est l'espace des suites d'éléments de E tendant vers zéro).

(4) Si μ est une loi telle que $\int |t|^p \, d\mu(t) < \infty$ $(0 < p \leq 1)$, on a $l^p(E) \subset \mathcal{C}^p_\mu(E)$.

(5) Si μ est une loi telle que $\int |t|^p \, d\mu(t) < \infty$ $(1 \leq p < \infty)$, on a $l^1(E) \subset \mathcal{C}^p_\mu(E)$.

(6) Si Supp μ est compact, pour tout $p \in [0, \infty[$ on a $\mathcal{C}^o_\mu(E) = \mathcal{C}^p_\mu(E)$ et $\mathcal{B}^o_\mu(E) = \mathcal{B}^p_\mu(E)$. En particulier, $\mathcal{C}^p_\beta(E)$ et $\mathcal{B}^p_\beta(E)$ ne dépendent pas de $p \in [0, \infty[$; on les notera par $\mathcal{C}_\beta(E)$ et $\mathcal{B}_\beta(E)$. $\mathcal{C}_\beta(E)$ est alors l'ensemble des suites $(x_n) \in E^{\mathbb{N}}$ telles que $\sum x_n \varepsilon_n$ converge presque sûrement.

(7) Si pour tout $n \in \mathbb{N}$, ξ_n^s désigne une symétrisée de ξ_n, alors

$$\mathcal{C}^p_{(\xi_n)}(E) \subset \mathcal{C}^p_{(\xi_n^s)}(E) \quad \text{et} \quad \mathcal{B}^p_{(\xi_n)}(E) \subset \mathcal{B}^p_{(\xi_n^s)}(E) .$$

(8) Si μ^{*n} désigne la $n^{\text{ième}}$ convoluée de μ $(n \geq 1)$, on a pour tout n :

$$\mathcal{C}^p_{\mu}(E) \subset \mathcal{C}^p_{\mu^{*n}}(E) \quad \text{et} \quad \mathcal{B}^p_{\mu}(E) \subset \mathcal{B}^p_{\mu^{*n}}(E).$$

(9) On a les implications suivantes :

$$(x_n) \in \mathcal{C}^p_{(\xi_n)}(E),\ (a_n) \in l^\infty \Rightarrow (a_n x_n) \in \mathcal{C}^p_{(\xi_n)}(E) ;$$

$$(x_n) \in \mathcal{B}^p_{(\xi_n)}(E),\ (a_n) \in l^\infty \Rightarrow (a_n x_n) \in \mathcal{B}^p_{(\xi_n)}(E) ;$$

$$(x_n) \in \mathcal{B}^p_{(\xi_n)}(E),\ (a_n) \in c_o \Rightarrow (a_n x_n) \in \mathcal{C}^p_{(\xi_n)}(E) .$$

Naturellement, si μ est dégénérée, $\mathcal{B}^p_\mu(E) = \mathcal{C}^p_\mu(E) = E^{\mathbb{N}}$.

Nous allons maintenant mettre une topologie sur $\mathcal{B}^p_\mu(E)$ et $\mathcal{C}^p_\mu(E)$, si <u>μ est non dégénérée.</u>

Soit $(x_n) \in E^{\mathbb{N}}$; $E^{\mathbb{N}}$ s'envoie dans $\left(L^p(E)\right)^{\mathbb{N}}$ par l'application

$$(x_n) \rightsquigarrow (\xi_n x_n)_{n\in\mathbb{N}}$$

où les (ξ_n) sont des copies indépendantes de loi μ.

Cette application est évidemment injective (μ étant non dégénérée !) et elle envoie $\mathcal{C}^p_\mu(E)$ dans $\mathcal{C}\left(L^p(E)\right)$ et $\mathcal{B}^p_\mu(E)$ dans $\mathcal{B}\left(L^p(E)\right)$.

On munira $\mathcal{B}^p_\mu(E)$ et $\mathcal{C}^p_\mu(E)$ de la topologie induite par celle de $\mathcal{B}\left(L^p(E)\right)$.

<u>LEMME 1</u> :

μ étant non dégénérée, $\mathcal{B}^p_\mu(E)$ et $\mathcal{C}^p_\mu(E)$ sont des F-espaces ; $\mathcal{C}^p_\mu(E)$ est l'adhérence dans $\mathcal{B}^p_\mu(E)$ des suites presques nulles.

<u>DEMONSTRATION</u> : $\mathcal{B}^p_\mu(E)$ pouvant être considéré comme un sous-espace de $\mathcal{B}\left(L^p(E)\right)$, et ce dernier étant un F-espace, il suffit de démontrer que $\mathcal{B}^p_\mu(E)$ est fermé dans $\mathcal{B}\left(L^p(E)\right)$.

Soit (ξ_n) une suite de copies indépendantes de loi   ; soit, pour tout k, $X^k = (x_n^k\ \xi_n)_{n\in\mathbb{N}}$, on suppose que (X^k) converge vers $X = (X_n) \in \mathcal{B}(L^p(E))$. Pour tout n, les $(x_n^k)_{n\in\mathbb{N}}$ forment une suite de Cauchy dans E, donc convergeant vers x_n. Il est maintenant facile de voir que $(x_n^k\ \xi_n)_n \xrightarrow[k\to\infty]{} (x_n\ \xi_n)_n$ pour la topologie de $\mathcal{B}(L^p(E))$.

Le reste de la démonstration est trivial.

<u>REMARQUE 2</u> : Si $x = (x_n) \in \mathcal{C}_\mu^p(E)$, posons

$$Tx = \sum \xi_n x_n \quad (\xi_n \text{ de loi } \mu) ;$$

on a vu que $\quad ||Tx||_{L^p(E)} \leq ||x_n||_{\mathcal{C}_\mu^p(E)} \quad$ (Chapitre I, n°6).

En outre, si μ est symétrique ($\mu(A) = \mu(-A)$ pour tout borélien A de $\mathbb{R}$), ou si μ est telle que $\int |t|^p\, d\mu(t) < \infty$ $(1 \leq p < \infty)$ et $\int t\, d\mu(t) = 0$, on a en outre, pour $p < \infty$,

$$\frac{1}{2}\, ||2x||_{\mathcal{C}_\mu^p(E)} \leq ||Tx||_{L^p(E)} \leq ||x||_{\mathcal{C}_\mu^p(E)} ;$$

(Donc pour $p \geq 1$, on a une isométrie de $\mathcal{C}_\mu^p(E)$ dans $L^p(\mu)$).

En effet, cela résulte immédiatement de la définition de $||x||_{\mathcal{C}_\mu^p(E)}$ et du lemme 3 du chapitre III.

On a un résultat analogue pour $p = \infty$ si μ est à support compact.

N° 2 : <u>COMPARAISON DES $\mathcal{C}_\mu^p(E)$ ET $\mathcal{C}_\nu^q(E)$.</u>

Dans toute la suite, toutes les lois que nous considérerons seront supposées non dégénérées (sauf mention expresse du contraire).

<u>PROPOSITION 1</u> : Soient μ et ν deux lois, p et q deux éléments de $[0, \infty]$. On suppose que $\mathcal{C}_\mu^p(E) \subset \mathcal{B}_\nu^q(E)$, alors

(a) $\mathcal{C}_\mu^p(E) \subset \mathcal{C}_\nu^q(E)$

(b) $\mathcal{B}_\mu^p(E) \subset \mathcal{B}_\nu^q(E)$.

DEMONSTRATION : Soit i l'injection canonique de $\mathcal{C}^p_\mu(E)$ dans $\mathcal{B}^q_\nu(E)$; alors

- i est continue : En effet, si $x_n \to x$ dans $\mathcal{C}^p_\mu(E)$ et si $i(x_n) \to y$ dans $\mathcal{B}^q_\nu(E)$, alors $y = i(x)$ car les topologies des espaces considérés sont plus fines que les topologies induites par la topologie produit sur $E^{\mathbb{N}}$. Le théorème du graphe fermé permet alors de conclure.

- i envoie des suites presque nulles sur les suites presque nulles. Or les suites presque nulles sont denses dans $\mathcal{C}^p_\mu(E)$; donc l'image de $\mathcal{C}^p_\mu(E)$ par i est contenue dans l'adhérence dans $\mathcal{B}^q_\nu(E)$ des suites presque nulles, qui est $\mathcal{C}^q_\nu(E)$.

Soit maintenant $x = (x_n) \in \mathcal{B}^p_\mu(E)$. Soit $\sigma \in \Phi(\mathbb{N})$ et soit $x_\sigma = (y_0, y_1, \ldots)$ avec $y_i = x_i$ si $i \in \sigma$, $y_i = 0$ si $i \notin \sigma$; alors l'ensemble x_σ est borné dans $\mathcal{C}^p_\mu(E)$; donc son image par i est bornée dans $\mathcal{B}^q_\nu(E)$. Donc $i(x) \in \mathcal{B}^q_\nu(E)$.

COROLLAIRE : Soient μ, ν, p, q, comme dans la proposition 1.

On a l'équivalence :

$$\mathcal{C}^p_\mu(E) = \mathcal{C}^q_\nu(E) \iff \mathcal{B}^p_\mu(E) = \mathcal{B}^q_\nu(E).$$

Si en outre on a $\mathcal{B}^p_\mu(E) = \mathcal{C}^q_\nu(E)$, alors

$$\mathcal{C}^p_\mu(E) = \mathcal{B}^p_\mu(E) = \mathcal{C}^q_\nu(E) = \mathcal{B}^q_\nu(E).$$

C'est en effet immédiat.

LEMME 2 :

Soit (X_n) une suite symétrique de v.a. à valeurs dans le Banach E. On a les équivalences :

(1) $\{\sum_{k\leq n} X_k \,;\, n \in \mathbb{N}\}$ est p.s. bornée $\iff (X_n(\omega))_{n\in\mathbb{N}} \in \mathcal{B}_\beta(E)$;

(2) $\sum_n X_n$ converge p.s. $\iff (X_n(\omega))_{n\in\mathbb{N}} \in \mathcal{C}_\beta(E)$ p.s.

Démontrons (1), (2) se démontrant de manière tout à fait analogue. Naturellement, d'après les résultats du chapitre III, dire que $\{\sum_{k\leq n} X_j \,;\, n \in \mathbb{N}\}$

est p.s. bornée, signifie que $(X_n) \in \mathscr{C}(L^0(E))$ ((X_n) étant symétrique).

Posons $S_n = \sum_{k\leq n} X_k$. Soit ε_n des Rademacher indépendantes des (X_n). On a vu que $(\varepsilon_n X_n)$ et (X_n) ont même répartition. Par conséquent :

$$P\left(\{S_n \,;\, n \in \mathbb{N}\} \text{ est bornée}\right) = \int_{E^{\mathbb{N}}} P\,(\omega \,;\, \{\sum_{k\leq n} \varepsilon_n(\omega)\, x_n \,;\, n\} \text{ est bornée}) \; P_X(dx_0, dx_1, \dots).$$

Maintenant, l'intégrant vaut 1 si $(x_n) \in \mathscr{B}^0_\beta(E)$ et zéro sinon.

Donc $\qquad P\,(\{S_n \,;\, n \in \mathbb{N}\} \text{ est bornée}) = P\;\{\omega \,;\, (X_n(\omega)) \in \mathscr{B}^0_\beta(E)\}$

et ceci démontre (1).

On déduit de ce résultat un résultat de comparaison :

<u>PROPOSITION 2</u> : Soit (X_n) une suite de v.a. à valeurs dans le Banach E, (ξ_n) une suite de v.a. numériques. On suppose que :

(a) les suites (X_n) et $(\xi_n X_n)$ sont symétriques ;

(b) $\sup_n |\xi_n| < \infty$ p.s.

Alors

- $\sum X_n$ converge p.s. $\Longrightarrow$ $\sum \xi_n X_n$ converge p.s.
- $\{\sum_{k\leq n} X_k \,;\, n \in \mathbb{N}\}$ est bornée en proba. $\Longrightarrow$ $\{\sum_{k\leq n} \xi_k X_k \,;\, n \in \mathbb{N}\}$ est bornée en probabilité.

Si l'on suppose en outre

(b') $\xi_n \to 0$ p.s., alors

$\{\sum_{k\leq n} X_k \,;\, n \in \mathbb{N}\}$ bornée en probabilité $\Longrightarrow$ $\sum \xi_n X_n$ converge p.s.

<u>DEMONSTRATION</u> : Comme (X_n) et $(\xi_n X_n)$ sont symétriques, on a

$$P\,\{\sum X_n \text{ converge}\} = P\,\{\omega \,;\, (X_n(\omega))_n \in \mathscr{C}_\beta(E)\}$$

$$P\,\{\sum \xi_n X_n \text{ converge}\} = P\{\omega \,;\, (X_n(\omega)\,\xi_n(\omega)) \in \mathscr{C}_\beta(E)\}\,.$$

D'après la propriété (9) du n° 1,

$$P\,\{\omega \,;\, (X_n(\omega))_n \in \mathscr{C}_\beta(E)\} = 1 \Longrightarrow P\,\{\omega \,;\, (\xi_n(\omega) X_n(\omega))_n \in \mathscr{C}_\beta(E)\} = 1 \,;$$

d'où le résultat.

PROPOSITION 3 : Soit $p \geq 1$ et $\xi_n \in L^p(\mathbb{R})$ $\forall n$; supposons $E(\xi_n) = 0$ $\forall n$.

Alors on a les égalités :

(1) $\mathcal{B}^p_{(\xi_n)}(E) = \mathcal{B}^p_{(\varepsilon_n \xi_n)}(E)$; et par conséquent

(2) $\mathcal{C}^p_{(\xi_n)}(E) = \mathcal{C}^p_{(\varepsilon_n \xi_n)}(E)$,

où $(\varepsilon_n)_n$ est indépendante de $(\xi_n)_n$.

DEMONSTRATION : Par le corollaire 1 de la proposition 1 du chapitre III, on a

$$\frac{1}{2} \, \|\inf_{k \leq n} |\varepsilon_k|\|_{L^p(E)} \, \|\sum_{k \leq n} \xi_k x_k\|_{L^p(E)} \leq \|\sum_{k \leq n} \varepsilon_k \xi_k x_k\|_{L^p(E)}$$

$$\leq 2 \, \|\sum_{k \leq n} |\xi_k|\|_{L^p(E)} \, \|\sum_{k \leq n} \xi_k x_k\|_{L^p(E)} ,$$

ce qui suffit pour démontrer (1).

(2) résulte alors trivialement de (1) et de la proposition 1.

PROPOSITION 4 : Soit (ξ_n) une suite de variables aléatoires indépendantes.

L'on suppose que :

(1) $\operatorname{Sup}|\xi_n| \in L^p(\mathbb{R})$ $(1 \leq p)$ et $\inf E\{|\xi_n|\} = a > 0$;

(2) $E(\xi_n) = 0$ $\forall n$.

Alors

- $\mathcal{C}^o_\beta(E) = \mathcal{C}^p_\beta(E) = \mathcal{C}^p_{(\xi_n)}(E)$;
- $\mathcal{B}^o_\beta(E) = \mathcal{B}^p_\beta(E) = \mathcal{B}^p_{(\xi_n)}(E)$.

DEMONSTRATION : Soit (ε_n) une suite de Bernoulli indépendante de la suite (ξ_n). Nous allons appliquer le théorème 7 du chapitre III deux fois :

- la première fois en remplaçant X_n par x_n, ξ_n par ε_n et η_n par ξ_n.

On en déduit alors :

$$\mathcal{C}^p_\beta(E) \subset \mathcal{C}^p_{(\xi_n)}(E) ;$$

- la seconde fois en remplaçant X_n par x_n, ξ_n par ξ_n et η_n par ε_n.

Alors $\mathcal{C}^p_{(\xi_n)}(E) \subset \mathcal{C}^p_\beta(E)$.

Donc on a $\mathcal{C}^p_{(\xi_n)}(E) = \mathcal{C}^p_\beta(E)$. Comme l'on sait que $\mathcal{C}^p_\beta(E) = \mathcal{C}^o_\beta(E)$, on a démontré la première partie.

La seconde partie se déduit trivialement de la première ; on peut d'ailleurs la démontrer directement à l'aide du théorème 7 du chapitre III.

Comme conséquence facile, on obtient alors le

THEOREME 1 :

Soit μ une loi non dégénérée à support compact et de moyenne nulle. Pour tout $p \in [0, \infty[$ on a :

(1) $\mathcal{B}^p_\mu(E) = \mathcal{B}_\beta(E)$

(2) $\mathcal{C}^p_\mu(E) = \mathcal{C}_\beta(E)$.

DEMONSTRATION : On sait déjà que $\mathcal{B}^p_\mu(E)$ ne dépend pas de p (μ étant à support compact). Soit (ξ_n) une suite de copies indépendantes de loi μ. La suite (ξ_n) satisfait aux conditions (1) et (2) de la proposition 4 pour tout $p \geq 1$.

On en déduit donc que pour tout $p \geq 1$, $\mathcal{B}^p_\mu(E) = \mathcal{B}^p_\beta(E)$. D'où le résultat pour tout $p \in [0, \infty[$.

Le théorème 1 signifie que $\mathcal{C}^p_\mu(E)$ et $\mathcal{B}^p_\mu(E)$ ne dépendent pas de μ pour autant qu'elle soit non dégénérée, à support compact et de moyenne nulle. Dans le cas général, on a le

THEOREME 2 :

Pour toute loi μ (non dégénérée !) on a les inclusions

- $\mathcal{C}^p_\mu(E) \subset \mathcal{C}^o_\beta(E)$ $\quad \forall\, p \in [0, \infty[$
- $\mathcal{B}^p_\mu(E) \subset \mathcal{B}^o_\beta(E)$ $\quad \forall\, p \in [0, \infty[$.

DEMONSTRATION : Il suffit de démontrer la 2ème inclusion. On peut, d'après les propriétés énoncées au n° 1, supposer μ symétrique.

Par ailleurs, on peut, pour tout $n \geq 1$ remplacer μ par μ^{*n}. Mais il est bien connu que pour tout intervalle borné I de $\mathbb{R}$, $\mu^{*n}(I)$ tend vers zéro (uniformément sur la longueur de l'intervalle). En définitive, on supposera :

- μ symétrique ;

- $\mu([-1, +1]) < \frac{1}{2}$.

Soit (η_n) une suite de v.a. réelles indépendantes et de loi μ. Si $(x_n) \in \mathcal{B}^p_\mu(E)$, alors d'après le lemme 2, pour presque tout ω,

$$(X_n(\omega))_n = (\xi_n(\omega)\, x_n)_n \in \mathcal{B}^c_\beta(E) = \mathcal{B}^1_\beta(E).$$

Il résulte alors des principes de contraction que si (ξ_n) est une famille de v.a. (sur le même espace probabilisé que celui sur lequel sont définis les (η_n)) telle que

$$|\xi_n| \leq |\eta_n| \text{ p.s. } \forall n ,$$

la famille $(x_n \xi_n)$ appartient à $\mathcal{B}^o_\beta(E)$ p.s.

Nous allons maintenant choisir les (ξ_n) satisfaisant à la condition ci-dessus. Pour cela, nous aurons besoin du

LEMME 3 :

Soit (ξ_n) une suite de v.a. indépendantes de loi $\frac{1}{2}\delta_{(0)} + \frac{1}{2}\delta_{(1)}$. Soit (x_n) une suite d'éléments de E. On suppose que p.s. $(\xi_n x_n) \in \mathcal{B}^o_\beta(E)$. Alors $x_n \in \mathcal{B}^o_\beta(E)$.

DEMONSTRATION : On peut supposer pour le moment que les (ξ_n) sont définis sur l'espace

$$(\mathbb{Z}_2^{\mathbb{N}}, \mathcal{B}_{\mathbb{Z}_2^{\mathbb{N}}}, m) ,$$

où $\mathbb{Z}_2$ désigne le groupe additif modulo 2, $\mathcal{B}_{\mathbb{Z}_2^{\mathbb{N}}}$ est la tribu borélienne du groupe compact $\mathbb{Z}_2^{\mathbb{N}}$ et m la probabilité de Haar : c'est la mesure produit des probabilités de Haar des composantes ; les (ξ_n) sont alors les composantes.

On désignera par $\oplus$ l'addition dans $\mathbb{Z}_2$ et dans $\mathbb{Z}_2^{\mathbb{N}}$.

Soit $A = \{\omega \in \mathbb{Z}_2^{\mathbb{N}} ; (\omega_n x_n)_n \in \mathcal{B}^o_\beta(E)\}$.

Par hypothèse, m(A) = 1. Donc, et c'est une propriété bien connue de "la" mesure de Haar d'un groupe localement compact, $A \oplus A$ contient un voisinage de zéro. On peut naturellement supposer que ce voisinage de zéro est un ouvert élémentaire de la forme

$$V = \{\omega = (\omega_n),\ \omega_0 = 0,\ \omega_1 = 0,\ \dots\ \omega_N = 0\} \text{ où } N \in \mathbb{N}.$$

Par conséquent, si $(\omega_n) \in V$, il existe (ω'_n) et (ω''_n) dans A tels que

$$\omega'_n \oplus \omega''_n = 1 \quad \forall\, n \geq N+1.$$

Mais si $\omega'_n \oplus \omega''_n = 1$, on a aussi $\omega'_n + \omega''_n = 1$ au sens de l'addition dans $\mathbb{Z}$.

Or, par hypothèse, $(\omega'_n x_n)$ et $(\omega''_n x_n)$ appartiennent à $\mathcal{B}^o_\beta(E)$; donc aussi $\big((\omega'_n \oplus \omega''_n)x_n\big) = (y_n)$.

Mais si $n \geq N+1$, $(\omega'_n + \omega''_n)\, x_n = x_n$. Donc si $(y_n) \in \mathcal{B}^o_\beta(E)$, on a également $(x_n) \in \mathcal{B}^o_\beta(E)$; et le lemme est démontré.

<u>REMARQUE 3</u> : On aurait pu seulement supposer que $(\xi_n x_n)$ appartient à $\mathcal{B}^o_\beta(E)$ pour un ensemble de ω de probabilité non nulle.

<u>REMARQUE 4</u> : La démonstration du lemme a utilisé un espace probabilisé particulier. Il est clair que la conclusion ne dépend pas de cet espace.

Pour achever la démonstration du théorème 2, il nous faut trouver des v.a. indépendantes (ξ_n) de loi $\frac{1}{2}\delta_{(0)} + \frac{1}{2}\delta_{(1)}$, et telles que $|\xi_n| \leq |\eta_n|$ p.s. $\forall$ n. Nous allons faire intervenir le fait que $\mu([-1, +1]) < \frac{1}{2}$.

Soit η une v.a. de loi μ. On peut supposer que η est définie sur l'espace probabilisé unité $([0, 1]\ ,\ \lambda = dx)$.

Il suffit en effet pour cela de poser pour $\omega \in [0, 1]$:

$$\eta(\omega) = \sup \{x \ ;\ \mu(]-\infty, x]) \leq \omega\}\ .$$

Puisque $\mu\,([-1, +1]) < \frac{1}{2}$, il existe $a > 0$ tel que

$$\mu\,\{|t| > a\} \leq \frac{1}{2} \leq \mu\,\{|t| \geq a\}\ .$$

Et par conséquent

$$\lambda\,\{\omega\ ;\ |\eta(\omega)| > a\} \leq \frac{1}{2} \leq \lambda\,\{\omega\ ;\ |\eta(\omega)| \geq a\}.$$

Mais, λ étant diffuse, il existe un borélien B de [0, 1] tel que :

- $\{\omega\ ;\ |\eta(\omega)| > a\} \subset B \subset \{\omega\ ;\ |\eta(\omega)| \geq a\}$;
- $\lambda(B) = \frac{1}{2}$.

Posons alors $\xi(\omega) = 1_B \circ \eta$. Il est clair que ξ a pour loi $\frac{1}{2}\delta_{(0)} + \frac{1}{2}\delta_{(1)}$ et que $|\xi| \leq |\eta|$.

Maintenant, à partir de là, il est facile de définir une suite (ξ_n) de v.a. indépendantes, de loi $\frac{1}{2}\delta_{(0)} + \frac{1}{2}\delta_{(1)}$ et telles que $|\xi_n| \leq |\eta_n|$ $\forall$ n.

Examinons maintenant où nous en sommes :

- D'après le principe de contraction :

$$\text{si } (x_n) \in \mathcal{B}^p_\mu(E) \quad \text{alors} \quad \text{p.s. } (\xi_n x_n) \in \mathcal{B}_\beta(E).$$

- Il résulte alors du lemme 3 que $(x_n) \in \mathcal{B}_\beta(E)$.

Le théorème est donc démontré.

COROLLAIRE : Soient $\mu \in \mathcal{P}(\mathbb{R})$ $(\mu \neq \delta_{(0)})$ et $p \in \,]0, \infty[$. Il existe une constante K (dépendant de μ et de p en général) telle que, pour tout n et pour toute famille finie $\{x_1, x_2, \ldots, x_n\}$ de vecteurs de E on ait :

$$E\{\|\sum_{j=0}^{n} \varepsilon_j\, x_j\|^p\} \leq K \max_{0 \leq k \leq n} E\{\|\sum_{j=0}^{k} \eta_j\, x_j\|^p\} ,$$

où (ε_n) est une suite de Bernoulli et les (η_n) des copies indépendantes de loi μ.

DEMONSTRATION : Soit (η^s_n) une symétrisée de (η_n) et soit $(x_n) \in E^{\mathbb{N}}$. Notons tout d'abord que :

$$\sup_n \|\sum_{j \leq n} \eta^s_j\, x_j\|_{L^p(E)} \leq 2 \sup_n \|\sum_{j \leq n} \eta_j x_j\|_{L^p(E)} .$$

Donc la quantité du 1er membre de cette inégalité est finie quand celle du 2è membre l'est.

Mais, à cause de la symétrie, dire que $\sup_n \|\sum_{j \leq n} \eta^s_j\, x_j\|_{L^p(E)} < \infty$ équivaut à dire que $(x_n) \in \mathcal{B}^p_{\mu^s}(E)$; par le théorème 2, on en déduit que $(x_n) \in \mathcal{B}^o_\beta(E)$.

Soit $\mathcal{E}$ l'espace vectoriel des suites $x = (x_n) \in E^{\mathbb{N}}$ telles que :

$$||x||_{\mathcal{E}} = \sup_n ||\sum_{j\leq n} \eta_j x_j||_{L^p(E)} < \infty .$$

Il est facile de voir, comme dans le lemme 1, que $\mathcal{E}$ est un F-espace muni de la F-norme ci-dessus.

En outre, la topologie de $\mathcal{E}$ est plus fine que celle induite par la topologie produit sur $E^{\mathbb{N}}$.

Maintenant, l'injection canonique de $\mathcal{E}$ dans $\mathcal{B}^p_\beta(E)$ étant continue pour la topologie produit, est, par le théorème du graphe fermé, continue pour les topologies de F-espaces. Mais c'est précisément équivalent à ce qu'il fallait démontrer.

<u>REMARQUE 5</u> : Les variables (ε_n) et (η_n) n'ont pas besoin d'être définies sur le même espace probabilisé. C'est ainsi que l'on peut écrire :

$$\int_0^1 ||\sum_{i\leq n} R_i(t)\; x_i||^p \,dt \;\leq\; K \sup_{0\leq k\leq n} \int ||\sum_{i\leq k} \eta_i(\omega)\; x_i||^p \,P(d\omega).$$

N° 3 : <u>ESPACES DE SUITES ASSOCIES A UNE LOI STABLE D'ORDRE</u> p

Soit $0 < p \leq 2$, on appelle γ_p la loi sur $\mathbb{R}$ dont la transformée de Fourier est la fonction $t \rightsquigarrow e^{-|t|^p}$. On dit que γ_p est la loi stable symétrique d'ordre p. Naturellement, si $p = 2$, γ_2 n'est autre que la loi gaussienne normale réduite.

Si $p = 2$, nous savons que γ_2 possède des moments absolus de tous ordres et nous connaissons la densité de γ_2 (par rapport à la mesure de Lebesgue).

<u>Aussi, supposons d'abord que $p < 2$</u>. Et nous étudierons ensuite directement le cas $p = 2$.

Posons $\psi(t) = \int_0^\infty (1 - \cos ut)\,\frac{du}{u^{1+p}}$ (cette intégrale ayant évidemment un sens pour $0 < p < 2$).

Alors, si $s > 0$, on a :

$$\psi(st) = s^p \psi(t) .$$

Puisque la fonction $t \rightsquigarrow \psi(t)$ est paire, on en déduit que :

$$\psi(t) = K'_p |t|^p \quad \text{où } K'_p \text{ est une constante.}$$

La formule (1) : $|t|^p = K_p \int_0^\infty (1 - \cos ut) \frac{du}{u^{p+1}}$ (où l'on pose $K_p = K'^{-1}_p$) a une conséquence importante : <u>Pour tout $q \in]0, p[$, γ_p a des moments absolus d'ordre q. En outre, γ_p n'a pas de moment absolu d'ordre p.</u>

En effet, soit X une v.a. réelle quelconque. On déduit de (1) que si $0 < q < 2$,

$$E\{|X|^q\} = K_q \times \int_0^\infty (1 - \operatorname{Re} \varphi_X(t)) \frac{dt}{t^{q+1}} \quad (\text{avec } \varphi_X(t) = E\{e^{itX}\})$$

En particulier, si X a une loi stable d'ordre p :

$$E\{|X|^q\} = K_q \int_0^\infty (1 - e^{-|t|^p}) \frac{dt}{t^{q+1}} .$$

Et la dernière intégrale n'est finie que si $0 < q < p$. D'où le résultat annoncé.

Pour $p = 1$, γ_p n'est autre que la loi de Cauchy, et l'on connaissait déjà le résultat relatif à l'existence des moments.

Dû au fait que $t \rightsquigarrow e^{-|t|^p}$ est intégrable pour la mesure de Lebesgue, γ_p a une densité, disons f_p (et naturellement, c'est un fait archi-connu si $p = 1$ ou 2 !).

Indiquons un résultat qui sera très important pour la suite :

Si $0 < p < 2$, $f_p(x)$ est équivalent à $\frac{1}{|x|^{p+1}}$ au voisinage de l'infini : c'est-à-dire, il existe a et $b > 0$:

$$\frac{a}{|x|^{p+1}} \leq f_p(x) \leq \frac{b}{|x|^{p+1}}$$

pour x suffisamment grand. (Ce qui permet de retrouver le résultat sur l'existence des moments). En effet :

- c'est bien connu si $p = 1$ (et c'est trivial !)

- pour $p \neq 1$, cela résulte de FELLER, Volume II, p. 583.

Naturellement, $f_2(x)$ est rapidement décroissante, donc la propriété ci-dessus est fausse pour $p = 2$.

On en déduit que pour $|x|$ suffisamment grand, on a :

$$\limsup_{|y| \to \infty} \frac{f_p(xy)}{f_p(y)} \simeq \frac{K}{|x|^{p+1}} . \qquad (2)$$

De (2), on déduit alors que si X est une v.a. stable symétrique d'ordre p, on a : il existe $K_1 > 0$ tel que

$$\text{pour tout } t \geq 1, \quad P\{|X| \geq st\} \leq K\,|t|^{-p}\, P\{|X| \geq s\} \qquad (3)$$

pour $|s| > K_1$.

REMARQUE 6 : Si $p = 2$, $f_2(x)$, densité de la loi γ_2, est a décroissance rapide ; en outre, pour tout $q \in]0, \infty[$ il existe une constante K_1 et une constante K_2, positives, telles que :

$$\forall\, t \geq 1, \qquad P\{|X| \geq st\} \leq K_1\,|t|^{-q}\, P\{|X| \geq s\}$$

pour tout $s > K_2$.

Dans la suite, nous utiliserons la convention suivante :

Si p est tel que $0 < p \leq 2$, on posera :

$$p^* = \begin{cases} p & \text{si } p < 2 \\ +\infty & \text{si } p = 2 \end{cases}$$

Le principal résultat est alors le suivant :

THEOREME 2 :

Soit E un Banach et $(\xi_n)_n$ une suite de v.a. réelles indépendantes de loi γ_p $(0 < p \leq 2)$. Soit (x_n) une suite de vecteurs de E. On pose :

$$S_n = \sum_{k=0}^{n} \xi_k\, x_k \quad \text{si } n \in \mathbb{N}. \text{ Alors :}$$

1) Si $\{S_n, n \in \mathbb{N}\}$ est bornée en probabilité dans $L^0(E)$, elle est bornée dans $L^q(E)$ pour tout $q \in]0, p^*[$.

2) Si (S_n) converge dans $L^0(E)$, elle converge dans $L^q(E)$ pour tout $q \in]0, p^*[$.

<u>DEMONSTRATION</u> : Soit $q \in]0, p^*[$, nous nous servirons du fait qu'il existe deux constantes positives C_1 et C_2 telles que :

$$P\{|\xi_1| > st\} \leq C_1 |t|^{-q} P\{|\xi_1| > s\}$$

pour tout $t \geq 1$ et tout $s \geq C_2$.

Démontrons 1). $\{S_n, n \in \mathbb{N}\}$ étant bornée en probabilité par le corollaire 1 du théorème 3 du chapitre III, il suffit de démontrer que $q \in]0, p^*[$ étant donné, $N = \sup_n ||\xi_n x_n|| \in L^{q'}(\mathbb{R})$ pour tout $q' < q$.

Ce qui revient à dire que $\sum_{k \in \mathbb{N}} P\{N \geq k^{\frac{1}{q'}}\} < \infty$. Il nous faut donc évaluer $P\{N \geq t\}$ pour $t > 0$.

Soit $t > 0$, on a $P\{\sup_n ||\xi_n x_n|| > t\} \leq \sum_n P\{||\xi_n x_n|| > t\}$.

Remarquons tout d'abord que, d'après la remarque (2) suivant le théorème 1 du chapitre III, $N < \infty$ presque sûrement.

La v.a. réelle $\limsup_n ||\xi_n x_n||$ appartient à la tribu asymptotique des v.a. (ξ_n). D'après la loi de zéro-un, elle est presque certaine, et finie, car $N < \infty$ p.s.

Posons donc $c = \limsup_n ||\xi_n x_n||$.

Soit $A_n = \{||\xi_n x_n|| \geq c+1\}$; les A_n sont indépendantes et $P\{\limsup_n A_n\} = 0$. Donc, par Borel-Cantelli, cela implique que $\sum P(A_n) < \infty$.

Pour tout $t \geq 0$, posons $F_p(t) = P\{|\xi_1| \geq t\}$.

Si l'on pose $a_n = \dfrac{c+1}{||x_n||}$, alors $P(A_n) = F_p(a_n)$ et par conséquent $\sum_n F_p(a_n) < \infty$.

Dû au fait que F_p est décroissante et que γ_p est non dégénérée, on a :

$$\inf_n a_n = b > 0 .$$

En effet, il existe $\alpha > 0$ tel que $F_p(\alpha) > 0$ et alors $a_n > \alpha$ pour n assez grand. D'où le résultat annoncé.

Naturellement :

$$\sum_n P\{||\xi_n x_n|| \geq t\} = \sum_n P\{|\xi_n| \geq \frac{t}{||x_n||}\}$$
$$= \sum_n P\{|\xi_n| \geq \frac{a_n}{c+1} t\} .$$

Remplaçant le cas échéant x_n par $\frac{b}{C_2} x_n$, on peut supposer que $a_n \geq C_2 \ \forall n$.
Alors, si $\frac{t}{c+1} \geq 1$:

$$P\{|\xi_n| \geq \frac{a_n}{c+1} t\} \leq C_1 (c+1)^q t^{-q} P\{|\xi_n| \geq a_n\}$$

$$\leq C_1 (c+1)^q t^{-q} F_p(a_n).$$

Mais, dû au fait que $\sum_n F_p(a_n) < \infty$, on déduit que :

$$P\{N \geq t\} \leq \sum_n P\{\|\xi_n x_n\| \geq t\} \leq K_1 t^{-q}$$

où K_1 est une constante. Si maintenant $q' < q$, on en déduit :

$$\sum_{k \geq k_o} P\{N \geq k^{\frac{1}{q'}}\} \leq K_1 \sum_{k \geq k_o} k^{-\frac{q}{q'}} + k_o < \infty \text{ pour un } k_o \text{ de } \mathbb{N}.$$

Donc, pour tout $q' < q$, $E\{N^{q'}\} < \infty$ et (1) est donc démontré.

2) se déduit alors immédiatement de 1) en utilisant le corollaire 1 du théorème 3 du chapitre III. Et le théorème est complétement démontré.

<u>REMARQUE 7</u> : Pour démontrer le théorème 2, nous n'avons pas utilisé toutes les propriétés de la loi γ_p, mais seulement le fait suivant :

"Il existe une constante $r \in]0, \infty]$ (en l'occurence, dans le théorème 2 $r = p^*$) tel que pour tout $q \in]0, r[$ on ait :

$$P\{|\xi_1| \geq st\} \leq C_1 t^{-q} P\{|\xi_1| > s\}, \ \forall t \geq 1, \forall s \geq C_2$$

(où C_1 et C_2 sont des constantes finies positives dépendant de q éventuellement)".

Par conséquent, si μ est une loi satisfaisant seulement à la condition ci-dessus, les conclusions du théorème 2 restent valables. Par exemple, si la loi de μ est à support compact, $r = +\infty$.

<u>COROLLAIRE 1</u> : Soit $p \in]0, 2]$ et E un Banach, alors :

- $\mathcal{B}^o_{\gamma_p}(E) = \mathcal{B}^q_{\gamma_p}(E) \quad \forall q \in]0, p^*[$;
- $\mathcal{C}^o_{\gamma_p}(E) = \mathcal{C}^q_{\gamma_p}(E) \quad \forall q \in]0, p^*[$;

C'est immédiat.

Dû au fait que l'on a affaire à des F-espaces, les isomorphismes algébriques ci-dessus sont des isomorphismes topologiques. Donc :

COROLLAIRE 2 : Si E est un Banach et si les (ξ_n) sont des copies réelles indépendantes de loi γ_p, pour tout q, $r \in]0, p^*[$ il existe une constante finie $C_{q,r}$ telle que pour toute famille $(x_j)_{j\in\sigma}$ finie de vecteurs de E on ait :

$$\|\sum_{j\in\sigma} x_j \xi_j\|_{L^r(E)} \leq C_{q,r}\|\sum_{j\in\sigma} x_j \xi_j\|_{L^q(E)} .$$

REMARQUE 8 : Les résultats ci-dessus peuvent s'interpréter autrement. Soit $(\xi_j)_{j\in J}$ une famille quelconque (donc pas forcément dénombrable !) de v.a. indépendantes réelles de loi γ_p $(0 < p \leq 2)$ (Naturellement, on a choisi $(\Omega, \mathcal{F}, P)$ de telle façon que cette famille existe !) et soit E un Banach.

On considère le sous-espace de $L^0(E)$ formé des v.a. de la forme :

$$X = \sum_{j\in\sigma} \xi_j x_j$$

où σ décrit l'ensemble des parties finies de J et où les x_j sont des vecteurs de E.

Soit $\mathcal{E}$ sa fermeture en probabilité. Alors, pour tout $q \in [0, p^*[$, on a $\mathcal{E} \subset L^p(E)$; en outre, les topologies induites sur $\mathcal{E}$ par les topologies des $L^q(E)$ coîncident.

En particulier si p > 1, la topologie de $\mathcal{E}$ est une topologie de Banach. Dans le cas de E = $\mathbb{R}$, ce résultat pouvait s'obtenir plus facilement en remarquant que :

1) Si ξ est une v.a. de loi γ_p et si $a \in \mathbb{R}$, $a\xi$ a pour loi la loi dont la fonction caractéristique est $t \rightsquigarrow \exp(-|a|^p |t|^p)$.

2) Si $\xi_1, \xi_2, \ldots, \xi_n$ sont des v.a. indépendantes de loi γ_p et si $a_1, a_2, \ldots, a_n$ sont des constantes réelles, $\sum a_i \xi_i$ a pour fonction caractéristique $t \rightsquigarrow \exp\left(-(\sum|a_i|^p) |t|^p\right)$. D'où l'on conclut que :

3) Si les $(\xi_n)_{n\in\mathbb{N}}$ sont des v.a. indépendantes de loi γ_p et si $(a_n) \in \mathbb{R}^{\mathbb{N}}$, on a l'équivalence :

$$(a_n) \in l^p \iff \sum_{n\in\mathbb{N}} a_n \xi_n \quad \text{existe en probabilité ; et}$$

$\sum a_n \xi_n$ a pour fonction caractéristique $t \rightsquigarrow e^{-\sum |a_n|^p |t|^p}$.

4) Si $r \in]0, p^*[$, il existe une constante finie $K_{r,p}$ telle que l'on ait :

$$(\sum |a_i|^p)^{\frac{1}{p}} = K_{r,p}\left(\int |\sum a_i \, \xi_i(\omega)|^r \, P(d\omega)\right)^{1/r}$$

pour toute $(a_i) \in \mathbb{R}_0^N$ (où les ξ_i sont indépendantes de loi γ_p). En effet, il suffit par raison d'homogénéité de supposer $\sum |a_i|^p = 1$. Mais alors $\sum a_i \xi_i$ est une v.a. stable symétrique d'ordre p. On trouve alors :

$$K_{r,p} = \|\gamma_p\|^{-1}_{L^r(\Omega, \mathcal{F}, P)} .$$

5) Pour tout $\varepsilon > 0$, si l'on pose, X étant une v.a. réelle :

$$J_\varepsilon(X \, ; \, P) = \inf \{A > 0, \, P\{|X| > A\} \leq \varepsilon\} , \quad \text{alors}$$

$$(\sum |a_i|^p)^{1/p} = K_{\varepsilon,p} \, J_\varepsilon(\sum a_i \, \xi_i \, ; \, P) ,$$

où les (a_i) et les (ξ_i) sont comme dans le point 4). Il suffit en effet de prendre $K_{\varepsilon,p} = J_\varepsilon(\gamma_p)^{-1}$.

Maintenant, le résultat annoncé se déduit, si $r > 0$, de la définition de la topologie de $L^r(\Omega, \mathcal{F}, P)$ et du fait que la topologie de $L^0(\Omega, \mathcal{F}, P)$ est définie par la famille de "jauges" $(J_\varepsilon)_{\varepsilon>0}$.

N° 4 : UNE NOUVELLE CARACTERISATION DES C-ESPACES DE BANACH

Jusqu'à présent, nous avons surtout composé les $\mathcal{D}^p_\mu(E)$ entre eux et les $\mathcal{C}^p_\mu(E)$ entre eux.

On peut naturellement se poser alors le problème : à quelles conditions portant sur μ et E a-t-on l'égalité suivante :

$$\mathcal{C}^p_\mu(E) = \mathcal{D}^p_\mu(E) \quad ?$$

Le seul résultat dans ce sens que l'on a obtenu est lorsque μ est la répartition de Bernoulli. Plus précisément :

THEOREME 3 :

Soit E un Banach ; on a l'équivalence de :

(1) E est un C-espace ;

(2) $\mathcal{B}_\beta(E) = \mathcal{C}_\beta(E)$;

(3) $\mathcal{B}_\beta(E) \subset C_o(E)$.

Pour démontrer ce résultat, il suffit de démontrer le

THEOREME 3' :

Soit E un Banach ; on a l'équivalence de :

(1) E n'est pas un C-espace (i.e. E contient C_o isomorphiquement) ;

(2) $\mathcal{B}_\beta(E) \neq \mathcal{C}_\beta(E)$;

(3) $\mathcal{B}_\beta(E) \not\subset C_o(E)$.

DEMONSTRATION :

(1) ⟹ (2) est trivial, en vertu du fait que la base canonique de C_o appartient à $\mathcal{B}_\beta(E)$ mais non à $\mathcal{C}_\beta(E)$ comme on le voit facilement.

(2) ⟹ (3) : Supposons donc que $\mathcal{B}_\beta(E) \neq \mathcal{C}_\beta(E)$ et soit $(x_n) \in \mathcal{B}_\beta(E) \setminus \mathcal{C}_\beta(E)$. Soit (ε_n) une suite de Bernoulli indépendantes ; alors la suite $(\sum_{k\leq n} \varepsilon_k x_k)_{n\in\mathbb{N}}$ n'est pas de Cauchy dans $L^1(E)$. Il existe donc $\alpha > 0$ et une suite $0 = n_o < n_1 < n_2 < \ldots$ telle que :

$$E\{\|\sum_{n_k\leq j<n_{k+1}} \varepsilon_j x_j\|\} \geq a \quad \forall k.$$

Posons $X_k = \sum_{n_k\leq j<n_{k+1}} \varepsilon_j x_j$; les X_k sont indépendantes et symétriques.

Alors $E\{\|X_k\|\} \geq a \quad \forall k$; en outre, dû au fait que $\mathcal{B}^o_\beta(E) = \mathcal{B}^1_\beta(E)$, il existe $M \in L^1(\mathbb{R})$ tel que :

$$\|X_k(\omega)\| \leq M(\omega) \quad \text{p.s.}$$

Il résulte alors du théorème de Lebesgue que :

$$P\{X_n \not\to 0\} > 0 .$$

Maintenant, grâce au fait que les (X_k) sont symétriques, il résulte du lemme 2 que $\left(X_n(\omega)\right)_n \in \mathcal{B}_\beta(E)$ presque sûrement.

En définitive, il existe $\omega \in \Omega$ tel que $\left(X_n(\omega)\right)_n \notin C_0(E)$ et $(X_n(\omega))_n \in \mathcal{B}_\beta(E)$. Donc (2) $\Rightarrow$ (3).

(3) $\Rightarrow$ (1) : Si (3) est vérifiée, on voit facilement qu'il existe une suite (x_n) de vecteurs telle que

$$\inf \|x_n\| > 0 \quad \text{et} \quad P\,\{\sup_n \|\sum_{i \leq n} \varepsilon_i x_i\| = \infty\} = 0$$

où les (ε_i) sont des v.a.de Bernoulli (définies sur un $(\Omega, \mathcal{F}, P)$).

Soit $\mathcal{G}$ la plus petite tribu sur Ω rendant les (ε_n) mesurables ; pour tout $B \in \mathcal{G}$ on a :

$$(1) \quad \lim_{n\to\infty} P\left(B \cap \{\varepsilon_n = 1\}\right) = \lim_{n\to\infty} P\left(B \cap \{\varepsilon_n = -1\}\right) = \frac{1}{2} P(B) .$$

En effet, cela est trivial si $B \in \sigma(\varepsilon_k \,;\, k \leq n)$; et le cas général en résulte facilement.

Soit $M(\omega) = \sup_n \|\sum_{k\leq n} \varepsilon_k x_k\|$; et soit $\infty > K > 0$ telle que :

$$P\,\{\omega \,;\, M(\omega) < K\} > \frac{1}{2} \quad ; \quad \text{Posons } A = \{M < K\} .$$

Alors $A \in \mathcal{G}$. Il résulte de (1) que l'on peut définir par récurrence eune suite croissante d'indices (n_k) telle que :

$$(2) \quad \begin{cases} - P\,(A \cap \{\varepsilon_{n_0} = 1\}) > \frac{1}{4} \quad \text{et} \quad P\,(A \cap \{\varepsilon_{n_1} = -1\}) > \frac{1}{4} \;; \\ - P\,(A \cap \bigcap_{i=0}^{k} \{a_i \varepsilon_{n_i} = 1\}) > \frac{1}{2^{k+2}} \quad \text{pour toute famille} \\ \quad (a_i)_{0\leq i\leq k} \in \{-1, +1\}^{(k+1)} \quad (k = 1, 2, \ldots) \end{cases}$$

Soit $\varepsilon_i' = \varepsilon_i$ si $i = n_k$ et $\varepsilon_i' = -\varepsilon_i$ autrement, et soit

$$A' = \{\omega \,;\, \sup_n \|\sum_{i\leq n} \varepsilon_i'(\omega)\, x_i\| \leq K\} \,.$$

Dû au fait que les suites (ε_n) et (ε_n') ont même loi, il résulte de (2) que :

$$(3) \quad P\left(A \cap \bigcap_{i=0}^{k} \{a_i \varepsilon_i = 1\}\right) = P\left(A \cap \bigcap_{i=0}^{k} \{a_i \varepsilon_i' = 1\}\right) > \frac{1}{2^{k+2}}$$

pour tout $(a_i)_{i\in\mathbb{N}} \in \{-1, +1\}^{\mathbb{N}}$ et tout $k \in \mathbb{N}$.

Fixons-nous $k \in \mathbb{N}$ et $a_o, a_1, \ldots, a_k$; alors, puisque

$$P\{a_o\varepsilon_{n_o} = 1, a_1\varepsilon_{n_1} = 1, \ldots, a_k\varepsilon_{n_k} = 1\} = \frac{1}{2^{k+1}},$$

il résulte de (2) et (3) qu'il existe $\omega_o \in A \cap A' \cap \bigcap_{i=0}^{k} \{a_i\varepsilon_{n_i} = 1\}$.

Ainsi donc :

$$\|\sum_{i=0}^{k} a_i x_{n_i}\| = \|\frac{1}{2}(\sum_{j=0}^{n_k} \varepsilon_j(\omega_o)x_j + \sum_{j=0}^{n_k} \varepsilon'_j(\omega_o)x_j)\| \leq K.$$

De par le principe de contraction, il résulte alors que (x_{n_k}) est une C-suite telle que $\inf \|x_{n_i}\| > 0$. Donc E n'est pas un C-espace. Et le théorème est démontré.

COROLLAIRE : Soit E un espace de Banach ; les propriétés suivantes sont équivalentes :

(1) E est un C-espace ;

(2) Pour toute suite (X_n) symétrique de v.a. à valeurs dans E, la bornitude p.s. des sommes partielles $S_n = \sum_{j\leq n} X_j$ implique la convergence presque sûre de (S_n).

DEMONSTRATION : Il suffit de remarquer que la suite (X_n) étant symétrique, la bornitude p.s. des S_n équivaut à $(S_n(\omega))_n \in \mathcal{B}_\beta(E)$ pour presque tout ω (grâce au lemme 2).

REMARQUE : Dans le cas $E = \mathbb{R}$ où plus généralement E est un Hilbert, ce résultat était connu.

CHAPITRE V

TYPES DES ESPACES DE BANACH

N° 1 : DEFINITIONS GENERALES

DEFINITION 1 : Soit E un espace de Banach, μ une probabilité sur $\mathbb{R}$ $(\mu \neq \delta_{(0)})$ et soient p et q deux nombres réels tels que $0 < p < \infty$ et $0 \leq q < \infty$. L'on dit que :

- E est de type (p, q, μ) si $l^p(E) \subset \mathcal{C}_\mu^q(E)$
- E est de cotype (p, q, μ) si $\mathcal{C}_\mu^q(E) \subset l^p(E)$.

Nous nous bornerons dans la suite à étudier le type.

En vertu du théorème du graphe fermé, dire que le Banach E est du type (p, q, μ) équivaut à dire que $l^p(E)$ s'injecte continuement dans $\mathcal{C}_\mu^q(E)$.

Dû au fait que les suites presque nulles sont denses dans $l^p(E)$, E est de type (p, q, μ) si et seulement si $l^p(E) \subset \mathcal{B}_\mu^q(E)$.

Si (η_n) est une suite de copies indépendantes de loi μ, la condition "E est de type (p, q, μ)" équivaut aux conditions suivantes :

- dans le cas $q \neq 0$:

A) Il existe une constante $K < \infty$ telle que pour tout $(x_n) \in E^{\mathbb{N}}$ on ait :

$$\left(\int \|\sum \eta_n(\omega)x_n\|^q P(d\omega)\right)^{1/q} \leq K \left(\sum \|x_n\|\right)^{1/p} ;$$

ou encore

A') Il existe une constante $K < \infty$ telle que pour toute famille finie $\{x_1, x_2, \ldots, x_n\}$ d'éléments de E on ait :

$$\left(\int \|\sum_{i=1}^n \eta_i(\omega)x_i\|^q P(d\omega)\right)^{1/q} \leq K \left(\sum_{i=1}^n \|x_i\|^p\right)^{1/p} .$$

- dans le cas $q = 0$:

B) Pour tout $\alpha \in]0, 1[$, il existe une constante $K_\alpha < \infty$ telle que pour toute suite $(x_n) \in E^{\mathbb{N}}$ on ait :

$$J_\alpha(\|\sum \eta_n x_n\|, P) \leq K_\alpha(\sum \|x_n\|^p)^{1/p} ;$$

ou encore

B') Pour tout $\alpha \in]0, 1[$, il existe une constante $K_\alpha < \infty$ telle que pour toute famille finie $\{x_1, x_2, \ldots, x_n\}$ d'éléments de E on ait :

$$J_\alpha(\|\sum_{i=1}^{n} \eta_i x_i\|, P) \leq K_\alpha(\sum_{i=1}^{n} \|x_i\|^p)^{1/p} .$$

L'on rappelle que :

$$J_\alpha(\|\sum \eta_n x_n\|, P) = \inf \{A > 0 ; P\{\|\sum \eta_n x_n\| > A\} \leq \alpha\} .$$

Sous cette forme, on peut généraliser la notion de type à un espace normé, ou même un espace r-normé $(0 < r \leq 1)$.

<u>DEFINITION 2</u> : Soit E un espace r-normé ; E est dit de type (p, q, μ) si l'une des conditions A') ou B') est satisfaite.

On peut également généraliser la notion de type comme suit :

<u>DEFINITION 3</u> : Soient E et F deux espaces quasi-normés et $u : E \to F$ linéaire continue. u est dite de type (p, q, μ) si elle satisfait à la condition suivante, <u>dans le cas $q \neq 0$</u> : Il existe une constante $K < \infty$ telle que pour toute famille finie $\{x_1, x_2, \ldots, x_n\}$ d'éléments de E on ait :

$$\left(\int \|\sum \eta_i(\omega) u(x_i)\|_F^q \, P(d\omega)\right)^{1/q} \leq K\left(\sum_{i=1}^{n} \|x_i\|_E^p\right)^{1/p}$$

(définition analogue dans le cas $q = 0$).

Dire que E est de type (p, q, μ) équivaut alors à dire que l'application identique de E dans E est de type (p, q, μ).

Soit E un espace normé de type (p, q, μ) :

- il est de type (p', q', μ) pour tout $p' \leq p$ et $q' \geq q$;
- tout sous-espace de E est de type (p, q, μ) ;
- tout quotient séparé de E est de type (p, q, μ).

Les principaux cas que nous examinerons par la suite sont :

- $\mu = \beta$ est la répartition de Bernoulli.
- $\mu = \gamma_p$ est la loi stable symétrique d'ordre p $(0 < p \leq 2)$.

REMARQUE : On peut encore généraliser comme suit la notion de type : soit (φ_n) une suite de v.a. sur un $(\Omega, \mathcal{F}, P)$; on dit que E est de type $(p, q, (\varphi_n))$ où p et q sont comme dans la définition 1 si sont satisfaites des conditions analogues aux conditions A') et B') en remplaçant les (η_n) par les (φ_n).

N° 2 : ESPACES DE TYPE RADEMACHER (OU BERNOULLI)

L'on sait que pour tout $q \in]0, \infty[$ on a :

$$\mathcal{C}^q_\beta(E) = \mathcal{C}^o_\beta(E) .$$

On en déduit alors que si le Banach E est de type (p, q_o, β) pour un $q_o \in [0, \infty[$, il est de type (p, q, β) pour tout $q \in [0, \infty[$.

On dira donc que E est de type p-Rademacher (ou simplement de type p) s'il est de type (p, q, β) pour un certain $q \in [0, \infty[$.

E est de type p-Rademacher s'il existe un $q \in [0, \infty[$ tel qu'une condition de type A) ou B) ci-dessus soit satisfaite. Les propriétés suivantes sont immédiates :

1) Tout espace normé est de type p-Rademacher pour tout $p \in]0, 1]$.

2) $\mathbb{R}$ est de type 2-Rademacher (et aussi de cotype 2-Rademacher). Cela résulte en effet de l'équivalence suivante :

$(x_n) \in l^2 \iff \sum x_n \varepsilon_n$ converge presque sûrement,

(où (ε_n) est une suite de Bernoulli).

3) Si $E \neq 0$ est de type p-Rademacher, nécessairement $p \leq 2$. En effet, soit $e \in E$ non nul et soit $(\lambda_n) \in \mathbb{R}^{\mathbb{N}}$ tel que $\sum (\lambda_n e)\varepsilon_n$ converge p.s. ; alors $\sum \lambda_n \varepsilon_n$ converge presque sûrement ; donc $\sum |\lambda_n|^2 < \infty$.

Si donc E est de type p, la condition $\sum |\lambda_n|^p < \infty$ implique $\sum |\lambda_n|^2 < \infty$; d'où le résultat.

EXEMPLE : Soit $(X, \mathcal{F}, \mu)$ un espace mesuré σ-fini et $p \in [1, \infty[$; l'espace $L^p(X, \mathcal{F}, \mu)$ est de type min $(p, 2)$-Rademacher. Plus généralement, si E est un Banach de type s-Rademacher, $L^p(X, \mathcal{F}, \mu, E)$ a le type min (p, s).

DEMONSTRATION : Soient $f_1, f_2, \dots, f_n$ n éléments de $L^p(E)$ et soit (ε_n) une suite de Bernoulli définie sur un $(\Omega, \mathcal{F}, P)$. Alors :

$$\int_\Omega \|\sum_{i=1}^n f_i \, \varepsilon_i(\omega)\|^p_{L^p(E)} \, P(d\omega) = \int_X d\mu(x) \int_\Omega \|\sum_{i=1}^n f_i(x) \, \varepsilon_i(\omega)\|^p_E \, P(d\omega).$$

Mais E étant de type s-Rademacher, pour tout x on a :

$$\int_\Omega \|\sum_{i=1}^n f_i(x) \, \varepsilon_i(\omega)\|^p_E \, P(d\omega) \leq \text{Cte} \, (\sum_{i=1}^n \| f_i(x)\|^s_E)^{p/s}$$

(on a noté par la même lettre un élément de $L^p(E)$ et un de ses représentants.)

Donc, en définitive :

$$\int_\Omega \|\sum_{i=1}^n f_i \varepsilon_i(\omega)\|^p_{L^p(E)} \, P(d\omega) \leq \text{Cte} \int_X (\sum_{i=1}^n \|f_i(x)\|^s_E)^{p/s} \, d\mu(x) .$$

Si donc $p \leq s$, $(\sum_{i=1}^n \|f_i(x)\|^s_E)^{p/s} \leq \sum \|f_i(x)\|^p$ et $L^p(X, \mathcal{F}, \mu, E)$ est de type p-Rademacher.

Si $p > s$, par Minkowski on obtient :

$$\left[\int_X (\sum_{i=1}^n \|f_i(x)\|^s_E)^{p/s} \, d\mu(x)\right]^{s/p} \leq \sum_{i=1}^n \left(\int_X \|f_i(x)\|^p_E \, d\mu(x)\right)^{s/p}$$

$$= \sum_{i=1}^n \|f_i\|^s_{L^p(E)}$$

et $L^p(E)$ est donc de type s-Rademacher.

REMARQUE 1 : On en déduit que E est de type p-Rademacher $(1 < p \leq 2)$ si et seulement s'il existe un espace $(X, \mathcal{F}, \mu)$ σ-fini tel que $L^p(X, \mathcal{F}, \mu, E)$ soit de type p-Rademacher.

En effet, il suffit de remarquer que E peut alors s'identifier à un sous-espace de $L^p(X, \mathcal{F}, \mu, E)$ par l'application $e \rightsquigarrow 1_A \cdot e$ de E dans $L^p(X, \mathcal{F}, \mu, E)$ où $A \in \mathcal{F}$ et $0 < \mu(A) < \infty$.

PROPOSITION 1 :

Soit $1 \leq p \leq 2$ et E un Banach ; on a équivalence de :

(1) E est de type p ;

(2) Il existe $C < \infty$ telle que l'on ait :

$$E\{\|\sum_{k=1}^{n} X_k\|^p\} \leq C \sum_{k=1}^{n} E\{\|X_k\|^p\} ,$$

quelle que soit la suite (X_k) de v.a. dans $L^p(E)$ indépendantes et centrées.

DEMONSTRATION :

(1) $\Rightarrow$ (2) : Ce n'est autre qu'une reformulation du corollaire 2 du lemme 8 du chapitre III.

(2) $\Rightarrow$ (1) : Il suffit de prendre $X_k = \varepsilon_k x_k$ où $(\varepsilon_1, \varepsilon_2, \ldots, \varepsilon_n)$ est une suite de Bernoulli et les x_k sont des vecteurs de E.

COROLLAIRE : Soit $1 \leq p \leq 2$, et μ une loi non dégénérée et symétrique sur $\mathbb{R}$. On suppose que $\int |t|^p d\mu < \infty$. Soit E un Banach. Les propriétés suivantes sont équivalentes :

(1) E est de type p ;

(2) $l^p(E) \subset \mathcal{C}_\mu^o(E)$;

(3) $l^p(E) \subset \mathcal{B}_\mu^o(E)$; et en particulier :

(4) $l^p(E) \subset \mathcal{B}_{\gamma_2}^o(E)$ (où γ_2 est la loi gaussienne réduite).

DEMONSTRATION :

(1) $\Rightarrow$ (2) : cela résulte de la proposition (1) en prenant les X_k de loi μ.

(2) $\Rightarrow$ (3) : cela a déjà été remarqué au n° 1.

(3) $\Rightarrow$ (1) : cela résulte de $\mathcal{B}_\mu^o(E) \subset \mathcal{B}_\beta^o(E)$ comme nous l'avons vu au chapitre IV.

On en déduit alors (1) $\Leftrightarrow$ (4) en faisant $\mu = \gamma_2$.

Le résultat suivant montre l'équivalence de la condition "E est de type p-Rademacher" avec la condition "E satisfait à une loi forte des grands nombres". Plus précisément :

THEOREME 1 : (HOFFMANN - JØRGENSEN)

Soit E un Banach et $p \in [1, 2]$. Les propriétés suivantes sont équivalentes :

(1) E est de type p ;

(2) Pour toute suite de v.a. de $L^p(E)$ indépendantes, centrées et vérifiant :

$$(*) \qquad \sum_{n\geq 1} \frac{E\{||X_n||^p\}}{n^p} < \infty \ ;$$

alors $\frac{1}{n} \sum_{1\leq k\leq n} X_k \longrightarrow 0$ presque sûrement ;

(3) Pour toute suite $(x_k) \in E^{\mathbb{N}}$ telle que $\sum_n \frac{||x_n||^p}{n^p} < \infty$, alors $\frac{1}{n} \sum_{1\leq k\leq n} \varepsilon_k x_k \longrightarrow 0$ presque sûrement.

DEMONSTRATION : Soit (X_n) une suite satisfaisant aux conditions de (2). Alors, d'après la proposition 1 :

$$E\{||\sum_{k=1}^{n} \frac{X_k}{k}||^p\} \leq C \sum_{k=1}^{n} \frac{E\{||X_k||^p\}}{k^p}$$

Donc $\sum_{k\geq 1} \frac{1}{k} X_k$ converge dans $L^p(E)$, et aussi presque sûrement. Il résulte alors du lemme de KRONECKER que $\frac{1}{n} \sum_{k=1}^{n} X_k \longrightarrow 0$ presque sûrement.

(2) ⟹ (3) est trivial.

(3) ⟹ (1) : Soit $\mathcal{E}$ le sous-espace de $E^{\mathbb{N}^*}$, formé des suites $(x_n)_{n\geq 1}$ telles que : $\sum_{n\geq 1} \frac{||x_n||^p}{n^p} < \infty$.

La fonction $(x_n) \rightsquigarrow \left(\sum_{n\geq 1} \frac{||x_n||^p}{n^p}\right)^{1/p}$ est alors une norme sur $\mathcal{E}$; et $\mathcal{E}$ est un Banach pour cette norme.

Soit $\mathcal{E}_1$ le sous-espace de $E^{\mathbb{N}^*}$ formé des suites $(x_n)_{n\geq 1}$ pour lesquelles

$$\sup_{n\geq 1} \frac{1}{n} \left(E(||\sum_{k\leq n} \varepsilon_k x_k||^p)\right)^{1/p} < \infty \ ;$$

$\mathcal{E}_1$ est un Banach pour la norme

$$(x_n) \rightsquigarrow \sup_n \frac{1}{n} ||\sum_{k\leq n} \varepsilon_k x_k||_{L^p(E)} .$$

Par hypothèse, $\mathcal{E} \subset \mathcal{E}_1$; il résulte alors du théorème du graphe fermé que l'injection $\mathcal{E} \longrightarrow \mathcal{E}_1$ est continue. Par conséquent, il existe une constante

$C < \infty$ telle que :

$$E\{\|\sum_{k=1}^{n} \varepsilon_k x_k\|^p\} \leq C\, n^p \sum_{k=1}^{n} \frac{\|x_k\|^p}{k^p} ,$$

pour tout $n \geq 1$ et toute famille $(x_1, x_2, \ldots, x_n)$ d'éléments de E.

Soient donc $x_1, x_2, \ldots, x_n$ donnés et soit un entier $N \geq 1$. Considérons la famille $\{y_1^N, y_2^N, \ldots, y_n^N, \ldots, y_{N+n}^N\}$ définie comme suit :

$$y_k^N = \begin{cases} 0 & \text{si } 1 \leq k \leq N \\ x_{k-N} & \text{si } N < k \leq N+n . \end{cases}$$

Alors
$$E\{\|\sum_{k=1}^{n} \varepsilon_k x_k\|^p\} = E\{\|\sum_{k=1}^{N+n} \varepsilon_k y_k^N\|^p\}$$

$$\leq C\,(N+n)^p \sum_{k=1}^{N+n} \frac{\|y_N^k\|^p}{k^p} = C\,(N+n)^p \sum_{k=1}^{n} \frac{\|x_k\|^p}{(N+k)^p}$$

$$\leq C\,(\frac{N+n}{N+1})^p\,(\sum_{k=1}^{n} \|x_k\|^p) .$$

Ceci étant vrai pour tout $N \geq 1$, on en déduit :

$$E\{\|\sum_{k=1}^{n} \varepsilon_k x_k\|^p\} \leq C(\sum_{k=1}^{n} \|x_k\|^p) ;$$

donc E est de type p et le théorème est démontré.

REMARQUE 1 : Dans le cas où E est un Hilbert, donc de type deux, ce résultat était déjà connu.

REMARQUE 2 : Si E est un Banach quelconque, E. MOURIER a démontré que la loi suivantes des grands nombres était valable : Si (X_n) est une suite de v.a. à valeurs dans E, indépendantes, intégrables, centrées et de même loi,

$$\frac{X_1 + X_2 + \ldots + X_n}{n} \longrightarrow 0 \text{ presque sûrement.}$$

Le théorème 1 n'implique pas ce résultat (compte tenu du fait qu'un Banach est de type un) et n'est pas impliqué par cette loi des grands nombres.

Nous reviendrons encore sur ce point au n° 4.

N° 3 : ESPACES DE TYPE p-STABLE

Soit $p \in]0, 2]$ et soit γ_p la loi stable symétrique d'ordre p. L'on sait que pour tout $q \in [0, p^*[$ on a :

$$\mathcal{C}^q_{\gamma_p}(E) = \mathcal{C}^0_{\gamma_p}(E) \quad \text{(algébriquement et topologiquement).}$$

Par conséquent :

E est de type (r, q, γ_p) $r \in]0, \infty[$, $0 < q < p^*$, $0 < p \le r$, si et seulement si E est de type $(r, 0, \gamma_p)$.

DEFINITION 3 : Soit E un Banach et $p \in]0, 2[$, E est dit p-stable si $l^p(E) \subset \mathcal{C}^0_{\gamma_p}(E)$.

Cela équivaut encore à : il existe $r \in]0, p^*[$ et il existe une constante $C < \infty$ telle que pour toute famille finie $x_1, x_2, \dots, x_n$ d'éléments de E on ait :

$$\left(\int_\Omega \left\|\sum_{k=1}^n \xi_k(\omega)\, x_k\right\|^r P(d\omega)\right)^{1/r} \le C \left(\sum_{k=1}^n \|x_k\|^p\right)^{1/p},$$

où les (ξ_k) sont indépendantes de loi γ_p.

Nous allons tout d'abord comparer la notion de type p-Rademacher et celle de type p-stable.

Pour cela, nous aurons besoin de lemmes.

LEMME 1 :

Soit E un espace normé et $p \in [1, \infty[$. Soit (φ_n) une suite symétrique de v.a. réelles intégrables sur un $(\Omega, \mathcal{F}, P)$ et soit (ε_n) une suite de Bernoulli définie sur un $(\Omega', \mathcal{F}', P')$. Soit $x = (x_n)$ une suite presque nulle d'éléments de E. Alors :

$$\inf_n \|\varphi_n\|_{L^1} \left(\int_{\Omega'} \left\|\sum \varepsilon_n(\omega')\, x_n\right\|^p P'(d\omega')\right)^{1/p} \le \left(\int_\Omega \left\|\sum \varphi_n(\omega) x_n\right\|^p P(d\omega)\right)^{1/p}.$$

DEMONSTRATION : Soit ω' fixé. Alors :

$$\left\| \int \left(\sum \varepsilon_n(\omega') \ |\varphi_n(\omega)| \ x_n\right) P(d\omega)\right\|$$

$$\leq \int \left\|\sum \varepsilon_n(\omega') \ |\varphi_n(\omega)| \ x_n\right\| P(d\omega) \ ;$$

et par conséquent,

$$\left\|\sum \varepsilon_n(\omega') \ x_n \ \|\varphi_n\|_{L^1}\right\|^p$$

$$\leq \int \left\|\sum \varepsilon_n(\omega') \ |\varphi_n(\omega)| \ x_n\right\|^p P(d\omega) \ .$$

Donc, en intégrant par rapport à ω' :

$$\int_{\Omega'} \left\|\sum \varepsilon_n(\omega') \ x_n \ \|\varphi_n\|_{L^1}\right\|^p P'(d\omega')$$

$$\leq \int_{\Omega\times\Omega'} \left\|\sum \varepsilon_n(\omega') \ |\varphi_n(\omega)| \ x_n\right\|^p dP \ dP' \ .$$

Mais, d'après le principe de contraction :

$$\inf_n \|\varphi_n\|_{L^1} \left(\int_{\Omega'} \left\|\sum \varepsilon_n(\omega') \ x_n\right\|^p P'(d\omega')\right)^{1/p}$$

$$\leq \left(\int_{\Omega'} \left\|\sum \varepsilon_n(\omega') \ x_n \|\varphi_n\|_{L^1}\right\|^p dP'\right)^{1/p}.$$

D'autre part, la suite (φ_n) étant symétrique :

$$\left(\int_{\Omega\times\Omega'} \left\|\sum \varepsilon_n(\omega')|\varphi_n(\omega)|x_n\right\|^p dPdP'\right)^{1/p} = \left(\int_{\Omega} \left\|\sum \varphi_n(\omega)x_n\right\|^p dP\right)^{1/p} .$$

Donc le lemme 1 est démontré.

LEMME 2 :

Soient p, q, r des nombres réels tels que $0 < r < q < p \leq 2$; soit $(\alpha_n) \in \mathbb{R}^{\mathbb{N}}$ et soit (f_n) une suite de v.a. réelles indépendantes de loi γ_q. Soit g_1 une v.a. de loi γ_p. On a alors

$$\|f_1\|_{L^2}\left(\sum |\alpha_n|^q\right)^{1/q} \leq \left[\int \left(\sum |\alpha_n|^p \ |f_n(\omega)|^p\right)^{r/p} P(d\omega)\right]^{1/r}$$

$$\leq \frac{\|f_1\|_{L^r} \ \|g_1\|_{L^q}}{\|g_1\|_{L^r}} \left(\sum |\alpha_n|^q\right)^{1/q} .$$

DEMONSTRATION : Soit (g_n) une suite stable d'ordre p et soit $(\beta_n) \in \mathbb{R}_o^N$

D'après ce que l'on a vu au chapitre IV, on a :

$$(1) \qquad \left(\sum |\beta_n|^p\right)^{1/p} ||g_1||_{L^r} = \left(\int |\sum \beta_n g_n(\omega)|^r P(d\omega)\right)^{1/r}$$

$$(2) \qquad \left(\sum |\beta_n|^q\right)^{1/q} ||f_1||_{L^r} = \left(\int |\sum \beta_n f_n(\omega')|^r P'(d\omega')\right)^{1/r} .$$

De (1) on déduit alors que

$$\int_{\Omega'} \left(\sum |\alpha_n|^p |f_n(\omega')|^p dP'\right)^{r/p} = \frac{1}{||g_1||_{L^r}} \left(\int_{\Omega\times\Omega'} |\sum \alpha_n f_n(\omega') g_n(\omega)|^r dP dP'\right)^{1/r}$$

Mais, par (2), l'intégrale du second membre vaut :

$$||f_1||_{L^1} \, ||g_1||_{L^r}^{-1} \left(\int_\Omega \left(\sum |\alpha_n|^q \, |g_n(\omega)|^q\right)^{r/q} dP\right)^{1/r}$$

Par ailleurs, on a :

$$\left(\int_\Omega \left(\sum |\alpha_n|^q \, |g_n(\omega)|^q\right)^{r/q} dP\right)^{1/r} \leq \left(\int \sum |\alpha_n|^q \, |g_n(\omega)|^q \, dP\right)^{1/q}$$

$$= (\sum |\alpha_n|^q)^{1/q} \, ||g_1||_{L^q} \quad ,$$

et d'autre part :

$$\left(\int \left(\sum |\alpha_n|^q \, |g_n(\omega)|^q\right)^{r/q} P(d\omega)\right)^{1/r} \geq \left[\sum \left(\int |\alpha_n|^r \, |g_n(\omega)|^r \, dP\right)^{q/r}\right]^{1/q}$$

$$= ||g_1||_{L^r} \, (\sum |\alpha_n|^q)^{1/q} \quad .$$

CQFD.

Comme conséquence, nous obtenons alors la :

PROPOSITION 2 : Soit E un Banach et $p \in]1, 2]$. Si E est de type p-stable, il est de type p-Rademacher.

DEMONSTRATION : Soit (φ_n) une suite stable d'ordre p. (φ_n) est symétrique et, du fait que $p > 1$, les φ_n sont intégrables. Puisque

$$\inf_n ||\varphi_n||_{L^1} = ||\varphi_1||_{L^1} > 0 \ ,$$

le lemme 1 montre que E est de type p-Rademacher.

REMARQUE 3 : La réciproque de la proposition 2 est vraie si p = 2 :

Donc type 2-stable $\iff$ type 2-Rademacher.

En effet, cela résulte trivialement du corollaire de la proposition 1.

REMARQUE 4 : Si $p < 2$, la réciproque de la proposition 2 est fausse, comme le montre l'exemple suivant :

Soit $E = l^p$ $(1 \leq p < 2)$; E est alors de type p-Rademacher comme on l'a vu plus haut. Je dis alors que E n'est pas de type p-stable.

En effet, supposons $E = l^p$ de type p-stable, c'est-à-dire $l^p(E) \subset \mathcal{C}^o_{\gamma_p}(E)$.

Soit (e_n) la base canonique de l^p et soit $(\lambda_n) \in l^p$. Posons $x_n \in \lambda_n e_n$. Alors $(x_n) \in l^p(E)$. On en déduit donc :

$$\sum |\lambda_n|^p < \infty \implies \sum |\lambda_n f_n|^p < \infty \text{ p.s. },$$

(f_n) désignant une suite stable d'ordre p.

Mais ce résultat contredit alors le

LEMME 3 :

Soit $0 < p < 2$ et (f_n) une suite stable d'ordre p. Soit $(\lambda_n) \in \mathbb{R}^{\mathbb{N}}$. On a l'équivalence de :

(a) $\sum |\lambda_n|^p \left(1 + \log \left|\frac{1}{\lambda_n}\right|\right) < \infty$;

(b) $\sum |\lambda_n f_n|^p < \infty$ presque sûrement.

DEMONSTRATION du lemme : Par le théorème des deux séries, on a l'équivalence :

$$\sum |\lambda_n f_n|^p < \infty \text{ p.s. } \iff \begin{cases} - \sum P\{|\lambda_n f_n| > 1\} < \infty \ ; \\ - \sum |\lambda_n|^p \int_{|\lambda_n f_n| \leq 1} |f_n|^p \, dP < \infty . \end{cases}$$

Mais, dû au fait que $\gamma_p \{u \in \mathbb{R} ; |u| \geq t\} \sim \frac{1}{t^p}$ $(t \to \infty)$, la première condition de droite équivaut à $\sum |\lambda_n|^p < \infty$. D'autre part, les conditions $(\lambda_n) \in c_o$ et la seconde condition de droite équivalent à :

$$\sum |\lambda_n|^p \int_1^{\frac{1}{|\lambda_n|}} \frac{dt}{t} < \infty.$$

Le lemme est donc démontré.

Toutefois, on a le :

THEOREME 2 :

Soit E un Banach et $p \in]0, 2]$. Si u est de type p-Rademacher, il est de type q-stable pour tout $q < p$.

DEMONSTRATION : Supposons que E est de type p-Rademacher. Alors pour tout $r \in]0, \infty[$ il existe $C_r < \infty$ tel que

$$||\sum x_n \varepsilon_n||_{L^r(E)} \leq C_r \left(\sum_n ||x_n||^p \right)^{1/p} .$$

Soit donc $q \in]0, p[$ et choisissons $r < p$. Si (f_n) est une suite stable d'ordre p, alors :

$$||\sum x_n f_n \varepsilon_n||_{L^r} = ||\sum f_n x_n||_{L^r} ,$$

si (ε_n) et (f_n) sont indépendantes. Et par conséquent,

$$||\sum f_n x_n||_{L^r} \leq C_r \left(\int \left(\sum_n |f_n(\omega)|^p \, ||x_n||^p \right)^{r/p} dP \right)^{1/r} .$$

Le lemme 2 donne immédiatement le résultat.

En particulier, tout Banach est de type q-stable pour tout $q < 1$; tout $L^p(X, \mathcal{F}, \mu)$ $(1 \leq p \leq 2)$ est de type q-stable pour tout $q < p$.

COROLLAIRE : Soit E un Banach et $p \in]0, 2]$; si E est de type p-stable, il est de type q-stable pour tout $q \in]0, p]$.

DEMONSTRATION : Le résultat est trivial si $p \leq 1$, car tout Banach est de type q-stable $(0 < q < 1)$.

Supposons donc $p > 1$. Par la proposition 2, E est de type p-Rademacher et le théorème 2 donne alors le résultat.

COROLLAIRE 2 : Si E est un espace de Banach de type p-Rademacher, $(p \in [1,2])$ pour tout $q \in]0, p[$ on a :

$$\mathcal{C}^{o}_{\gamma_q}(E) = l^q(E) .$$

<u>DEMONSTRATION</u> : On sait déjà par le théorème 2 que

$$l^q(E) \subset \mathcal{C}^o_{\gamma_q}(E) .$$

D'autre part, si E est un Banach et si $r \in]0, 2[$ on a :

$$\mathcal{C}^o_{\gamma_r}(E) \subset l^r(E) .$$

En effet, si les (f_n) sont des copies indépendantes de loi γ_p et si $(x_n) \in \mathcal{C}^o_{\gamma_r}(E)$, $\sum f_n x_n$ converge presque sûrement. Et par conséquent,

$$\sup_n ||f_n x_n|| < \infty \text{ p.s.}$$

Il existe donc $C > 0$ tel que $\sum P \{||f_n x_n|| \geq C\} < \infty$. D'où résulte que $(x_n) \in l^r(E)$. Et cela démontre le corollaire 2.

En particulier pour tout Banach E et tout $q \in]0, 1[$ on a

$$l^q(E) = \mathcal{C}^o_{\gamma_q}(E) .$$

<u>REMARQUE 5</u> : Si $E = \mathbb{R}$, on peut voir directement grâce aux propriétés des lois stables énoncées au chapitre IV que $\forall\ p \in]0, 2]$,

$$l^p = l^p(\mathbb{R}) = \mathcal{C}^o_{\gamma_p}(\mathbb{R}) .$$

Indiquons sans les démontrer quelques propriétés des espaces de type p-stable :

1) Soit $p \in [1, 2[$; tout Banach de type p-stable est de type $(p+\varepsilon)$-stable pour un $p \in]0, 2-p]$ (PISIER).

2) Soit $p \in]0, 2]$ et soit E un Banach de type p-stable. Soit $(\Omega, \mathcal{F}, P)$ un espace probabilisé et soit $q \in [0, p[$, soit $u : E \to L^q(\Omega, \mathcal{F}, P)$ linéaire continue. Soit r tel que $\frac{1}{q} = \frac{1}{r} + \frac{1}{p}$ (r = 0 si q = 0). u admet la factorisation suivante :

$$\begin{array}{ccc} E & \xrightarrow{\quad u \quad} & L^q(\Omega, \mathcal{F}, P) \\ & {\scriptstyle v}\searrow \quad \nearrow{\scriptstyle M_h} & \\ & L^p(\Omega, \mathcal{F}, P) & \end{array}$$

où v est linéaire continue et M_h est définie par $\varphi \rightsquigarrow h\varphi$, où $h \in L^r(\Omega, \mathcal{F}, P)$. Ce résultat est dû à MAUREY.

3) Soit $p \in [1, 2[$ et soit S un sous-espace d'un $L^p(\Omega, \mathcal{F}, P)$ (où $(\Omega, \mathcal{F}, P)$ est un espace probabilisé). Les propriétés suivantes sont équivalentes :

a) S est de type p-stable ;

b) S ne contient pas l^p isomorphiquement ;

c) S ne contient pas de sous-espace complémenté isomorphe à l^p ;

d) $L^p(\Omega, \mathcal{F}, P)$ et $L^0(\Omega, \mathcal{F}, P)$ induisent sur E la même topologie ;

En outre, <u>si p = 1</u>, les propriétés ci-dessus équivalent à :

c) S est réflexif.

(Ce résultat est dû à KADEC-PELCZYNSKI-PISIER, voir Studia Mathematica 21, pp. 161-176).

N° 4 : <u>ESPACES DE BANACH DE TYPE RADEMACHER</u>

Nous avons vu que tout Banach est de type p-Rademacher pour tout $p \in]0, 1]$. D'autre part, il existe des Banach qui ne sont pas de type p-Rademacher pour un $p \in]1, 2]$: c'est le cas pour l^1. En effet, si l^1 était de type p-Rademacher pour $p > 1$, il serait de type 1-stable, ce qui ne peut être.

Par ailleurs, il existe des Banach <u>réflexifs</u> qui ne sont pas de type p-Rademacher pour $p \in]1, 2]$. Par exemple, si

$$E = (\oplus_n l^1_n)_2 = \{y = (y_n)_n \quad y_n \in l^1_n \text{ et } ||y|| = (\sum ||y_n||^2_{l^1_n})^{1/2} < \infty\} ,$$

on voit que E muni de la norme $||.||$ est réflexif et ne peut être de type p-Rademacher pour un $p > 1$.

Aussi, la définition suivante présente a priori un intérêt :

<u>DEFINITION 4</u> : Un espace de Banach E est dit <u>de type Rademacher</u> s'il existe $p \in]1, 2]$ tel que E soit de type p-Rademacher.

Nous nous proposons de donner différentes caractérisations du type Rademacher. Pour cela, nous aurons besoin de nouvelles définitions.

DEFINITION 5 : Un Banach E est dit B-convexe s'il existe un entier $N \geq 2$ et $\varepsilon \in]0, 1[$ tel que pour tout $(x_1, x_2, \ldots, x_N) \in E^N$ on ait :

$$\inf_{\varepsilon_i = \pm 1} \left\| \sum_{k=1}^{N} \varepsilon_k x_k \right\| \leq N (1-\varepsilon) \sup_{k \leq N} \|x_k\| .$$

Cette condition a été introduite par A. BECK. Nous verrons plus loin son intérêt en liaison avec la loi des grands nombres.

DEFINITION 6 : Un espace de Banach E est dit uniformément convexe si pour tout $\varepsilon \in]0, 2[$ le nombre

$$\delta(\varepsilon) = \inf \{1 - \frac{\|x+y\|}{2} \ ; \ \|x\| \leq 1 \ ; \ \|y\| \leq 1 \ ; \ \|x-y\| \geq \varepsilon\}$$

est strictement positif.

On démontre qu'un espace uniformément convexe est réflexif. Les espaces $(L^p(X, \mathcal{F}, \mu)$ $(1 < p < \infty)$ sont uniformément convexes. Il existe des espaces réflexifs non uniformément convexes ; c'est le cas pour l'espace E ci-dessus.

DEFINITION 7 : Soit E un Banach et $\lambda \geq 1$; on dit que E contient les l^1_n λ-uniformément si pour tout $n \geq 1$ il existe $(x_1, x_2, \ldots, x_n)$ dans E tels que pour tous réels $\alpha_1, \alpha_2, \ldots, \alpha_n$ on ait la double inégalité :

$$\frac{1}{\lambda} \sum_{k=1}^{n} |\alpha_k| \leq \left\| \sum_{k=1}^{n} \alpha_k x_k \right\| \leq \sum_{k=1}^{n} |\alpha_k| .$$

Cela signifie encore que pour tout $n \geq 1$ il existe un sous-espace E_n de E de dimension n et un isomorphisme T_n de l^1_n sur E_n tel que :

$$\|T_n\| \cdot \|T_n^{-1}\| \leq \lambda \ ;$$

On dit que E_n et l^1_n sont λ-isomorphes.

REMARQUE : Les définitions ci-dessus nous conduisent à définir des constantes liées à E, exprimant certaines propriétés géométriques de la norme.

Si N est un entier, $\lambda_N(E)$ est définie comme la plus petite constante positive λ vérifiant pour tout N-uple $(x_1, x_2, \ldots, x_N)$ dans E :

$$\inf_{\varepsilon_i=\pm 1} \|\sum_{i=1}^{N} \varepsilon_i x_i\| \leq \lambda N \sup_{1\leq i\leq N} \|x_i\| .$$

L'on voit facilement que :

- $0 \leq \lambda_N(E) \leq 1 \quad \forall N$;

- $\lambda_N(E) \geq \frac{1}{N}$ si $E \neq \{0\}$; $\lambda_N(\{0\}) = 0 \quad \forall N$;

- $(N + N')\, \lambda_{N+N'}(E) \leq N\, \lambda_N(E) + N'\lambda_{N'}(E) \quad \forall N, N'$.

- $N < N' \Rightarrow N\lambda_N(E) \leq N'\lambda_{N'}(E)$.

De même, si N est un entier, on définit $\rho_N(E)$ par :

$$\frac{1}{\inf \|T_N\|\ \|T_N^{-1}\|}$$

où l'infimum est pris sur les isomorphismes linéaires de l_N^1 sur un sous-espace de dimension N de E. On voit facilement que

- $\rho_1(E) = 1$ et $N \rightsquigarrow \rho_N(E)$ est décroissante ;

- $\rho_N(E)=\inf\{\delta>0\ \forall (x_1,x_2,\ldots x_N) \in E^{\mathbb{N}} \inf_{\sum_{k=1}^{N}|\alpha_k|=1} \|\sum_1^N \alpha_k x_k\| \leq \delta \sup_{k\leq N}\|x_k\|\}$;

REMARQUE : Toutes les définitions ci-dessus gardent évidemment un sens si E est seulement supposé normé.

Indiquons une autre propriété des $\lambda_N(E)$:

PROPOSITION 3 :

Soit E un espace normé.

$\forall$ k, n $\in \mathbb{N}$, on a $\lambda_{nk}(E) \leq \lambda_n(E)\, \lambda_k(E)$ (autrement dit, $N \to \lambda_N(E)$ est sous-multiplicatif).

DEMONSTRATION : Soit $(x_j)_{1 \leq j \leq nk}$ un nk-uple d'éléments de E. Soit $i \in [1, n]$ et soit

$(\varepsilon_j^i)_{(i-1)k< j\leq ik}$ tel que l'on ait :

$$\Big\|\sum_{(i-1)k<j\le ik} \varepsilon_j^i x_j\Big\| = \inf_{\varepsilon_j=\pm 1} \Big\|\sum_{(i-1)k<j\le ik} \varepsilon_j x_j\Big\| .$$

Posons alors

$$X^i = \sum_{(i-1)k<j\le ik} \varepsilon_j^i x_j \quad (i \in [1, n]).$$

Alors il est clair que :

$$\|X_i\| \le \lambda_k(E)k \sup_{(i-1)k<j\le ik} \|x_j\| .$$

En outre,

$$\inf_{\varepsilon_i=\pm 1} \Big\|\sum_{i=1}^{n} \varepsilon_i X_i\Big\| \le n\,\lambda_n(E) \sup_{1\le i\le n} \|X_i\|$$

$$\le n\,\lambda_n(E)\, k\,\lambda_k(E) \sup_{1\le j\le nk} \|x_j\| .$$

En définitive,

$$\inf_{\varepsilon_j=\pm 1} \Big\|\sum_{j=1}^{nk} \varepsilon_j x_j\Big\| \le \inf_{\varepsilon_i=\pm 1} \Big\|\sum_{i=1}^{n} \varepsilon_i X_i\Big\|$$

$$\le nk\,\lambda_n(E)\,\lambda_k(E) \sup_{1\le j\le nk} \|x_j\| ;$$

Et ceci n'est autre que la conclusion.

On peut alors énoncer (et démontrer) le

THEOREME 3 :

Soit E un espace normé ; les propriétés suivantes sont équivalentes :

a) Pour tout $\lambda \ge 1$, E ne contient pas l_n^1 λ-uniformément ;

b) Il existe $\lambda > 1$ tel que E ne contienne pas de l_n^1 λ-uniformément ;

c) E est B-convexe.

DEMONSTRATION :

a) $\Longrightarrow$ b) : c'est trivial.

b) $\Longrightarrow$ c) : supposons que E ne soit pas B-convexe. Pour tout n et tout $\varepsilon \in]0, 1[$ il existe un n-uple $(x_1, \dots, x_n)$ d'éléments de E tel que :

$$\inf_{\varepsilon_i=\pm 1} \Big\|\sum \varepsilon_i x_i\Big\| \ge n-\varepsilon \quad \text{et} \quad \sup \|x_i\| \le 1 .$$

Soit $(\alpha_i)_{1\le i\le n} \in \mathbb{R}^{\mathbb{N}}$ tel que $\sum_{i=1}^{n} |\alpha_i| = 1$ et soit $\varepsilon_i = \operatorname{sgn} \alpha_i$. Alors :

$$n-\varepsilon \leq \left\|\sum_{i=1}^{n} \varepsilon_i x_i\right\| = \left\|\sum_{i=1}^{n} [\varepsilon_i (1-|\alpha_i|) + \alpha_i]\, x_i\right\| ;$$

et par conséquent :

$$n-\varepsilon \leq \left\|\sum_{i=1}^{n} \varepsilon_i (1 - |\alpha_i|)\, x_i\right\| + \left\|\sum_{i=1}^{n} \alpha_i x_i\right\|$$

$$\leq \sum_{i=1}^{n} |(1 - |\alpha_i|)| + \left\|\sum_{i=1}^{n} \alpha_i x_i\right\| .$$

Donc, on a démontré que

$$\sum |\alpha_i| = 1 \Longrightarrow 1-\varepsilon \leq \left\|\sum_{i=1}^{n} \alpha_i x_i\right\| \leq 1 ,$$

c'est-à-dire, par raison d'homogénéité :

$$(1-\varepsilon) \sum_{i=1}^{n} |\alpha_i| \leq \left\|\sum_{i=1}^{n} \alpha_i x_i\right\| \leq \sum_{i=1}^{n} |\alpha_i| \qquad \forall\, (\alpha_i) \in \mathbb{R}^n .$$

Mais cela signifie que les (x_i) engendrent un sous-espace de E qui est $(\frac{1}{1-\varepsilon})$-isomorphe à l_n^1 .

En définitive, si E n'est pas B-convexe, pour tout $\varepsilon \in]0, 1[$ E contient l_n^1 $(\frac{1}{1-\varepsilon})$-uniformément, et b) ne peut être vérifiée.

c) $\Longrightarrow$ a) : Remarquons tout d'abord que si E est B-convexe, la proposition 3 implique que $\lambda_n(E) \xrightarrow[n \to \infty]{} 0$.

Supposons alors que a) n'est pas vérifiée : il existe λ tel que $\lambda_n(E) \geq \frac{1}{\lambda}$ $\forall\, n$, donc E ne peut être B-convexe.

Le théorème est donc démontré.

Nous indiquerons sans démonstration le principal résultat de ce chapitre (Voir PISIER, exposé VII du Séminaire MAUREY-SCHWARTZ, 1973-1974).

<u>THEOREME 4</u> :

Soit E un Banach ; les propriétés suivantes sont équivalentes :

a) E est de type Rademacher ;

b) E est de type p-stable pour un $p \in]1, 2]$;

c) E est de type 1-stable ;

d) E est B-convexe.

(Les implications a) ⟹ b) ⟹ c) ont déjà été démontrées).

Ce théorème a d'intéressantes conséquences que nous allons démontrer.

COROLLAIRE 1 : Un Banach E de type Rademacher ne peut contenir les l_n^∞ C-uniformément.

DEMONSTRATION : Supposons que E contient l_n^∞ C-uniformément. Par analogie avec la définition 7, on dit que E contient les l_n^∞ C-uniformément s'il existe une suite (E_n) de sous-espaces de E de dimension n et une suite d'isomorphismes T_n de E_n sur l_n^∞ tels que

$$\sup_n \|T_n\| \, \|T_n^{-1}\| < C .$$

Cela signifie encore que pour tout $n \geq 1$ il existe un n-uple de E $(x_1^n, x_2^n, \ldots, x_n^n)$ tel que pour tout $(\lambda_k)_{1\leq k\leq n} \in \mathbb{R}^n$ on ait :

$$\sup_{1\leq k\leq n} |\lambda_k| \leq \Big\| \sum_{1\leq k\leq n} \lambda_k x_k \Big\| \leq C \sup_{1\leq k\leq n} |\lambda_k| .$$

Soit $\{\sigma_1^n, \sigma_2^n, \ldots, \sigma_{2^n}^n\}$ une énumération de l'ensemble $\{-1, +1\}^n$, et pour tout $l \in \{1, 2, \ldots, n\}$ posons

$$y_l^n = \sum_{k=1}^{2^n} \sigma_k^n(l)\, x_k^{2^n} .$$

En remarquant que :

$$\sum_{k\leq n} |\lambda_k| = \sup_{1\leq k\leq 2^n} \Big| \sum_{l=1}^{n} \sigma_k^n(l)\, \lambda_l \Big|$$

on obtient

$$\sum_{k\leq n} |\lambda_k| \leq \Big\| \sum_{k\leq n} \lambda_k y_k^n \Big\| \leq C \sum_{k\leq n} |\lambda_k| .$$

Donc, par le théorème 2, E ne peut être B-convexe ; et par le théorème 3, E ne peut être de type Rademacher.

COROLLAIRE 2 : Un espace de Banach de type Rademacher est un C-espace, et ne peut contenir de sous-espace isomorphe à l^1.

DEMONSTRATION : Il suffit de remarquer que si E n'est pas un C-espace, c'est à dire si E contient isomorphiquement C_o, il contient les $(l^\infty_n)_n$ λ-uniformément.

On déduit de ce corollaire que si E est de type Rademacher, $\mathcal{C}_\beta(E) = \mathcal{B}_\beta(E)$. En outre, $L^\infty([0, 1], dx)$, $L^1([0, 1], dx)$, $\mathcal{C}([0, 1])$ qui ne sont pas des C-espaces ne sont pas de type Rademacher.

COROLLAIRE 3 : Tout Banach uniformément convexe est de type Rademacher.

DEMONSTRATION : Supposons donc le Banach E uniformément convexe. Démontrons que E est B-convexe ; cela sera fait si l'on démontre que

$$\{\inf (||x+y||, ||x-y||)\} < 2 .$$

Soit δ(1) la quantité correspondant à 1 comme il est dit dans la définition 6 (en fait $\delta(1) \leq \frac{1}{2}$ comme on le voit facilement).

Soient x et y dans E tels que $\sup (||x||, ||y||) \leq 1$. Si $||x-y|| \geq 1$, alors $||x+y|| \leq 2(1-\delta(1))$ et $\inf (||x-y||, ||x+y||) < 2$.

Si $||x-y|| < 1$, a fortiori $\inf (||x-y||, ||x+y||) < 1 \leq 2(1-\delta(1))$. Donc le corollaire est démontré.

COROLLAIRE 4 : Soient E un Banach et E' son dual. E est de type Rademacher si et seulement si E' est de type Rademacher.

DEMONSTRATION : Supposons que E n'est pas de type Rademacher, c'est-à-dire d'après les théorèmes 4 et 3 ci-dessus, il existe λ > 1 tel que E contient les (l^1_n) λ-uniformément. Démontrons que E' contient alors les (l^1_n) 2λ-uniformément.

Il existe par hypothèse, pour tout n, un sous-espace E_n de E, de dimension n, et un isomorphisme T_n de E_n sur l^1_n tel que :

$$||T_n|| \; ||T_n^{-1}|| \leq \lambda \; , \; \forall n .$$

La transposée T'_n de T_n envoie l^∞_n sur $E'_n = E'/E^\perp_n$ ($E^\perp_n$ orthogonal de E_n relativement à la dualité entre E et E'), et :

$$||T'_n|| \; ||T'^{-1}_n|| \leq \lambda \; , \; \forall n.$$

On en déduit que, pour tout n, il existe un n-uple $(y_1^n, y_2^n, \ldots, y_n^n)$ dans E' tel que :

$$\sup_{1\leq k\leq n} |\alpha_k| \leq \|\sum_{k\leq n} \alpha_k y_k^n\|_{E'} \leq 2\lambda \sup_{k\leq n} |\alpha_k|$$

pour tout n-uple $(\alpha_1, \alpha_2, \ldots, \alpha_n) \in \mathbb{R}^n$. Mais, d'après ce que l'on a vu dans la démonstration du corollaire 1, cela implique que E' contient les (l_n^1) 2λ-uniformément, et E' n'est pas de type Rademacher.

On démontrerait de même que si E' n'est pas de type Rademacher, E n'est pas de type Rademacher. CQFD.

Donnons enfin une caractérisation des espaces de type Rademacher (ou des espaces B-convexes) au moyen d'une loi des grands nombres.

THEOREME 5 :

Soit E un Banach ; les propriétés suivantes sont équivalentes :

a) E est de type Rademacher ;

b) Pour toute suite (X_n) de v.a. indépendantes, à valeurs dans E telles que :

$$\sup_n E\{\|X_n\|^2\} < \infty \quad \text{et} \quad E\{X_n\} = 0 \quad \forall n ,$$

la suite

$$\left(\frac{1}{n} \sum_{1\leq k\leq n} X_k\right)_n \quad \text{converge vers zéro en probabilité ;}$$

c) Pour toute suite bornée (x_n) d'éléments de E et toute suite (ε_n) de Bernoulli,

$$\frac{1}{n} \sum_{1\leq k\leq n} \varepsilon_k x_k \xrightarrow{\text{Probabilité}} 0 .$$

DEMONSTRATION :

a) ⟹ b) : Si E est de type Rademacher, E est de type p pour un $p \in]1,2]$. Si (X_n) vérifie la condition de b), elle vérifie aussi

$$\forall n \quad E\{X_n\} = 0 \quad \text{et} \quad \sum_{n\geq 1} \frac{E\{\|X_n\|^p\}}{n^p} < \infty .$$

Mais alors, par le théorème 1, la conclusion de b) est vraie.

b) ⟹ c) : C'est trivial.

c) ⟹ a) : Supposons donc que c) est vérifiée et que a) ne soit pas vraie. Alors pour tout entier N on a :

$$\sup_{\sup_{1\le k\le N}||x_k|| \le 1} \left(\inf_{\sum_{k\le N} |\alpha_k|=1} ||\sum_{k\le N} \alpha_k x_k||\right) = 1 .$$

Par conséquent, pour tout entier N ≥ 1 on a également :

$$\sup_{\sup_{k\le N}|| x_k|| \le 1} \frac{1}{N} \int || \sum_{k\le N} \varepsilon_k x_k|| \, dP = 1 .$$

(où (ε_n) désigne une suite de Bernoulli sur un $(\Omega, \mathcal{F}, P)$).

Alors, pour tout entier n ≥ 1, il existe n^n éléments de E, soient $x_1^n, x_2^n, \ldots, x_{n^n}^n$ tels que :

$$\sup_{1\le k\le n^n} ||x_k^n|| = 1 \text{ et } \frac{1}{n^n} \int || \sum_{k=1}^{n^n} \varepsilon_k x_k^n|| \, dP > \frac{1}{2} .$$

Posons $k_n = 1 + 2^2 + \ldots + n^n$ et $y_i = x^l_{i-k_l}$ si $k_l < i \le k_{l+1}$. Nous obtenons :

$$\frac{1}{k_n} \int || \sum_{k=1}^{k_n} \varepsilon_k y_k|| \, dP \ge \frac{1}{k_n} (\frac{1}{2} n^n - k_{n-1})$$

$$\ge \frac{1}{2n^n} (\frac{1}{2} n^n - (n-1)^n) \ge \frac{1}{2n^n} (\frac{1}{2} n^n) \frac{1}{2} (1 - \frac{1}{n})^n$$

$$\to \frac{1}{4} - \frac{1}{2e} > 0 . \text{ Ce qui est contradictoire.}$$

Et le théorème 5 est donc démontré.

Enfin, indiquons un dernier résultat démontré par PISIER et HOFFMANN-JØRGENSEN :

Un Banach E est de type r-Rademacher si et seulement si dans E un "théorème central limite" est vrai.

Ce sera l'objet du cours 1976 de St-Flour.

CHAPITRE VI

ESPACES DE BANACH POSSÉDANT LA PROPRIÉTÉ DE RADON-NYKODYM

Dans ce chapitre, nous utiliserons constamment les notations et résultats du chapitre II. Les remarques qui suivent aideront à la compréhension de ce chapitre. Aussi les mettrons-nous en tête.

$(X, \mathcal{F}, \mu)$ désignera un espace mesuré σ-fini (ou bien de Radon). On identifiera une fonction avec sa classe dans $L^p(X, \mu, E)$, ou $L^p_s(X, \mu, E)$ si E est un Banach.

1) Soit F un Banach et $v : F \to L^p(X, \mathcal{F}, \mu, R)$ $(0 \leq p \leq \infty)$ linéaire et p-latticiellement bornée. Soit $\varphi : X \to F'$ une décomposition de v : $\varphi \in L^p_*(X, \mu, F')$. Comme d'habitude, on posera $\Phi = \bigvee_{||e|| \leq 1} |v(e)| = \bigvee_{||e|| \leq 1} |\widehat{\langle \varphi, e \rangle}|$.

Par définition même, $\Phi \in L^p(X, \mu, \mathbb{R})$. Alors, pour tout $B \in \mathcal{F}$ $\Phi\mu$-intégrable, $\int_B \varphi \, d\mu$ existe et $\int_B \varphi \, d\mu \in E'$.

En effet, $\int_B \langle \varphi, e \rangle \, d\mu$ existe pour tout $e \in F$; donc $\int_B \varphi \, d\mu$ existe. A priori $\int_B \varphi \, d\mu \in F^*$ (dual algébrique de F). Mais, dû au fait que $|\langle \varphi, e \rangle| \leq \Phi ||e||$ μ-presque partout, il résulte du théorème de Lebesgue que $e \rightsquigarrow \int \langle \varphi, e \rangle \, d\mu$ est continue sur F ; donc $\int_B \varphi \, d\mu \in F'$.

On démontrerait de même que, plus généralement, si $f : X \to \mathbb{R}$ est $\Phi\mu$-intégrable, $\int f \varphi \, d\mu$ existe et appartient à F'.

2) Soit E un Banach, E' son dual et $\varphi \in L^p_s(X, \mu, E)$. Soit v l'application linéaire $E' \to L^p(X, \mu, \mathbb{R})$ correspondante : elle est p-latticiellement bornée.

Alors, si $B \in \mathcal{F}$ est $\Phi\mu$-intégrable (Φ ayant été définie comme plus haut), on a $\int_B \varphi \, d\mu \in E''$.

De même, si $f : X \to \mathbb{R}$ est $\Phi\mu$-intégrable, on a $\int f\varphi d\mu \in E''$. (Il suffit en effet d'appliquer le résultat précédent à $F = E'$).

3) Soit E un Banach ; on a vu que l'on a un isomorphisme algébrique, entre $L^{\infty}_{*}(X, \mu, E')$ et $\mathcal{L}(E, L^{\infty}(X, \mu, \mathbb{R}))$.

Je dis que cet isomorphisme est également une isométrie. En effet, soit $v : E \to L^{\infty}(X,\mu, \mathbb{R})$ et soit φ_M une décomposition de Maharam de v. Alors, pour μ-presque tout x :

$$||\varphi_M(x)|| = |\Phi(x)| \le ||v|| .$$

Donc

$$||\varphi_M||_{L^{\infty}_{*}(X, \mu, E')} = ||\Phi||_{L^{\infty}(\mu, \mathbb{R})} \le ||v|| .$$

D'autre part :

$$||v(e)||_{L^{\infty}(\mu, \mathbb{R})} \le ||\Phi||_{L^{\infty}(\mu, \mathbb{R})} \quad \forall\ ||e|| \le 1.$$

Donc

$$||v|| \le ||\Phi||_{L^{\infty}(\mu, \mathbb{R})}$$. D'où le résultat.

En conséquence, $\mathcal{L}(E, L^{\infty}(\mu, \mathbb{R}))$, l'espace des applications bilinéaires de $L^1(X, \mu, \mathbb{R}) \times E$ dans $\mathbb{R}$ $\mathcal{L}(L^1(\mu, \mathbb{R}), E')$, $L^1(X, \mu, E)'$ et $L^{\infty}_{*}(X, \mu, E')$ sont isométriques.

N° 1 : LES ESPACES $L^p_{**}(X, \mu, E)$ ET LES ESPACES DE RADON-NIKODYM

DEFINITION 1 : Soit E un Banach et $p \in [0,\infty]$. On désigne par $L^p_{**}(X, \mu, E)$ le sous-espace de $L^p_{*}(X, \mu, E'')$ formé des $\varphi : X \to E''$ telles que pour toute $B \in \mathcal{R}$ $\Phi\mu$-intégrable on ait $\int_B \varphi d\mu \in E$.

(On a posé $\Phi = \bigvee_{\substack{||e'||\le 1 \\ e' \in E'}} |\overset{\circ}{\overbrace{<\varphi, e'>}}|$).

Cette notation est ambigue, car on appelle $L^p_{**}(X, \mu, E)$ un espace de fonctions à valeurs dans E'', mais elle est commode.

PROPOSITION 1 :

Soit $\varphi \in L^p_*(X, \mu, E'')$; on a l'équivalence de :

a) $\varphi \in L^p_{**}(X, \mu, E)$;

b) Pour tout $B \in \mathfrak{F}$ à la fois μ et $\Phi\mu$-intégrable, on a $\int_B \varphi d\mu \in E$.

DEMONSTRATION : Il est clair que a) $\Rightarrow$ b).

Supposons donc b) vérifiée et soit $B \in \mathfrak{F}$, $\Phi\mu$-intégrable. On a évidemment $B = B \cap \{\Phi = 0\} \cup B \cap \{\Phi > 0\}$, et $B \cap \{\Phi = 0\}$ est $\Phi\mu$-intégrable. En outre, $\int_{B\cap\{\Phi=0\}} d\mu = 0$. D'autre part, $B \cap \{\Phi > 0\} = \bigcup_n B \cap \{\Phi \geq \frac{1}{n}\}$.

Posons $B_n = B \cap \{\Phi \geq \frac{1}{n}\}$; B_n est $\Phi\mu$-intégrable ; en outre, B_n est μ-intégrable, car

$$\int_B \Phi \, d\mu \geq \int_{B_n} \frac{1}{n} \, d\mu = \frac{1}{n} \mu(B_n) .$$

Donc $\int_{B_n} \varphi \, d\mu \in E$ pour tout n. Mais alors

$$\left|\left|\int_B \varphi d\mu - \int_{B_n} \varphi d\mu\right|\right|_{E''} \leq \int_{B \cap \{\Phi<\frac{1}{n}\}} \Phi d\mu .$$

(En effet, on peut sans changer l'intégrale vectorielle "faible" remplacer φ par une décomposition de Maharam de l'application $e' \rightsquigarrow \widehat{\langle \overset{\circ}{\varphi}, e' \rangle}$)

B étant $\Phi\mu$-intégrable, ce dernier terme tend vers zéro quand n tend vers l'infini, d'après le théorème de Lebesgue. Et par conséquent $\int_B \varphi \, d\mu \in E$.

REMARQUE 1 : $L^\infty_{**}(X, \mu, E)$ peut encore être défini comme le sous-espace de $L^\infty_*(X, \mu, E'')$ formé des φ telles que pour toute $B \in \mathfrak{F}$ <u>μ-intégrable</u> , on ait $\int_B \varphi \, d\mu \in E$.

En effet, si B est μ-intégrable, elle est $\Phi\mu$-intégrable car Φ est μ-essentiellement bornée.

Naturellement, $L^p_{**}(X, \mu, E)$ sera muni de la topologie induite par celle de $L^p_*(X, \mu, E'')$. On voit facilement qu'il est fermé dans cet espace.

PROPOSITION 2 :

L'espace $\mathcal{L}(L^1(X, \mu, R), E)$ est isométrique à $L^\infty_{**}(X, \mu, E)$. L'isomorphisme entre ces espaces associe à $\varphi \in L^\infty_{**}(X, \mu, E)$, l'application linéaire continue de $L^1(X, \mu, R)$ dans E : $f \rightsquigarrow \int \varphi f \, d\mu \in E$.

DEMONSTRATION : On sait déjà que $L^\infty_*(X, \mu, E")$ est isométrique à $\mathcal{L}(L^1(X,\mu),E")$. D'autre part, $\mathcal{L}(L^1(X, \mu), E)$ est le sous-espace de $\mathcal{L}(L^1(X, \mu, \mathbb{R}), E")$ envoyant $L^1(X, \mu)$ dans E.

Maintenant une application linéaire continue de $L^1(X, \mu)$ dans E" envoie $L^1(X, \mu)$ dans E si et seulement si elle envoie les fonctions étagées μ-intégrables dans E, donc si elle envoie les 1_B($B \in \mathcal{F}$, B μ-intégrable) dans E. D'où le résultat.

On peut maintenant poser la

DEFINITION 2 : Soit E un Banach ; on dit que E est de Radon-Nikodym (en abrégé E est R.N.) si toute $\varphi \in L^0_{**}(X, \mu, E)$ admet un représentant $\varphi_1 \in L^0(X,\mu,E)$, quel que soit l'espace mesuré $(X, \mathcal{F}, \mu)$.

Dire que E est de Radon-Nikodym signifie encore que l'application naturelle de $L^0(X, \mu, E)$ dans $L^0_*(X, \mu, E")$ est une bijection de $L^0(X, \mu, E)$ sur $L^0_{**}(X, \mu, E)$, ou encore en abrégé $L^0(X, \mu, E) = L^0_{**}(X, \mu, E)$.

Il est clair que si E est de Radon-Nikodym, quel que soit l'espace mesuré $(X,\mathcal{F}, \mu)$ et quel que soit $p \in [0, \infty]$, toute φ appartenant à $L^p_{**}(X,\mu,E)$ admet un représentant φ_1 dans $L^p(X, \mu, E)$.

REMARQUE 2 : Si E est de Radon-Nikodym et si $\varphi \in L^p_{**}(X, \mu, E)$, son représentant $\varphi_1 \in L^p(X, \mu, E)$ réalise le module minimum de la classe φ, car on a vu que $||\varphi_1||_{L^p(X,\mu,E")} = ||\varphi||_{L^p_*(X,\mu,E")}$.

THEOREME 1 :

Soit E un Banach ; E est de Radon-Nikodym si et seulement si pour tout $(X, \mathcal{F}, \mu)$ σ-fini, on a :

$$L^\infty_{**}(X, \mu, E) = L^\infty(X, \mu, E) .$$

DEMONSTRATION : La partie "seulement si" a déjà été remarquée. Supposons alors que E satisfait la condition du théorème et soit $\varphi \in L^0_{**}(X, \mu, E)$, à laquelle est associée $\Phi \in L^0(X, \mu, \mathbb{R})$.

Si $f \in L^1(X, \Phi\mu, \mathbb{R})$, on a vu que $\int f\varphi \, d\mu \in E$. En outre l'application $f \leadsto \int f\varphi \, d\mu$ de $L^1(X, \Phi\mu, \mathbb{R})$ dans E est continue, et de norme ≤ 1.

Elle est donc définie par une fonction de $L^\infty_{**}(X, \Phi\mu, E)$. Mais E vérifiant la condition du théorème, il existe $\psi \in L^\infty(X, \Phi\mu, E)$, $||\psi|| \leq 1$ telle que

$$\int \varphi f \, d\mu = \int \psi f \Phi \, d\mu \qquad \forall f \in L^1(X, \Phi\mu, \mathbb{R}) .$$

On peut évidemment remplacer ψ par 0 sur $\{\Phi = 0\}$ et par conséquent supposer que $\psi \in L^\infty(X, \mu, E)$. Posons $\bar{\varphi} = \Phi.\psi$. Alors $\int \varphi f \, d\mu = \int \bar{\varphi} f \, d\mu$ $\forall f \in L^1(X, \Phi\mu, \mathbb{R})$ et $||\bar{\varphi}|| \leq \Phi$; en outre, $\bar{\varphi} \in L^0(X, \mu, E)$. Donc $\bar{\varphi} = \varphi$ scalairement $\Phi\mu$-partout. D'autre part, $\bar{\varphi} = 0$ sur $\{\Phi = 0\}$ et $\psi = 0$ scalairement μ-presque partout (par la dualité entre E" et E'). Donc $\varphi = \bar{\varphi}$ scalairement μ-presque partout. Et par conséquent $L^0(X, \mu, E)$ s'envoie surjectivement sur $L^0_{**}(X, \mu, E)$ (et l'on savait déjà qu'il s'envoie injectivement).

Le théorème est donc démontré.

REMARQUE 3 : Dans la condition de la définition 2, on peut supposer que $(X, \mathcal{F}, \mu)$ est un espace mesuré fini (ou même un espace probabilisé).

En effet, le cas général (i.e. μ σ-finie, ou μ de Radon) se ramène au cas"fini"par concassage.

On en déduit alors la :

PROPOSITION 3 :

Si E est de Radon-Nikodym, pour tout $p \in [0, \infty]$ on a $L^p(X, \mu, E) = L^p_{**}(X, \mu, E)$, quel que soit $(X, \mathcal{F}, \mu)$.
Inversement, s'il existe $p \in [0, \infty]$ tel que pour tout espace probabilisé $(X, \mathcal{F}, \mu)$ on ait l'égalité, E est de Radon-Nikodym.

DEMONSTRATION : La première partie a été remarquée.

Si pour tout espace probabilisé on a $L^p(X, \mu, E) = L^p_{**}(X, \mu, E)$, on a $L^\infty(X, \mu, E) = L^\infty_{**}(X, \mu, E)$, donc E est de Radon-Nikodym.

Nous verrons plus loin des exemples d'espaces de Radon-Nikodym. Remarquons au préalable que, E étant un Banach, on a

$$L^\infty_{**}(X, \mu, E') = L^\infty_*(X, \mu, E')$$

(algébriquement et topologiquement). (Naturellement, $L^\infty_{**}(X, \mu, E')$ est un espace de fonctions à valeurs dans E''').

En effet, $L^\infty_{**}(X, \mu, E') \simeq \mathcal{L}(L^1(\mu, R), E')$

$$\simeq \mathcal{L}(E, L^\infty(\mu, R)) \simeq L^\infty_*(X, \mu, E').$$

D'où le résultat.

On en déduit, avec l'aide du théorème 1, que E' est de Radon-Nikodym si et seulement si

$$L^\infty_*(X, \mu, E') = L^\infty(X, \mu, E').$$

Mais alors :

PROPOSITION 4 :

E' est de Radon-Nikodym si et seulement si, pour tout $(X, \mathcal{T}, \mu)$ on a :

$$L^0(X, \mu, E') = L^0_*(X, \mu, E') .$$

DEMONSTRATION : Il est clair que si E' satisfait à la condition de la proposition 4, on a $L^\infty(X, \mu, E') = L^\infty_*(X, \mu, E')$; donc E' est de Radon-Nikodym.

Réciproquement, supposons que E' est de Radon-Nikodym, ou, ce qui revient au même, que $L^\infty_*(X, \mu, E') = L^\infty(X, \mu, E')$. Soit $\varphi \in L^0_*(X, \mu, E')$, à laquelle est associée $\Phi \in L^0(\mu, \mathbb{R})$.

Alors l'application $e \rightsquigarrow \dfrac{<\varphi, e>}{\Phi}$ envoie continuement E dans $L^\infty(X,\mu,\mathbb{R})$ et est de norme ≤ 1 ; elle est donc définie par un élément de $L^\infty_*(X, \mu, E')$. Mais E' étant R.N., il existe $\psi \in L^\infty(X, \mu, E')$ tel que $||\psi|| \leq 1$ et $<\psi, e> = <\frac{\varphi}{\Phi}, e> \forall e \in E$. Donc $\varphi = \psi\Phi$ scalairement presque partout. Puisque $\psi\Phi \in L^0(X, \mu, E')$, on a le résultat annoncé.

REMARQUE 4 : Comme plus haut, on démontrerait que si E' est de Radon-Nikodym et si $p \in [0, \infty]$ on a

$$L^p_*(X, \mu, E') = L^p(X, \mu, E')$$

pour tout $(X, \mathcal{F}, \mu)$.

Réciproquement, si pour un $p \in [0, \infty]$ on a l'égalité ci-dessus pour tout espace probabilisé, E' est de Radon-Nikodym.

On en déduit alors :

THEOREME 2 :

E' est de Radon-Nikodym si et seulement si il existe $p \in [1, \infty[$ tel que pour tout $(X, \mathcal{F}, \mu)$ on ait $L^p(X, \mu, E)' = L^{p'}(X, \mu, E')$ (avec $\frac{1}{p} + \frac{1}{p'} = 1$). Et l'égalité ci-dessus est vraie pour tout p tel que $1 \leq p < \infty$.

DEMONSTRATION : On sait en effet (chapitre II) que $L^p(X, \mu, E)' = L^{p'}_*(X, \mu, E')$ et l'on applique alors la remarque 4.

N° 2 : MESURES A VALEURS DANS UN ESPACE DE RADON-NIKODYM

Soit $(X, \mathcal{F})$ un espace mesurable et soit E un Banach.

On appelle mesure vectorielle à valeurs dans E une application $\vec{\nu} : \mathcal{F} \to E$ dénombrablement additive, c'est à dire :

Si (A_n) est une suite d'éléments de $\mathcal{F}$ deux à deux disjoints et si $A = \cup A_n$, on a

$$\vec{\nu}(A) = \sum_n \vec{\nu}(A_n)$$ (la série du second membre étant d'ailleurs commutativement convergente).

DEFINITION 3 : Soit $(X, \mathcal{F}, \mu)$ un espace mesuré ; $\vec{\nu}$ une mesure vectorielle sur $(X, \mathcal{F})$ à valeurs dans E. On dit que $\vec{\nu}$ est majorée par μ si :

$$||\vec{\nu}(A)|| \leq \mu(A) \qquad \forall\ A \in \mathcal{F}.$$

On démontre que l'ensemble des majorantes de $\vec{\nu}$ possède un plus petit élément : c'est la variation de $\vec{\nu}$, désignée par $||\vec{\nu}||$

La mesure positive $||\vec{\nu}||$ n'est pas forcément bornée : elle l'est si E est de dimension finie.

On peut même démontrer que E est de dimension finie si et seulement si toute mesure vectorielle à valeurs dans E est à variation bornée. Donc, dans un Banach, il existe des mesures non-majorables par une mesure finie.

Si $\vec{\nu}$ est majorée par μ , on écrira $\vec{\nu} << \mu$

Une des formes du théorème de Radon-Nikodym classique affirme que si $\vec{\nu}$ est une mesure complexe majorée par μ <u>finie</u>, alors $\vec{\nu} = f \cdot \mu$ avec $f \in L^{\infty}(X, \mathscr{F}, \mu, \mathbb{C})$. On va chercher ce que devient ce résultat si on remplace $\mathbb{C}$ par un Banach.

En première étape, on a le :

<u>THEOREME 3</u> :

Soit $(X, \mathscr{F}, \mu)$ un espace mesuré fini et E un Banach. Soit $\varphi \in \mathscr{L}^{\infty}_{**}(X, \mu, E)$ de norme ≤ 1 ; l'application

$$A \rightsquigarrow \int_A \varphi \, d\mu = \vec{\nu}_{\varphi}(A)$$ définit une mesure à valeurs dans E majorée par μ .

φ et φ' définissent la même mesure vectorielle si et seulement si elles sont égales scalairement μ-presque partout (par la dualité entre E' et E").

Inversement, si $\vec{\nu}$ est une mesure vectorielle majorée par μ il existe $\varphi \in L^{\infty}_{**}(X, \mu, E)$ unique telle que $\vec{\nu} = \vec{\nu}_{\varphi}$. Donc la boule unité de $L^{\infty}_{**}(X, \mu, E)$ est en correspondance bijective avec les mesures vectorielles à valeurs dans E majorées par μ.

<u>DEMONSTRATION</u> : Soit $\vec{\nu} << \mu$, on définit une application u de l'ensemble des fonctions étagées, définies sur X et à valeurs réelles dans E par :

$$u\left(\sum 1_{B_i} C_i\right) = \sum C_i \vec{\nu}(B_i) .$$

En outre, $||u(\sum 1_{B_i} C_i)||_E \leq \sum |C_i| \mu(B_i)$; donc $||u(f)||_E \leq \int |f| d\mu$ $\forall$ f μ-étagée à valeurs dans E. u se prolonge donc de manière unique en une application de $L^1(X, \mathcal{F}, \mu)$ dans E, linéaire et continue, de norme ≤ 1. Inversement, une telle application u, définit une mesure vectorielle sur $(X, \mathcal{F})$ par

$$B \rightsquigarrow \vec{\nu}(B) = u(1_B) \qquad B \in \mathcal{F}$$

(Il est facile de prouver la σ-additivité).

En outre :
$$||\vec{\nu}(B)|| = ||u(1_B)|| \leq ||u|| \, ||1_B||_{L^1(X,\mu,R)}$$
$$\leq \mu(B) \text{ (car } ||u|| \leq 1).$$

En résumé, la boule unité $\mathcal{L}(L^1(X, \mu, R), E)$ est en bijection avec l'ensemble des mesures majorées par μ .

Comme $\mathcal{L}(L^1(X, \mu, R), E)$ est en bijection isométrique avec $L^\infty_{**}(X,\mu, E)$, le théorème est complétement démontré.

REMARQUE 5 : On démontrerait de même que si E est un Banach (ou même un espace normé), l'espace des mesures vectorielles à valeurs dans E', majorées par μ est en bijection avec la boule-unité de $L^\infty_*(X,\mu, E') = \mathcal{L}(L^1(X, \mu, R), E')$.

On en déduit immédiatement le :

THEOREME 4 :

Soit E un Banach ; E est de Radon-Nikodym si et seulement si pour tout espace mesuré fini $(X, \mathcal{F}, \mu)$, toute mesure vectorielle $\vec{\nu}$ à valeurs dans E, majorée par μ se met sous la forme suivante

$$\vec{\nu}(A) = \int_A \varphi \, d\mu \qquad \text{avec } \varphi \in L^\infty(X, \mu, E).$$

On notera par $\varphi . \mu$ les mesures figurant dans les énoncés des théorèmes 3 et 4.

DEFINITION 4 : Soit $(X, \mathcal{F}, \mu)$ un espace mesuré <u>fini</u> ; E un Banach et $\vec{\nu}$ une mesure sur $(X, \mathcal{F})$ à valeurs dans E.

1) On dit que $\vec{\nu}$ est dominée par μ et on écrit $\vec{\nu} < \mu$ si $B \in \mathcal{F}$ et

$$\mu(B) = 0 \Longrightarrow \vec{\nu}(B) = 0 .$$

2) On dit que $\vec{\nu}$ est de base μ si $\vec{\nu} = \varphi . \mu$, avec $\varphi \in \mathcal{L}^1(X, \mu, E)$ (c'est-à-dire $\vec{\nu}(A) = \int_A \varphi \, d\mu \quad \forall A \in \mathcal{F}$).

Dans le cas où $\vec{\nu}$ est une mesure complexe, les conditions 1) et 2) de la définition 4 sont équivalentes. En outre, toute mesure majorée par μ a pour base μ.

C'est faux si E est un Banach quelconque, comme nous le verrons.

REMARQUE 6 : $\vec{\nu}$ est dominée par μ si et seulement si pour tout $e' \in E'$, $e'(\vec{\nu})$ (mesure à valeurs dans $\mathbb{R}$) est dominée par μ (ou, ce qui revient au même, est de base μ).

En effet, $\vec{\nu}(B) = 0$ équivaut à $< \vec{\nu}(B), e' > = 0 \quad \forall e' \in E' \quad (B \in \mathcal{F})$.

Nous allons examiner certaines implications mutuelles entre ces définitions.

THEOREME 5 :

Soit $(X, \mathcal{F}, \mu)$ un espace mesuré <u>fini</u>.

1) Si $\varphi \in L^1_{**}(X, \mu, E)$, la mesure $\varphi . \mu$ est majorable par une mesure finie (donc est à variation totale bornée). Sa variation (ou plus petite majorante finie) est égale à $\Phi\mu$, où Φ est associée à φ comme il a été dit.

2) Si $\vec{\nu}$ est une mesure sur $(X, \mathcal{F})$ à valeurs dans E <u>majorée</u> par μ et <u>ayant une base finie</u> σ, elle s'écrit $\vec{\nu} = \varphi . \mu$ avec $\varphi \in L^\infty(X, \mu, E)$ de norme ≤ 1 (ou encore $\vec{\nu}$ a μ pour base).

3) Si $\vec{\nu}$ est dominée par μ et si elle admet une mesure de base finie σ, elle admet μ pour base, μ étant finie.

DEMONSTRATION :

1) Il est facile de voir que $B \rightsquigarrow \int_B \varphi \, d\mu$ est une mesure à valeurs dans E si $\varphi \in L^1_{**}$. On peut supposer que $||\varphi|| = \Phi$ car la mesure $\varphi . \mu$ ne dépend que de la classe de scalaire équivalence de φ.

Il est clair alors que $\Phi\mu$ est une majorante finie de $\varphi . \mu = \vec{\nu}$. Je dis que c'est la plus petite majorante.

En effet, $||\vec{\nu}|| \leq \Phi\mu$ donc par Radon-Nikodym classique $||\nu|| = f . \Phi\mu$ avec $f \geq 0$, $\Phi\mu$-mesurable et $0 \leq f \leq 1$. On peut supposer que $f = 0$ si $\Phi = 0$, donc f est μ-mesurable.

Soit $e' \in E'$ tel que $||e'|| \leq 1$; alors $e'(\vec{\nu}) = \langle \varphi, e' \rangle \mu$ est une mesure réelle majorée par $f \, \Phi\mu = ||\vec{\nu}||$.

Donc $|\langle \varphi, e' \rangle| \leq f \, \Phi$, μ-presque partout.

Par conséquent,

$$\bigvee_{||e'|| \leq 1} |\widehat{\langle \varphi, e' \rangle}| = \Phi$$

est majorée par $f.\Phi$, μ-presque partout. Donc $f = 1$ μ-presque partout et

$$||\vec{\nu}|| = \Phi \, \mu. \qquad \text{CQFD.}$$

2) Soit $\vec{\nu}$ une mesure majorée par μ finie et ayant une base finie σ. Alors $\vec{\nu} = \psi \, \sigma$ avec $\psi \in L^1(X, \sigma, E)$. D'après 1), $||\nu|| = ||\psi|| . \sigma$. Donc $||\psi|| \, \sigma \leq \mu$, et $||\psi|| \, \sigma = f\mu$ avec f, μ-mesurable et $0 \leq f \leq 1$.

Cela étant,

$$\vec{\nu} = \psi . \sigma = \frac{\psi}{||\psi||} ||\psi|| \, \sigma = \frac{\psi}{||\psi||} f\mu = \varphi . \mu .$$

En outre, $\varphi \in L^\infty(X, \mu, E)$ et $||\varphi|| \leq 1$. D'où le résultat.

3) Soit $\vec{\nu}$ dominée par μ et ayant une base finie σ: $\nu = \psi . \sigma$ avec $\psi \in L^1(X, \sigma, E)$.

$\vec{\nu}$ est alors dominée par μ et σ donc par $\inf(\mu, \sigma) = f . \sigma$ (f borélienne, $0 \leq f \leq 1$).

Je dis que $\psi(x) = 0$ σ-presque partout sur $\{f = 0\}$.

En effet, $\{f = 0\}$ est inf (μ, σ) négligeable, donc si $B \in \mathcal{F}$, $B \subset \{f = 0\}$, on a : $\vec{\nu}(B) = 0$.

Comme pour $e' \in E'$, $< \vec{\nu}, e' > = < \psi, e' > .\sigma$, si $B \in \mathcal{F}$ est telle que $B \subset \{f = 0\}$ on a $\int_B < \psi, e' > d\sigma = 0$. Donc $< \psi, e' > = 0$ μ-presque partout sur $\{f = 0\}$ pour tout $e' \in E'$. Mais ψ étant σ-mesurable, $\psi = 0$ σ-presque partout sur $\{f = 0\}$.

On peut définir $\frac{\psi}{f}$ (en posant $\frac{0}{0} = 0$). Il est clair alors que

$$\vec{\nu} = \psi \, . \, \sigma = \frac{\psi}{f} f\sigma \, .$$

D'autre part, inf $(\mu, \sigma) = g\mu$ $(0 \leq q \leq 1$, μ-mesurable$)$. Donc

$$\vec{\nu} = \frac{\psi}{f} \inf(\mu,\sigma) = \frac{\psi}{f} g\mu = \varphi\mu$$

avec φ μ-mesurable. En outre :

$$\int \frac{||\psi||}{f} g \, d\mu = \int \frac{||\psi||}{f} d(\inf(\mu, \sigma)) = \int \frac{||\psi||}{f} f \, d\sigma$$

$$= \int ||\psi|| \, d\sigma < .\infty$$

Donc $\vec{\nu} = \varphi . \mu$ et $\varphi \in L^1(X, \mu, E)$.

Le théorème est donc démontré.

A partir de là, on obtient le :

<u>THEOREME 6</u> :

Soit E un Banach, les propriétés suivantes sont équivalentes :

a) E est de Radon-Nikodym ;

b) Toute mesure à valeurs dans E à variation bornée possède une mesure de base ;

c) Quel que soit l'espace mesuré $(X, \mathcal{F}, \mu)$, toute mesure sur $(X, \mathcal{F})$ à valeurs dans E, à variation bornée et dominée par μ admet μ pour base.

<u>DEMONSTRATION</u> :

a) $\Rightarrow$ b) : Soit $\vec{\nu} << \mu$ (μ mesure finie), alors $\vec{\nu} = \varphi . \mu$, $\varphi \in L^\infty(X,\mu,E)$ et $||\varphi|| \leq 1$ d'après le théorème 4 ; donc $\vec{\nu}$ est basée sur μ.

b) ⟹ a) : Supposons que toute $\vec{\nu}$ à valeurs dans E, à variation bornée admet une base ; et soit $\vec{\nu} \ll \mu$. Alors $\vec{\nu}$ admet une base et d'après le théorème (5, 2), elle a une densité par rapport à μ :

$$\vec{\nu} = \varphi . \mu \quad \varphi \in L^{\infty}(X, \mu, E), \ \|\varphi\| \leq 1.$$

Donc E est de Radon-Nikodym d'après le théorème 4.

a) ⟹ c) : Supposons que E est de Radon-Nikodym, et soit $\vec{\nu}$ à variation bornée, dominée par μ. D'après ce qui précède, $\vec{\nu}$ admet une mesure de base ; mais d'après le théorème 5, 3), $\vec{\nu}$ admet μ pour base.

c) ⟹ a) : Supposons que toute mesure à variation bornée $\vec{\nu}$ dominée par une mesure finie μ ait μ pour base ; alors toute mesure à variation bornée admet une base, et E vérifie R.N. d'après l'implication b) ⟹ a).

Le théorème est donc complétement démontré.

<u>REMARQUE 7</u> : Si $\vec{\nu} = \varphi . \mu$ on dit que la mesure $\vec{\nu}$ a pour densité φ par rapport à μ.

Par exemple, le théorème 3 dit que $\vec{\nu}$ a une densité $\varphi \in L^{\infty}_{**}(X, \mu, E)$ si et seulement si $\vec{\nu}$ est majorée par μ.

Cela étant, le théorème 6 s'énonce sous la forme suivante :

Soit E un Banach ; les propriétés suivantes sont équivalentes :

a) E est de Radon-Nikodym

b) Quel que soit l'espace mesurable $(X, \mathcal{F})$ et la mesure $\vec{\nu}$ définie sur $(X, \mathcal{F})$ à valeurs dans E, et à variation totale bornée, $\vec{\nu}$ a une densité par rapport à $\|\vec{\nu}\|$ dans $L^{\infty}(X, \|\vec{\nu}\|, E)$.

c) Quel que soit l'espace mesuré fini $(X, \mathcal{F}, \mu)$ et quelle que soit la mesure vectorielle $\vec{\nu}$ à <u>variation bornée</u>, à valeurs dans E définie sur $(X, \mathcal{F})$, si $\vec{\nu}$ est dominée par μ, elle admet une densité dans $L^1(X, \mu, E)$.

N° 3 : EXEMPLES D'ESPACES DE RADON-NIKODYM

PROPOSITION 5 :

Soit E un espace de Radon-Nikodym. Tout sous-espace fermé F de E est de Radon-Nikodym.

DEMONSTRATION : Soit $\varphi \in L^{\infty}_{**}(X, \mu, F)$; φ définit un élément de $L^{\infty}_{**}(X, \mu, E)$ que l'on désignera par φ_1.

Il existe alors dans la classe de scalaire équivalence de φ_1 un élément $\varphi'_1 \in L^{\infty}(X, \mu, E)$.

Par hypothèse, pour tout A μ-intégrable, $\int_A \varphi'_1 \, d\mu \in F$; Donc par le théorème de la moyenne, φ'_1 prend p.p. ses valeurs dans F ; donc définit un élément φ' dans $L^1(X, \mu, F)$. Ceci démontre que F est de Radon-Nikodym.

On en déduit que si un Banach E ne possède pas la propriété de Radon-Nikodym, tout Banach E_1 contenant E isomorphiquement ne peut être de Radon-Nikodym.

EXEMPLE 1 : L'espace C_0 n'est pas de Radon-Nikodym.

En effet, soit $([0, 1], \mathcal{B}_{[0, 1]}, dt)$ l'espace mesuré "unité". Soit $\vec{\nu} : \mathcal{B}_{[0, 1]} \to C_0$ définie par

$$\vec{\nu}(A) = \left(\int_A \cos 2\pi \, nt \, dt \right)_{n \in \mathbb{N}} \qquad A \in \mathcal{B}_{[0,1]} .$$

$\vec{\nu}$ envoie $\mathcal{B}_{[0,1]}$ dans C_0 par le théorème de Riemann-Lesbesgue. Par ailleurs, on voit facilement que $\vec{\nu}$ est une mesure. Il est facile de voir que $\vec{\nu}$ est majorée par dt et est dominée par dt.

Je dis que $\vec{\nu}$ ne peut posséder une densité $\vec{\varphi} : [0, 1] \to C_0$. En effet, si $\vec{\varphi}(t) = (\varphi_n(t))_{n \in \mathbb{N}}$ était une densité de $\vec{\nu}$, on aurait pour tout n et tout A

$$\int_A \cos 2\pi nt \, dt = \int_A f_n(t) \, dt$$

Donc $f_n(t) = \cos(2\pi \, nt)$ dt presque partout. Or ceci est absurde, car $(\cos(2\pi \, nt))_n$ n'appartient à C_0 pour aucune valeur de t.

Comme conséquence, on obtient :

PROPOSITION 6 :

Un Banach qui n'est pas un C-espace ne peut être de Radon-Nikodym.

DEMONSTRATION : Il suffit de remarquer que E n'est pas un C-espace si et seulement si E contient C_o isomorphiquement.

En d'autres termes, un espace de Radon-Nikodym est un C-espace. Toutefois, un C-espace peut satisfaire ou non la propriété de Radon-Nikodym, comme le montrent les exemples suivants :

EXEMPLE 2 : l^1 (qui est un C-espace) vérifie la propriété de Radon-Nikodym.

Soit en effet $(X, \mathcal{F}, \mu)$ espace mesuré fini et soit $\vec{\nu} : \mathcal{F} \to l^1$ mesure à variation bornée $v = ||\vec{\nu}||$. $\vec{\nu}$ est donnée par une suite de mesures scalaires ν_n v-absolument continues. Donc $\nu_n = \varphi_n . v$ $\forall n$ (φ_n mesurables).

Il nous reste à montrer que la fonction $x \to (f_n(x))_n = \vec{f}(x)$ peut être choisie à valeurs dans l^1.

Commençons par remarquer que la boule unité B_1 de l^1 est un disque compact de l'espace $\mathbb{R}^{\mathbb{N}}$.

La fonction $\vec{f} : X \to \mathbb{R}^{\mathbb{N}}$ est mesurable. Pour tout $z' = (z'_n) \in \mathbb{R}^{\mathbb{N}}_o$ (qui est le dual de $\mathbb{R}^{\mathbb{N}}$) et pour tout $A \in \mathcal{F}$ on a :

$$\frac{1}{v(A)} \int_A < \vec{f}, z' > dv = < \frac{\vec{\nu}(A)}{v(A)} , z' > \in z'(B_1)$$

(puisque $||\vec{\nu}(A)|| \leq v(A)$). Comme $z'(B_1)$ est un intervalle compact, le théorème de la moyenne dit que $< \vec{f}(x), z' > \in z'(B_1)$ v-presque partout. Donc $\vec{f} : X \to R^N$ est v-scalairement presque partout dans B_1. Dû au fait que $\mathbb{R}^{\mathbb{N}}$ est métrisable, donc limite projective d'une famille dénombrable de Banach, on en déduit que $\vec{f}$ est v-presque partout dans l^1. (En effet, si E est un Banach et si $\varphi : X \to E$ est μ-mesurable et scalairement presque partout à valeurs dans un disque fermé, elle prend presque partout ses valeurs dans ce disque fermé.

Reste à démontrer que $\vec{f} : X \to l^1$ est v-mesurable pour la topologie de l^1. Or, cela résulte du fait que l^1 est séparable et du fait que $\vec{f} : X \to l^1$

est scalairement v-mesurable (cas pour tout $\xi = (\xi_n) \in l^\infty$, la série $\sum \xi_n f_n(x)$ est convergente pour tout x).

Donc en définitive on a démontré que l^1 est de Radon-Nikodym.

EXEMPLE 3 : Soit K un espace compact. Pour toute mesure de Radon ρ diffuse sur K, l'espace $E = L^1(K, \rho)$ ne vérifie pas la propriété de Radon-Nikodym (or on sait que $L^1(K, \rho)$ est un C-espace).

(Nous admettrons ce résultat).

THEOREME 7 :

Tout espace de Banach réflexif possède la propriété de Radon-Nikodym.

SCHEMA DE LA DEMONSTRATION : Soit $\vec{\nu} : \mathcal{F} \to E$ une mesure sur $(X, \mathcal{F})$ à valeurs dans E et de variation totale $v = ||\vec{\nu}||$ bornée. Pour tout $B \in \mathcal{F}$, l'ensemble $\{\frac{\vec{\nu}(A)}{v(A)}, A \in \mathcal{F}\}$ est contenu dans la boule unité de E, qui est faiblement compacte (vu la réflexivité).D'après un théorème de MOEDOMO-UHL, cela suffit pour affirmer que $\vec{\nu}$ est de base v. Donc E est de Radon-Nikodym.

Par exemple, un Hilbert, ou un $L^p(X, \mathcal{F}, \mu)$ $(1 < p < \infty)$ est de Radon-Nikodym.

THEOREME 8 :

Un dual séparable de Banach a la propriété de Radon-Nikodym.

DEMONSTRATION : Soit en effet $\vec{\nu} : \mathcal{F} \to E'$ une mesure sur $(X, \mathcal{F})$ à valeurs dans E' et de variation bornée $v = ||\vec{\nu}||$. Il existe donc $\varphi \in L^\infty_{**}(X, v, E') = L^\infty_*(X, v, E')$ (voir théorème 3 et remarque 5 suivant le théorème) telle que $\vec{\nu} = \varphi . v$. Dû au fait que E' est séparable, on en déduit que φ est v-mesurable : $\varphi \in L^\infty(X, v, E')$ et E' satisfait à Radon-Nikodym.

On redémontre ainsi que l^1, dual de C_0 et séparable est de Radon-Nikodym.

<u>COROLLAIRE</u> : C_o et L^1 ([0, 1], dx) ne peuvent être isomorphes à un dual d'espace de Banach (ou même à un sous-espace d'un dual séparable).

<u>DEMONSTRATION</u> : On a vu en effet que ces espaces ne satisfont pas à Radon-Nikodym et sont séparables. Le théorème 8 implique qu'ils ne peuvent être des duals.

D'autre part, si C_o ou $L^1([0, 1], dx)$ étaient isomorphes à un sous-espace d'un dual séparables, ils satisferaient à la propriété de Radon-Nikodym.

Le corollaire est donc démontré.

N° 4 : MARTINGALES A VALEURS DANS UN ESPACE DE RADON-NIKODYM

Nous aurons besoin ici de considérer des martingales indéxées par un ensemble filtrant quelconque. Rappelons-en la définition :

Soit $(\Omega, \mathcal{F}, P)$ un espace probabilisé ; soit T un ensemble ordonné filtrant, $(\mathcal{F}_t)_{t\in T}$ une famille filtrante croissante de sous-tribus de $\mathcal{F}$. On pose $\mathcal{F}_\infty = \bigvee_{t\in T} \mathcal{F}_t$.

Soit d'autre part E un Banach <u>quelconque</u> (pour le moment, E n'est pas de Radon-Nikodym) : Si $X \in L^1(\Omega, \mathcal{F}, P, E)$ et si $\mathcal{G}$ est une sous-tribu de $\mathcal{F}$ on notera $E^{\mathcal{G}}(X)$ l'espérance conditionnelle de X relativement à $\mathcal{G}$.

<u>DEFINITION 5</u> : T, $(\mathcal{F}_t)_{t\in T}$ étant comme plus haut, on appelle martingale à valeurs dans E, et indexée par T, une famille $(X_t)_{t\in T}$ d'éléments de $L^1(\Omega, \mathcal{F}, P, E)$ telle que :

1) $X_t \in L^1(\Omega, \mathcal{F}_t, P, E) \qquad \forall\, t \in T$
2) $\forall \; s < t , \quad E^{\mathcal{F}_s}(X_t) = X_s$.

Par exemple, si $X \in L^1(\Omega, \mathcal{F}, P, E)$ et si l'on pose pour tout $t \in T$ $X_t = E^{\mathcal{F}_t}(X)$, $(X_t)_{t\in T}$ forme une martingale. Dans ce cas, les $||X_t||$ sont uniformément intégrables, c'est-à-dire :

$$\text{(A)} \quad \sup_{t\in T} \int_{\{||X_t||\geq a\}} ||X_t|| \, dP \xrightarrow[a \to \infty]{} 0 .$$

En outre, $(X_t) \to X_\infty = E^{\mathcal{F}_\infty}(X)$ pour la topologie de $L^1(\Omega, \mathcal{F}, P, E)$.

En effet, cela est bien connu si $T = \mathbb{N}$ et $E = \mathbb{R}$.

La partie "uniforme intégrabilité" se généralise très facilement si T est quelconque et E est un Banach.

La partie "convergence" se généralise facilement au cas $T = N$ et E Banach. On passe ensuite au cas T quelconque en remarquant que pour toute suite(t_n)d'éléments de T telle que $t_n < t_{n+1}$ $\forall$ n, la famille $(X_{t_n})_{n\in N}$ est une martingale relative aux $(\mathcal{F}_{t_n})$. En outre :

- Si $T = N$, E Banach, $X_n \to X_\infty$ presque sûrement.
- Si $T = R_+$, E Banach, il existe pour tout t un représentant $\tilde{X}_t$ dans la classe des X_t et $\tilde{X}_\infty$ dans la classe des X_∞ tels que
  - les $t \rightsquigarrow X_t$ sont continues à droite et possèdent une limite à gauche ;
  - $X_t \xrightarrow[t \to \infty]{} X_\infty$ presque sûrement.

REMARQUE 8 : En général, si T est un ensemble ordonné filtrant quelconque, on n'a pas de convergence presque sûre vers X_∞, même si $E = \mathbb{R}$, comme DIEUDONNE l'a remarqué.

Un des problèmes fondamentaux de la théorie des martingales est le suivant :

"Soit $(X_t)_{t\in T}$ une martingale à valeurs dans le Banach E. Existe-t-il une application $X_\infty : \Omega \to E$ telle que $X_t \to X_\infty$ (dans un sens à préciser) ? Identifier ensuite cette X_∞".

Le résultat suivant est un résultat fondamental de la théorie des martingales réelles.

THEOREME 9 :

Soit $(\Omega, \mathcal{F}, P)$ un espace probabilisé ; soit $(\mathcal{F}_t)_{t\in\mathbb{R}_+}$ une famille croissante de sous-tribus de $\mathcal{F}$. On suppose que :

- les $(\mathcal{F}_t)$ sont continues à droite i.e. $\mathcal{F}_t = \bigcap_{s>t} \mathcal{F}_s$;
- $\mathcal{F}$ et les $\mathcal{F}_t$ sont P-complètes, c'est-à-dire contiennent tous les ensembles de Ω, P-négligeables.

Soit $(X_t)_{t\in\mathbb{R}_+}$ une martingale réelle relative aux $(\mathcal{F}_t)$. Alors :

1) La martingale (X_t) admet une modification continue à droite et ayant des limites à gauche ; on la notera encore (X_t).

2) Si $\sup_{t\in\mathbb{R}^+} E\{|X_t|\} < \infty$, les v.a. X_t convergent p.s. vers une v.a. intégrable X_∞, quand $t \to \infty$.

3) Si en outre les $||X_t||$ sont uniformément intégrables, (X_t) converge au sens de $L^1(\Omega, P, \mathbb{R})$ vers X_∞, X_∞ est mesurable par rapport à la tribu complétée de $\mathcal{F}_\infty$ relativement à P, et $(X_t)_{t\in\mathbb{R}_+\cup\{\infty\}}$ est une martingale.

Pour la démonstration de ce résultat, nous renvoyons à MEYER (Probabilités et Potentiel).

Dans le cas où E est un Banach, on a le résultat général suivant :

THEOREME 10 :

Soit E un Banach ; $(\Omega, \mathcal{F}, P)$, $(\mathcal{F}_t)_{t\in\mathbb{R}^+}$ une famille croissante de sous-tribus de $\mathcal{F}$, P-complètes et continues à droite. Soit (X_t) une martingale relative aux $(\mathcal{F}_t)$ et régulière (i.e. ayant une modification, c.a.d.l.a.g.).

Alors :

1) Si $\sup_t E\{||X_t||\} < \infty$, il existe $X_\infty \in L^1_*(\Omega, P, E")$ telle que pour tout $e' \in E'$, $< X_t, e'>$ converge vers $< X_\infty, e'>$ presque sûrement, quand t tend vers l'infini.

2) Si les $||X_t||$ sont en outre uniformément intégrables, pour tout $e' \in E'$, les $< X_t, e'>$ convergent au sens de $L^1(\Omega, P, \mathbb{R})$ vers $< X_\infty, e'>$ et la famille $(< X_t, e'>)_{t\in \mathbb{R}_+\cup\{\infty\}}$ est une martingale relative aux $\mathcal{F}_t$ et à la tribu P-complète de $\mathcal{F}_\infty$. En outre, $X_\infty \in L^1_{**}(\Omega, P, E)$.

DEMONSTRATION :

1) Les $||X_t||$ forment une sous-martingale dans L^1 à trajectoires c.a.d.l.a.g. et à intégrales uniformément bornées.

D'après le théorème de convergence des sous-martingales, analogue au théorème de convergence des martingales (9. 1), $||X_t||$ converge p.s. vers une v.a. $\Phi \geq 0$ et P-intégrable.

Soit maintenant $e' \in E'$, alors les $< X_t, e'>$ forment une martingale c a d l a g telle que $\sup_t E\{|< X_t, e'>|\} < \infty$; donc elle converge p.s. vers une limite $X_\infty^{e'}$ P-intégrable. En outre :

$$< X_t, e'> \leq ||X_t||\ ||e'||$$

donc
$$|X_\infty^{e'}| \leq \Phi \ . \ ||e'|| \ .$$

Par conséquent, l'application $e' \to \dfrac{X_\infty^{e'}}{\Phi}$ envoie E' continue dans $L^\infty(\Omega, \mathcal{F}, \mathbb{R})$ et est de norme ≤ 1.

Elle est donc définie par une $\varphi \in \mathcal{L}^\infty_*(\Omega, \mathcal{F}, E")$ telle que

$$<\varphi, e'> = \frac{X_\infty^{e'}}{\Phi} \quad \text{presque sûrement.}$$

Donc, si l'on pose $X_\infty = \varphi\,\Phi$, on a pour tout $e' \in E'$ $< X_\infty, e'> = X_\infty^{e'}$ presque sûrement et $||X_\infty|| \leq \Phi$. Donc $X^\infty \in \mathcal{L}^1_*(\Omega, \mathcal{F}, E")$.

On peut considérer P comme une probabilité sur la tribu P-complétée de $\mathcal{F}_\infty$, soit $\hat{\mathcal{F}}_\infty$ par rapport à laquelle sont mesurables les X_t, $||X_t||$, Φ ; donc on peut prendre $\varphi \in \mathcal{L}^\infty_*(\Omega, \hat{\mathcal{F}}_\infty, P, E")$ et $X \in \mathcal{L}^1_*(\Omega, \hat{\mathcal{F}}_\infty, P, E")$.

2) Si on suppose en plus les $||X_t||$ uniformément intégrables, pour tout $e' \in E'$, il en est de même des $|< X_t, e'>|$. Et, par conséquent,

$< X_t, e'> \xrightarrow[t \to \infty]{} <X_\infty, e'>$ dans $L^1(\Omega, P, \mathbb{R})$ et les $(< X_t, e'>)_{0 \le t \le \infty}$ forment une martingale relative aux $(\mathscr{F}_t)_{t<\infty}$ et $\hat{\mathscr{F}}_\infty$.

D'autre part, on peut toujours supposer que X_∞ réalise le module minimum de sa classe : $||X_\infty||$ est alors P-mesurable, donc P-intégrable.

Il nous reste à montrer que $X_\infty \in \mathscr{L}^1_{**}(\Omega, P, E)$, c'est-à-dire que pour toute fonction $\mathscr{F}$-mesurable réelle bornée, on a :

$$\int X_\infty \, f \, dP \in E .$$

Or X_∞ étant scalairement $\hat{\mathscr{F}}_\infty$-mesurable, on a :

$$\int X_\infty \, f \, dP = \int X_\infty \, E^{\hat{\mathscr{F}}_\infty}(f) \, dP \,;$$

donc on peut supposer f mesurable par rapport à $\hat{\mathscr{F}}_\infty$.

Maintenant,

$$||\int (X_\infty - X_t) \, f \, dP|| = || \int (X_\infty f - X_t f_t) \, dP||$$

$$= || \int X_\infty (f - f_t) \, dP|| \le \int ||X_\infty|| \, |f - f_t| \, dP$$

(si l'on pose $f_t = E^{\mathscr{F}_t}(f)$).

D'autre part, pour tout $t \in \mathbb{R}_+$, $\int X_t \, f \, dP \in E$; en outre, f_t converge p.s. vers f en restant bornée d'après ce que l'on a vu au début de ce numéro. Il résulte alors de Lebesgue que :

$$||\int (X_\infty - X_t) \, f \, dP|| \to 0 \,;$$

donc $\int X_\infty \, f \, dP \in E$

et le théorème est démontré.

On en déduit immédiatement le

<u>THEOREME 11</u> :

Supposons, en plus des hypothèses générales du théorème 10, que E est de Radon-Nikodym. Supposons en outre que les $||X_t||$ soient uniformément intégrables. Alors la v.a. X_∞ du théorème 10 peut être choisie appartenant à $L^1(\Omega, P, E)$. En outre, $X_t \to X_\infty$ dans $L^1(\Omega, P, E)$ et $(X_t)_{0 \le t \le \infty}$ forme une martingale relative aux $(\mathscr{F}_t)_{t \in \mathbb{R}_+}$ et $\hat{\mathscr{F}}_\infty$.

DEMONSTRATION : Par le théorème 10, il existe $X_\infty \in L^1_{**}(\Omega, P, E)$ tel que pour tout $e' \in E'$

$$< X_t, e'> \xrightarrow[t \to \infty]{} < X_\infty, e'> \quad \text{dans } L^1(\Omega, P, \mathbb{R}).$$

E étant de Radon-Nikodym, on peut supposer que $X_\infty \in L^1(\Omega, P, \mathbb{R})$. Mais alors, dû au fait que $(X_t, \mathcal{F}_t)$ est une "martingale scalaire", on a $X_t = E^{\mathcal{F}_t}(X_\infty)$ $\forall t \in \mathbb{R}_+$. Il suffit alors d'appliquer les remarques du début.

REMARQUE 9 : On peut démontrer que si E est de Radon-Nikodym, la fonction X_∞ du théorème (10. 1) (X_t satisfaisant aux hypothèses de ce théorème) peut être choisie P-intégrable à valeurs dans E et que $X_t \to X_\infty$ P-presque sûrement. (Voir par exemple SCHWARTZ, Séminaire MAUREY-SCHWARTZ 1974-75). La démonstration en est assez longue. Nous l'omettrons, car nous n'utiliserons pas ce résultat.

COROLLAIRE : Soit E un espace de Radon-Nikodym ; $(X_n, \mathcal{F}_n)_{n\in\mathbb{N}}$ une martingale à valeurs dans E telle que les $||X_n||$ soient uniformément intégrables. Cette martingale converge dans $L^1(\Omega, P, E)$ quand n tend vers l'infini.

En effet, à partir de $(X_n, \mathcal{F}_n)$ il est facile de construire une martingale $(X_t, \mathcal{F}_t)_{t\in\mathbb{R}_+}$ ayant les propriétés du théorème 10, en interpolant

$$X_t = X_n \quad \text{si} \quad n \le t < n+1$$
$$\mathcal{F}_t = \mathcal{F}_n \quad \text{si} \quad n \le t < n+1 .$$

Remarquons que, dans le cas $E = \mathbb{R}$, on démontre d'abord le théorème 9 dans le cas dénombrable, et on en déduit ensuite le théorème de convergence pour les $(X_t)_{t\in\mathbb{R}_+}$ presque immédiatement.

PROPOSITION 7 :

Soit E un Banach ; les propriétés suivantes sont équivalentes :

1) Toute martingale dénombrable $(X_n, \mathcal{F}_n)$ à valeurs dans E telle que $||X_n||$ soient uniformément bornées converge dans $L^1(\Omega, P, E)$.

2) Toute martingale $(X_t, \mathcal{F}_t)_{t \in T}$ indexée par un ensemble filtrant quelconque et uniformément intégrable, converge dans $L^1(\Omega, P, E)$.

DEMONSTRATION : La seule chose à démontrer est 1) $\Rightarrow$ 2).

Supposons que $(X_t)_{t \in T}$ ne converge pas dans $L^1(\Omega, P, E)$; les (X_t) ne forment donc pas un filtre de Cauchy. Par conséquent, il existe $\varepsilon > 0$ et une suite $t_0 < t_1 < t_2 < \dots$ telle que

$$E\{\|X_{t_{n+1}} - X_{t_n}\|\} > \varepsilon \quad \forall n .$$

Mais alors les $(Y_n) = (X_{t_n})$ formeraient une martingale uniformément intégrable et non convergente. La logique s'écroule !

On en déduit que si E est de Radon-Nikodym, toute martingale uniformément intégrale converge dans $L^1(\Omega, P, E)$.

THEOREME 12 :

Soit E un Banach ; les propriétés suivantes sont équivalentes :

a) E est de Radon-Nikodym ;

b) Toute martingale à valeurs dans E uniformément intégrable est convergente dans $L^1(\Omega, P, E)$.

DEMONSTRATION : La seule chose qui reste à démontrer est l'implication b) $\Rightarrow$ a).

Supposons b) vérifiée ; il nous suffit de démontrer que toute

$$u \in \mathcal{L}(L^1(\Omega, P, \mathbb{R}), E)$$

peut être définie par une $\varphi \in L^\infty(\Omega, P, E)$ de façon à ce que

$$u(f) = \int f \varphi \, dP \qquad \forall f \in L^1(\Omega, P, \mathbb{R}).$$

Soit $\mathcal{G}$ une sous-tribu finie de $\mathcal{F}$, engendrée par la partition finies $(A_i)_{i \in I}$. L'application $u^{\mathcal{G}}$ de $L^1(\Omega, P, \mathbb{R})$ dans E définie par

$u^{\mathcal{G}}(f) = u(E^{\mathcal{G}}(f))$, est associée à une fonction

$\varphi_{\mathcal{G}} \in L^\infty(\Omega, P, E)$. En effet :

$$u^{\mathcal{G}}(f) = \sum_{i\in I} \frac{1}{P(A_i)} \int_{A_i} f\, dP\ u(1_{A_i})$$

$$= \int \varphi_{\mathcal{G}}\, f\, dP$$

avec

$$\varphi_{\mathcal{G}} = \sum_{i\in I} \frac{1}{P(A_i)}\ u\,(1_{A_i}) \cdot 1_{A_i} .$$

En outre,

$$\|\varphi_{\mathcal{G}}(\omega)\| \leq \|u\| ;$$

En effet, si $\omega \in A_i$:

$$\varphi_{\mathcal{G}}(\omega) = \frac{u(1_{A_i})}{P(A_i)} \leq \|u\| \frac{1}{P(A_i)} \|1_{A_i}\|_{L^1(\Omega,\, P,\, \mathbb{R})} = \|u\| .$$

Je dis que, quand $\mathcal{G}$ décrit l'ensemble ordonné filtrant des sous-tribus finies de $\mathcal{F}$, les $\varphi_{\mathcal{G}}$ forment une martingale. En effet, dû au fait que si $f \in L^1(\Omega, P, \mathbb{R})$ est $\mathcal{G}$-mesurable, on a $E^{\mathcal{G}}(f) = f$, on en déduit :

$$u(f) = u^{\mathcal{G}}(f) = \int \varphi_{\mathcal{G}} \cdot f\, dP \quad \forall f,\ \mathcal{G}\text{-mesurable et intégrable.}$$

Si donc $\mathcal{G}_1 \subset \mathcal{G}_2$ et f est intégrable $\mathcal{G}_1$-mesurable, on a

$$\int \varphi_{\mathcal{G}_1} f\, dP = \int \varphi_{\mathcal{G}_2} f\, dP = u(f) \quad ;$$

donc

$$E^{\mathcal{G}_1}(\varphi_{\mathcal{G}_2}) = \varphi_{\mathcal{G}_1} .$$

Les $(\varphi_{\mathcal{G}}, \mathcal{G})$ (quand $\mathcal{G}$ décrit l'ensemble des sous-tribus finies de $\mathcal{F}$) forment une martingale bornée donc uniformément intégrable . D'autre part, $\mathcal{F}$ est la tribu engendrée par ses sous-tribus finies.

Cela étant, $E^{\mathcal{G}}(f) \to f$ dans $L^1(\Omega, P, \mathbb{R})$, donc $u^{\mathcal{G}}(f) = u\,(E^{\mathcal{G}}(f))$ converge vers u(f) dans E. D'autre part, d'après l'hypothèse $\varphi_{\mathcal{G}}$ converge vers φ dans $L^1(\Omega, P, E)$. Mais la boule unité de $L^\infty(\Omega, P, E)$ étant un fermé de $L^1(\Omega, P, E)$ (car de toute suite convergente dans L^1 on peut extraire une sous-suite convergeant p.s.), on a $\|\varphi(\omega)\| \leqq 1 \ \ \forall \omega$.

Il reste à démontrer que $u(f) = \int f \varphi\, dP \quad \forall f \in L^1(\Omega, P, \mathbb{R})$.

Mais si f est bornée, du fait que $\varphi_{\mathcal{G}} \to \varphi$ dans $L^1(\Omega, P, E)$, on a :

$$\int \varphi f\, dP = \lim_{\mathcal{G}} \int \varphi_{\mathcal{G}} f\, dP = \lim_{\mathcal{G}} u^{\mathcal{G}}(f) = u(f) \quad ;$$

Donc $f \rightsquigarrow u(f)$ et $f \rightsquigarrow \int f \varphi \, dP$ sont deux applications linéaires continues de $L^1(P, \mathbb{R})$ dans E coïncidant sur $L^\infty(P, \mathbb{R})$. Elles coïncident donc partout, et a) est vérifiée.

REMARQUE 10 : On peut encore dire que E vérifie Radon-Nikodym si et seulement si toute martingale dénombrable à valeurs dans E uniformément intégrable converge dans $L^1(\Omega, P, E)$.

COROLLAIRE 1 :

Si E vérifie Radon-Nikodym, tous ses sous-Banach aussi.

DEMONSTRATION : On a déjà démontré ce résultat d'une autre façon. Mais on peut la redémontrer autrement :

Soit F un sous-Banach de E, et soit une martingale à valeurs dans F uniformément intégrable : c'est aussi une martingale à valeurs dans E uniformément intégrable, donc convergente dans $L^1(\Omega, P, E)$. Mais alors la limite est nécessairement dans $L^1(\Omega, P, E)$.

COROLLAIRE 2 :

E est de Radon-Nikodym si et seulement si tous ses Banach séparables le sont.

Cela résulte immédiatement de ce qu'une martingale dénombrable prend p.s. ses valeurs dans un sous-espace séparable.

REMARQUE 11 : On peut encore retrouver qu'un espace de Radon-Nikodym est un C-espace.

En effet, E est un C-espace si et seulement si $\mathcal{C}_\beta(E) = \mathcal{B}_\beta(E)$ par le théorème 3 du chapitre IV.

Maintenant, si $(x_n) \in \mathcal{B}_\beta(E)$, (x_n) définit une martingale uniformément intégrable $X_n = (x_n \varepsilon_n)$, donc convergente si E est de Radon-Nikodym (les tribus $\mathcal{F}_n$ sont celles rendant mesurables les ε_i $0 \leq i \leq n$).

BIBLIOGRAPHIE

- BUCCHIONI - BUCHWALTER : Intégration vectorielle et théorème de Radon-Nikodym - Université Claude-Bernard Lyon I - Mathématiques - 1975

- CHEVET S. : Notes manuscrites impubliées -

- HOFFMANN - JØRGENSEN :

  [1] Sums of independent Banach Space valued random variables - Preprint Series n° 15 - 1972-73 - AARHUS -

  [2] Sums of independent Banach Space valued random variables - Studia Mathematica - T. III (1974) pp. 159-186 -

- KWAPIEN

  [1] Complément au Théorème de SAZONOV-MINLOS - C.R. Acad. Sci. Paris 267 (1968) pp. 698-700 -

  [2] On Banach space containing C_0 - Studia Mathematica T LII - 1974 - pp. 187-188 -

- MAUREY - SCHWARTZ : Séminaires 1972-73, 1973-74 et 1974-75 - Ecole Polytechnique, Paris -

- MOEDOMO - UHL : Radon-Nikodym Theorem for Bochner and Pettis Integrales : Pacific Journal of Mathematics 38 (1971) -

- PISIER G. : Exposés du Séminaire SCHWARTZ-MAUREY

- ROLEWICZ S. : Metric Linear Spaces - Varsovie, 1972

- RYLL-NARDZENSKI C. et WOYCZINSKI W.A. : Bounded Multiplier convergence in measure of Random Series - Proceedings A.M.S. 53 (1) (1973) pp. 96-98 -

- SCHWARTZ L. :

  [1] Un théorème de convergence dans les L^p $(0 \leq p < \infty)$ - C.R. Acad. Sci. Paris 268 (1969) pp. 704-706 -

[2] Exposés 4, 5, 6 du Séminaire Maurey-Schwartz 1974-75

URPIN : Exposé 6bis du Séminaire Maurey-Schwartz 1974-75 -

SUBADDITIVE PROCESSES

PAR J.F.C. KINGMAN

CHAPTER 1

THE ERGODIC THEOREM

1.1 - Why study subadditive processes ?

The concept of a subadditive random process first arose in a paper by Hammersley and Welsh [8] on problems of percolation in networks. They observed that certain families of random variables satisfied inequalities which, could they be replaced by equalities, would enable the classical laws of large numbers to be applied. They boldly conjectured, and went some way towards proving, that these laws may be applied to the families they encountered.

In the decade since their paper was published, it has become clear that their axioms are satisfied by random variables arising in a number of different contexts, and I will describe these in chapter 2. Moreover I was able to establish the Hammersley-Welsh conjecture by proving an ergodic theorem for subadditive processes [15] which is a complete generalisation of the Birkhoff - Von Neumann theorem for stationary random sequences. Thus a law of large numbers is available for a variety of problems to which the usual results do not apply.

The definition then is motivated by the problems to which the theory is to be applied, and is as follows. A subadditive process is a family of real random variables

$$X_{st} \quad (s, t \in \mathbb{Z}, s < t), \tag{1.1.1}$$

all defined of course on some underlying probability space, indexed by two integer-valued variables s and t, with s < t, and satisfying three axioms :

(S_1) whenever $s < t < u$,

$$X_{su} \leqslant X_{st} + X_{tu} ; \tag{1.1.2}$$

(S_2) <u>the joint distribution of</u> (X_{st}) <u>are the same as those of</u> $(X_{s+1,t+1})$;

(S_3) X_{ot} <u>has a finite expectation</u>

$$g_t = \mathbb{E}(X_{ot}) , \qquad (1.1.3)$$

<u>which satisties</u>

$$g_t \geq -At \qquad (1.1.4)$$

<u>for some constant A and all</u> $t > 1$.

The assumption (S_1) is the one which gives the whole theory its characteristic flavour, (S_2) is a condition of stationarity, and (S_3) brings the random variables into L_1, where ergodic theory may be supposed to operate. The condition (1.1.4) may appear a little odd, but the reason for it is as follows.

From (S_2),

$$\mathbb{E}(X_{st}) = \mathbb{E}(X_{s+1,t+1}),$$

from which it follows that

$$\mathbb{E}(X_{st}) = g_{t-s} \qquad (1.1.5)$$

for all $s < t$. Taking expectations in (1.1.2), we therefore have

$$g_{u-s} \leq g_{t-s} + g_{u-t} ,$$

or

$$g_{m+n} \leq g_m + g_n \quad (m, n \geq 1) \qquad (1.1.6)$$

Fix a positive integer k. Then for any $r \geq 1$ and $1 \leq s \leq k$, induction on (1.1.6) yields

$$g_{rk+s} \leq rg_k + g_s .$$

Hence

$$\limsup_{r\to\infty} \frac{g_{rk+s}}{rk+s} \leq \limsup_{r\to\infty} \frac{rg_k + g_s}{rk + s} = \frac{g_k}{k} .$$

Since this is true for $s = 1, 2, \ldots, k$, we have

$$\limsup_{n\to\infty} \frac{g_n}{n} \leq \frac{g_k}{k} \qquad (1.1.7)$$

This holds for all k, and taking lower limits as $k \to \infty$,

$$\limsup_{n\to\infty} \frac{g_n}{n} \leq \liminf_{k\to\infty} \frac{g_k}{k} .$$

Hence g_n/n converges to a limit, which by (1.1.7) cannot be $+ \infty$. Condition (1.1.4) is just what is needed to ensure that the limit is not $- \infty$. In fact, we have established the following result.

Theorem 1.1

The finite limit

$$\gamma = \lim_{n\to\infty} g_n/n \tag{1.1.8}$$

exists, and

$$\gamma = \inf_{n\geq 1} g_n/n \tag{1.1.9}$$

The constant γ turns out to be of central importance, as can be see by considering two very special cases. Suppose first that the random variables X_{st} are not in fact random, but that each takes only one value. Then (1.1.5) shows that

$$X_{st} = g_{t-s}$$

In particular, (1.1.8) can then be written

$$\lim_{t\to\infty} X_{ot}/t = \gamma \tag{1.1.10}$$

Secondly, suppose that the inequality (1.1.2) is actually satisfied with equality :

$$X_{su} = X_{st} + X_{tu} \tag{1.1.11}$$

Then

$$X_{st} = \sum_{j=s+1}^{t} Y_j , \tag{1.1.12}$$

when the random variables

$$Y_j = X_{j-1,j} \tag{1.1.13}$$

form a stationary sequence because of (S_2). Hence

$$g_n = n\,\mathbb{E}(Y_1),$$

so that

$$\gamma = \mathbb{E}(Y_1)$$

In this case the pointwise ergodic theorem (or strong law of large numbers) for stationary sequences ensures that the limit

$$\xi = \lim_{t\to\infty} X_{ot}/t \qquad (1.1.14)$$

$$= \lim_{t\to\infty} t^{-1} \sum_{j=1}^{t} Y_j$$

exists with probability one, and that

$$\mathbb{E}(\xi) = \gamma \qquad (1.1.15)$$

Moreover, if there is any way of proving that ξ is degenerate, (1.1.15) shows that $\xi = \gamma$, so that (1.1.10) holds with probability one.

Thus two quite different special cases lead to the conclusion that the limit (1.1.14) exists with probability one, and this suggests the possibility that it may exist for all subadditive processes. This is indeed true, and the main purpose of this chapter is to prove it and to establish the properties of the limit ξ.

1.2. The maximal ergodic lemma

In classical ergodic theory the usual route to the Birkhoff-Von Neumann theorem goes by way of this lemma, of which a beautifully simple proof was given by Garsia [6]. It turns out that Garsia's argument applies, almost without change, to subadditive processes.

As a matter of notation, we shall in this chapter consistently use primes to represent a shift of the parameter set $\mathbb{Z}$. Thus we write

$$X'_{st} = X_{s+1,t+1} \qquad (1.2.1)$$

and use a similar notation for random variables defined as functions of the X_{st}.

Theorem 1.2

Let (X_{st}) be a subadditive process, and write

$$A = \{X_{ot} \geqslant 0 \quad \text{for some } t \geqslant 1\} \tag{1.2.2}$$

Then

$$\int_A X_{01} \, d\mathbb{P} \geqslant 0. \tag{1.2.3}$$

(Here A is of course a measurable subset of the underlying probability space, and $\mathbb{P}$ is the probability measure).

Proof

Write $M_t = \max_{1 \leqslant s \leqslant t} X_{os}$

Then

$$\begin{aligned} M_t \leqslant M_{t+1} &= \max (X_{01}, \max_{1 \leqslant s \leqslant t} X_{0,s+1}) \\ &\leqslant \max (X_{01}, \max_{1 \leqslant s \leqslant t} (X_{01} + X_{1,s+1})) \\ &= X_{01} + \max (0, \max_{1 \leqslant s \leqslant t} X'_{0s}) \\ &= X_{01} + \max (0, M'_t). \end{aligned}$$

Therefore, if I_t is the indicator of the event $\{M_t \geqslant 0\}$, we have

$$\begin{aligned} \mathbb{E} (M_t I_t) &\leqslant \mathbb{E} \{[X_{01} + \max (0, M'_t)] I_t\} \\ &\leqslant \mathbb{E} (X_{01} I_t) + \mathbb{E} \{\max (0, M'_t)\} \\ &= \mathbb{E} (X_{01} I_t) + \mathbb{E} \{\max (0, M_t)\}, \text{ using } (S_2), \\ &= \mathbb{E} (X_{01} I_t) + \mathbb{E} (M_t I_t). \end{aligned}$$

Hence $\mathbb{E} (X_{01} I_t) \geqslant 0$.

As $t \to \infty$, I_t increases to the indicator function of A, so that (1.2.3) follows and the proof is complete.

This result has some interest in its own right ; it can for instance be used as in [3] to set bounds on

$$\sup_{t\geq 1} X_{ot}/t$$

But it does not seem possible to use it directly to establish the ergodic theorem, because it is essentially one-sided and has no companion result with the inequalities reversed.

Having get so far, and to keep the analysis self-contained, we will take the opportunity to remind the reader how, in the additive case (1.1.12), the maximal ergodic lemma leads to the proof of the strong law of large numbers.

Theorem 1.3 (Birkhoff-Von Neumann)

Let (Yn ; n = 1, 2, ...) be a stationary sequence of random variables with finite expectation. Then the limit

$$\eta = \lim_{n\to\infty} n^{-1} \sum_{j=1}^{n} Y_j \qquad (1.2.4)$$

exists with probability one and in L_1 norm and

$$\mathbb{E}(\eta) = \mathbb{E}(Y_1) \qquad (1.2.5)$$

Proof

Write $S_n = \sum_{j=1}^{n} Y_j$,

and $\eta_* = \liminf_{n\to\infty} S_n/n$, $\eta^* = \limsup_{n\to\infty} S_n/n$ (1.2.6)

Note that η_* and η^* are invariant, in the sense that if $Y'_j = Y_{j+1}$, then

$$\eta'_* = \eta_* \ , \ \eta^{*\prime} = \eta^*$$

It follows that, if B is any event of positive probability defined in terms of (η_*, η^*), then (Y_j) remains stationary if $\mathbb{P}$ is replaced by the conditional probability measure

$$\mathbb{P}_B(.) = \mathbb{P}(. \mid B).$$

We exploit this fact in two ways.

(i) Let $B = \{\eta^* > b\}$, where b is any constant such that $\mathbb{P}(B) > 0$.

Apply Theorem 1.2 with

$$X_{st} = \sum_{j=s+1}^{t} (Y_j - b),$$

and $\mathbb{P}$ replaced by $\mathbb{P}_B$, noting that $\mathbb{P}_B(A) = 1$ by (1.2.6). Then

$$\int (Y_1 - b)\, d\,\mathbb{P}_B \geqslant 0$$

or

$$\int_B Y_1\, d\mathbb{P} \geqslant b\, \mathbb{P}(B)$$

Since the left hand side is bounded by $\mathbb{E}|Y_1|$, this shows that

$$\mathbb{P}(\eta^* > b) \to 0$$

as $b \to \infty$, whence

$$\mathbb{P}(\eta^* < \infty) = 1.$$

An exactly similar argument, applied to

$$X_{st} = \sum_{j=s+1}^{t} (a - Y_j),$$

shows that $\mathbb{P}(\eta_* > -\infty) = 1$.

(ii) Now repeat the process, but with $B = \{\eta_* < a, \eta^* > b\}$, supposing that $\mathbb{P}(B) > 0$. Then we obtain

$$\int_B Y_1\, d\mathbb{P} \geqslant b\,\mathbb{P}(B), \quad a\,\mathbb{P}(B) \geqslant \int_B Y_1\, d\mathbb{P},$$

whence we deduce that $\mathbb{P}(\eta_* < a, \eta^* > b) = 0$ whenever $a < b$. Thus, with probability one, $-\infty < \eta_* = \eta^* < \infty$, so that the finite limit (1.2.4) exists.

To prove convergence in the L_1 norm $||.|| = \mathbb{E}|.|$, let $\varepsilon > 0$ and choose C so large that

$$||\tilde{Y}_1 - Y_1|| < \frac{1}{4}\varepsilon,$$

where

$$\tilde{Y}_n = Y_n \text{ if } |Y_n| \leqslant C$$
$$= 0 \text{ if } |Y_n| > C.$$

The sequence $\tilde{Y}_n$ is also stationary, so that

$$\tilde{\eta} = \lim_{n\to\infty} \tilde{S}_n/n$$

exists, and $||n^{-1}\tilde{S}_n - \tilde{\eta}|| \to 0$

by bounded convergence. Moreover,

$$||n^{-1}\tilde{S}_n - n^{-1}S_n|| \leqslant n^{-1}\sum_{j=1}^{n}||\tilde{Y}_j - Y_j|| = ||\tilde{Y}_1 - Y_1|| < \frac{1}{4}\varepsilon$$

Since $\lim (n^{-1}\tilde{S}_n - n^{-1}S_n) = \tilde{\eta} - \eta$,

Fatou's lemma shows that

$$||\tilde{\eta} - \eta|| \leqslant \frac{1}{4}\varepsilon$$

Thus, if N is chosen so that

$$||n^{-1}\tilde{S}_n - \tilde{\eta}|| < \frac{1}{2}\varepsilon \qquad (n \geqslant N),$$

we have $||n^{-1}S_n - \eta|| < \frac{1}{4}\varepsilon + \frac{1}{2}\varepsilon + \frac{1}{4}\varepsilon = \varepsilon$

for $n \geqslant N$. This shows that

$$S_n/n \to \eta$$

in L_1 norm. In particular,

$$\mathbb{E}(\eta) = \lim_{n\to\infty}\mathbb{E}(S_n/n) = \mathbb{E}(Y_1),$$

and the proof is complete.

1.3 - The easy half of the ergodic theorem

It turns out that Theorem 1.3 is all that is needed to establish most of the ergodic properties of a general subadditive process. In this section we get as much as possible from these relatively simple arguments, leaving till the next section the difficult final step.

Theorem 1.4

Let (X_{st}) be a subadditive process. Then

$$\xi = \limsup_{t\to\infty} X_{ot}/t \qquad (1.3.1)$$

<u>is almost surely finite, and satisfies</u>

$$\lim_{t\to\infty} \mathbb{E}\,|t^{-1} X_{ot} - \xi| = 0 \tag{1.3.2}$$

<u>and</u>

$$\mathbb{E}\,(\xi) = \gamma, \tag{1.3.3}$$

<u>where γ is given by (1.1.8)</u>.

Equation (1.3.2) asserts that ξ is the L_1 limit of X_{ot}/t. Thus all that is needed to make theorem 1.4 into a complete generalisation to subadditive processes of theorem 1.3 is to complement (1.3.1) by showing that

$$\xi = \liminf_{t\to\infty} X_{ot}/t \;;$$

This is the surprisingly difficult result.

<u>Proof</u>

Fix $k \geq 1$ and write $N(t)$ for the integral part of t/k. Then repeated application of (1.1.2) gives

$$X_{st} \leq \sum_{r=1}^{N(t)} X_{(r-1)k,rk} + X_{N(t)k,t}$$

$$\leq \sum_{r=1}^{N(t)} Y_r + W_{N(t)} ,$$

where $Y_r = X_{(r-1)k,\, rk}$

and $$W_N = \sum_{j=1}^{k-1} |X_{Nk,Nk+j}|.$$

By (S_2) the distribution of W_N is the same for all N, and $\mathbb{E}\,(W_1) < \infty$. Hence, for $\varepsilon > 0$,

$$\sum_{N=1}^{\infty} \mathbb{P}\,(W_N \geq \varepsilon N) = \sum_{N=1}^{\infty} \mathbb{P}\,(W_1 \geq \varepsilon N) \leq \varepsilon^{-1}\,\mathbb{E}\,(W_1) < \infty ,$$

and the Borel-Cantelli lemma shows that

$$\lim_{N\to\infty} W_N / N = 0$$

with probability one. Moreover, the sequence (Y_r) is stationary, with $\mathbb{E}\,(Y_1) = g_k$, and hence

$$\xi_k = \lim_{N\to\infty} (Nk)^{-1} \sum_{r=1}^{N} Y_r$$

exists, and $\mathbb{E}(\xi_k) = g_k / k$. Therefore,

$$\xi = \limsup_{t\to\infty} X_{ot} / t \leqslant \limsup_{N\to\infty} (Nk)^{-1} \{\sum_{r=1}^{N} Y_r + W_N\}$$

$$\leqslant \xi_k .$$

We have thus shown that there are random variables ξ_k for k = 1, 2, 3...

such that

$$\xi \leqslant \xi_k , \quad \mathbb{E}(\xi_k) = g_k / k \tag{1.3.4}$$

Letting $k \to \infty$,

$$\mathbb{E}(\xi) \leqslant \gamma \tag{1.3.5}$$

Now consider the non-positive subadditive process

$$Z_{st} = X_{st} - \sum_{j=s+1}^{t} X_{j-1,j}$$

and write

$$\zeta_n = \sup_{t\geqslant n} Z_{ot} / t$$

Then ζ_n decreases with n to the limit

$$\limsup_{t\to\infty} Z_{ot} / t = \xi - \xi_1$$

By monotone convergence,

$$\lim_{n\to\infty} \mathbb{E}(\zeta_n) = \mathbb{E}(\lim_{n\to\infty} \zeta_n) = \mathbb{E}(\xi - \xi_1) = \mathbb{E}(\xi) - g_1,$$

but on the other hand

$$\lim_{n\to\infty} \mathbb{E}(\zeta_n) \geqslant \liminf_{n\to\infty} \mathbb{E}(Z_{on}/n) = \liminf_{n\to\infty} (n^{-1} g_n - g_1) = \gamma - g_1,$$

so that $\mathbb{E}(\xi) - g_1 \geqslant \gamma - g_1$,

which with (1.3.5) implies (1.3.3).

Examining the above chain of inequalities, we see that

$$\lim_{n\to\infty} \mathbb{E}(\zeta_n) = \lim_{n\to\infty} \mathbb{E}(Z_{on}/n),$$

so that

$$\|\zeta_n - Z_{on}/n\| \to 0.$$

By monotone convergence

$$||\zeta_n - \xi + \xi_1|| \to 0.$$

Applying Theorem 1.3 with $Y_n = X_{n-1,n}$ we have

$$\left|\left| \frac{X_{on} - Z_{on}}{n} - \xi_1 \right|\right| \to 0.$$

These three results combine to give

$$\left|\left| \frac{X_{on}}{n} - \xi \right|\right| \to 0,$$

proving (1.3.2). This implies also that

$$\mathbb{E}\,(\xi) = \lim_{n\to\infty} \mathbb{E}\,(X_{on}/n) = \gamma,$$

and the proof is complete.

It may perhaps be important for some applications to note that this proof does not use the full force of the axiom (S_2). In their original formulation, Hammersley and Welsh used a weaker axiom

(S_{2a}) <u>the distribution of</u> X_{st} <u>depends only on</u> $t-s$.

At one time there seemed to be an interesting process (arising from the problem to be described in Section 3.3) which satisfied (S_1), (S_{2a}) and (S_3) but not (S_2). When Hammersley [10] raised this possibility, I remarked that the proof of Theorem 1.4 depended only on the property

(S_{2b}) <u>the sequence</u> $(X_{(r-1)k,rk}\ ;\ r = 1,2,\ldots)$ <u>is stationary for all</u> $k \geq 1$.

which was also enjoyed by his example. In the event, this turned out to be a futile discussion, since it was pointed out by Joshi (quoted in [12]) that Hammersley's example does not even satisfy (S_1). But the possibility remains open that there may be interesting processes satisfying (S_{2b}) but not (S_2), and for these Theorem 1.4 is available.The axiom (S_{2a}) (which is neither weaker nor stronger than (S_{2b})) seems now to be of merely historical interest.

In the additive case, the limit η of the theorem 1.3 is classically identified as a conditional expectation of Y_1 relative to a certain σ-field. An analogous description of the corresponding limit ξ of Theorem 1.4 can likewise be given. To state the result, we denote by $\underline{\underline{I}}$ the completion of the σ-field of events defined in terms of the random variables X_{st} and invariant under the shift (1.2.1).

Theorem 1.5

The limit ξ in (1.3.1) may be written

$$\xi = \lim_{t\to\infty} t^{-1} \; \mathbb{E}\,(X_{ot} \mid \underline{\underline{I}}). \tag{1.3.6}$$

In particular, if $\underline{\underline{I}}$ is trivial, then

$$\lim_{t\to\infty} X_{ot} \,/\, t = \gamma \tag{1.3.7}$$

in L_1 norm.

Proof

Using primes as before to denote quantities defined with respect to the shifted process $(X_{s+1,t+1})$, we have

$$\xi' = \limsup_{t\to\infty} X_{1t}/t \geqslant \limsup_{t\to\infty} (X_{ot} - X_{01}) \,/\, t = \xi,$$

using (1.1.2). But, by (S_2) and (1.3.3),

$$\mathbb{E}\,(\xi') = \gamma = \mathbb{E}\,(\xi),$$

so that $\quad \mathbb{P}\,(\xi' = \xi) = 1,$

and ξ is $\underline{\underline{I}}$-measurable.

Now let Φ_t be a version of the conditional expectation $\mathbb{E}\,(X_{ot} \mid \underline{\underline{I}})$. By (S_2),

$$\mathbb{E}\,(X_{st} \mid \underline{\underline{I}}) = \Phi_{t-s}$$

and (1.1.2) shows that

$$\Phi_{m+n} \leqslant \Phi_m + \Phi_n \tag{1.3.8}$$

with probability one. Hence, as in theorem 1.1,

$$\phi = \lim_{n\to\infty} \phi_n / n$$

exists with probability one.

If $I \in \underline{\underline{I}}$ has $P(I) > 0$, then (X_{st}) is still a subadditive process if $\mathbb{P}$ is replaced by the conditional probability measure $\mathbb{P}_I$, and $X_{ot}/t \to \xi$ in L_1 norm relative to $\mathbb{P}_I$. Hence

$$\mathbb{E}(\xi \mid I) = \lim_{t\to\infty} \mathbb{E}(t^{-1} X_{ot} \mid I) = \lim_{t\to\infty} \mathbb{E}(t^{-1} \phi_t) \mid I).$$

Taking m = n in (1.3.8) we see that the convergence of ϕ_t/t to ϕ is monotone if t is restricted to power of 2. With this restriction, we can use monotone convergence to deduce that

$$\mathbb{E}(\xi \mid I) = \mathbb{E}(\phi \mid I).$$

Hence

$$\int_I \xi \; d\mathbb{P} = \int_I \phi \; d\mathbb{P}$$

for all $I \in \underline{\underline{I}}$, and since ξ and ϕ are both $\underline{\underline{I}}$-measurable, this means that

$$\mathbb{P}(\xi = \phi) = 1,$$

and the theorem is proved.

We shall see later that, in all the applications so far explored, there is available a "zero-one law" showing that $\underline{\underline{I}}$ is trivial. Thus (1.3.7) is the typical situation, and emphasises the importance of the constant γ.

1.4 - The difficult half of the ergodic theorem

A subadditive process is called additive if (1.1.2) is satisfied in the stronger form of the equality

$$X_{su} = X_{st} + X_{tu} \qquad (s < t < u) \tag{1.4.1}$$

It is already been remarked that additive processes admit a representation (1.1.12) in terms of a stationary sequence, and that their ergodic properties are therefore classical. In order to complete the analysis of the ergodic properties of subadditive processes which are not additive, we show that there always exists an additive process, lying below the subadditive process, and <u>with the same value of</u> γ.

At the time of preparing these lectures, this was the only known approach to the ergodic theorem, but during the <u>Ecole</u>, M. Yves Derrienic produced an ingenious proof of Theorem 1.7, replacing the appeal to compactness by an application of the maximal ergodic lemma to (1.4.13). This argument, which will be published elsewhere, does not establish Theorem 1.6, but it does give a more constructive approach to the fundamental ergodic result.

<u>Theorem 1.6</u>

<u>Let</u> (X_{st}) <u>be a subadditive process. Then there is an additive process</u> (A_{st}) <u>satisfying</u>

$$A_{st} \leqslant X_{st} \qquad (s < t) \tag{1.4.2}$$

<u>and</u>

$$\mathbb{E}\,(A_{01}) = \gamma.$$

<u>Proof</u>

Denote by Σ the collection of functions

$$x \ : \ \{(s,\, t)\ ;\ s,\, t \in \mathbb{Z},\ s < t\} \to \mathbb{R}$$

which satisfy

$$x\,(s,\, u) \ \leqslant \ x\,(s,\, t) + x\,(t,\, u) \qquad (s < t < u) \tag{1.4.4}$$

and make Σ a measurable space in the usual way. Because of (1.1.2), the collection of real random variables X_{st} can be regarded as a random variable $\underline{X}$ taking values in Σ, and we denote by P the distribution of $\underline{X}$, so that P is a probability measure on Σ defined

$$P\,(B) = \mathbb{P}\,(\underline{X}^{-1}\,B) \tag{1.4.5}$$

The coordinate functions $j_{st} : x \to x\,(s,\, t)$ belong to the Banach space

$$L = L_1\,(\Sigma,\, P), \tag{1.4.6}$$

consisting of (equivalence classes of) functions $f\,:\ \Sigma \to \mathbb{R}$ with

$$||f|| = \int_\Sigma |f|\, dP = \mathbb{E}\ \ |f\,(\underline{X})| \tag{1.4.7}$$

finite.

A function $\Theta : \Sigma \to \Sigma$ is defined by

$$(\Theta x)(s, t) = x(s+1, t+1) \tag{1.4.8}$$

By (S_2), Θ preserves the measure P, and in particular it induces an isometry $T : L \to L$ by the recipe

$$(Tf)(x) = f(\Theta x) \tag{1.4.9}$$

The key to the proof is to establish the existence of a function $f \in L$ such that, for all $n \geq 1$,

$$f + Tf + \dots + T^{n-1} f \leq j_{on} \tag{1.4.10}$$

and such that

$$\int_\Sigma f \, dP = \gamma \tag{1.4.11}$$

This is essentially a "linear programming" result : γ is the largest value that may be attained for $\int f\,dP$ if f satisfies (1.4.10) since integration of (1.4.10) gives

$$n \int f\,dP \leq \int j_{on}\,dP = g_n$$

If the existence of f is proved, the assertion of the theorem follows at once, since the random variables

$$A_{st} = \sum_{i=s+1}^{t} f(\Theta^{i-1} \underline{\underline{X}}) \tag{1.4.12}$$

satisfy (1.4.2) and (1.4.3) as a consequence of (1.4.10) and (1.4.11), and form an additive process because $\Theta\underline{\underline{X}}$ has the same distribution as $\underline{\underline{X}}$.

To show that f exists, consider the function f_m in L defined for $m \geq 1$ by

$$f_m = m^{-1} \sum_{r=1}^{m} (j_{or} - j_{1r}),$$

with the convention that $j_{rr} = 0$. For any $n \geq 1$,

$$\begin{aligned} f_m + T f_m + \dots + T^{n-1} f_m &= m^{-1} \sum_{k=0}^{n-1} \sum_{r=1}^{m} (j_{k,k+r} - j_{k+1,k+r}) \\ &= m^{-1} \sum_{t=1}^{m+n-1} \sum_{s=a}^{b-1} (j_{st} - j_{s+1,t}) \\ &= m^{-1} \sum_{t=1}^{m+n-1} (j_{at} - j_{bt}), \end{aligned}$$

where $a = \max(t-m, 0)$, $b = \min(t, n)$. Hence we have, applying (1.4.4),

$$f_m + T f_m + \dots + T^{n-1} f_m \leq m^{-1} \sum_{t=1}^{m+n-1} j_{ab} \tag{1.4.13}$$

Notice that, as $m \to \infty$ for fixed n, the right hand side of (1.4.13) converges to j_{on}. Moreover

$$\int_\Sigma f_m \, dP = m^{-1} \sum_{r=1}^{m} (g_r - g_{r-1}) = g_m / m \to \gamma.$$

Accordingly, if the sequence (f_m) has a limit point in a suitable topology for L, this limit point will satisfy (1.4.10) and (1.4.11).

Thus we have to use some sort of compactness argument. Burkholder [1] has shown how an elegant treatment of this point can be given using a theorem of Komlos [19]. But Komlos's theorem is itself a deap result, and I will use my original argument, which depends on the compactness of the unit ball in the second dual of a Banach space. (For the necessary results from Banach space theory, see for instance [4]).

The dual of L is the space $L^* = L_\infty(\Sigma, P)$ of equivalence classes of bounded measurable functions $\phi : \Sigma \to R$, acting on L by the formula

$$(\phi, f) = \int_\Sigma \phi f \, dP$$

The dual L^{**} of L^* is the space of bounded finitely additive set functions μ on Σ which vanish on P-null sets, and L^{**} acts on L^* by

$$(\mu, \phi) = \int_\Sigma \phi \, d\mu.$$

The natural embedding $\kappa : L \to L^{**}$ is represented by

$$(\kappa f)(A) = \int_A f \, dP$$

From (1.4.13) with $n = 1$,

$$||f_m|| \leqslant ||j_{01}|| + ||j_{01} - f_m|| = ||j_{01}|| + \int (j_{01} - f_n) \, dP$$

$$= ||j_{01}|| + g_1 - g_m/m \leqslant ||j_{01}|| + g_1 - \gamma = M,$$

say. Hence the elements κf_m of L^{**} form a bounded sequence, which therefore has a limit point μ (say) in the weak* topology (the weakest topology on L^{**} which makes $\mu \to (\mu, \phi)$ continuous for all $\phi \in L^*$).

With this topology, the function $S : L^{**} \to L^{**}$ defined by $S_\mu = \mu(\Theta^{-1}.)$ is continuous, and $S\kappa = \kappa T$ on L. Hence,

$$\kappa f_m + S\kappa f_m + \ldots + S^{n-1} \kappa f_m \leq m^{-1} \sum_{t=1}^{m+n-1} \kappa j_{ab},$$

and since the right hand side converges as $m \to \infty$ to κj_{on}, we have

$$\mu + S\mu + \ldots + S^{n-1} \mu \leq \kappa j_{on}. \tag{1.4.14}$$

Moreover, since

$$(\kappa f_n, 1) = (1, f_n) = \int f_n \, dP = g_m / m \to \gamma$$

we have $(\mu, 1) = \gamma$,

or $$\mu (\Sigma) = \gamma. \tag{1.4.15}$$

From (1.4.14) with $n = 1$, $(\kappa j_{01} - \mu)$ is a non-negative finitely additive set function, and hence by a theorem of Yosida and Hewitt (Theorem 1.2.3 of [24]) can be decomposed as the sum of a measure and a non-negative purely finitely additive set function. Hence

$$\mu = \lambda - \pi,$$

where λ is a signed measure and π is non-negative and purely finitely additive. From (1.4.14),

$$\lambda + S\lambda + \ldots + S^{n-1} \lambda \leq \kappa j_{on} + \pi_n,$$

where

$$\pi_n = \pi + S\pi + \ldots + S^{n-1} \pi.$$

Now $S\pi$ is purely finitely additive, for suppose ν is a measure $\leq S\pi$, there $S^{-1} \nu \leq \pi$ so that $S^{-1} \nu = 0$ and $\nu = 0$. Hence π_n being the sum of purely finitely additive functions, is also purely finitely additive (Theorem 1.17 of [24]) and therefore

$$\lambda + S\lambda + \ldots + S^{n-1} \lambda \leq \kappa j_{on}.$$

Evaluating these set functions on Σ,

$$n \lambda (\Sigma) \leq (\kappa j_{on}) (\Sigma) = g_n$$

and letting $n \to \infty$,

$$\lambda (\Sigma) \leq \gamma = \mu (\Sigma) = \lambda (\Sigma) - \pi (\Sigma).$$

Hence $\pi (\Sigma) = 0$, showing that $\mu = \lambda$ is a signed measure. Moreover, since $\mu \in L^{**}$, it vanishes on P-null sets, and the Radon-Nikodyn theorem shows that $\mu = \kappa f$ for some $f \in L$. Then (1.4.14) and (1.4.15) translate into (1.4.10) and (1.4.11) and the proof is complete.

Having proved Theorem 1.6, the ergodic theorem follows at once. If we use the notation of Theorem 1.4, and write ξ_A for the corresponding limit for the additive process A, then by (1.4.2) and Theorem 1.3,

$$\liminf_{t \to \infty} X_{ot} / t \geq \liminf_{t \to \infty} A_{ot} / t = \xi_A .$$

Hence $\xi \geq \xi_A$, but

$$\mathbb{E}(\xi) = \gamma = \mathbb{E}(\xi_A).$$

Therefore $\mathbb{P}(\xi = \liminf_{t \to \infty} X_{ot}) = 1$.

as required. Collecting together all these results, we therefore have the basic ergodic theorem.

<u>Theorem 1.7</u>

<u>Let (X_{st}) be a subadditive process. Then the finite limit</u>

$$\xi = \lim_{t\to\infty} X_{ot} / t \tag{1.4.16}$$

<u>exists with probability one and in L_1 norm, and is given by (1.3.6).</u>

1.5 - <u>Some complementary remarks</u>

(i) It should be stressed that the additive process whose existence follows from Theorem 1.6 is not usually unique. For instance, suppose that, for each i in the finite index set, A^i_{st} denotes an additive process with trivial invariant σ-field, and suppose that

$$a = \mathbb{E}(A^i_{01})$$

is independent of i. Then it is easy to check that

$$X_{st} = \max_i A^i_{st}$$

is a subadditive process. Since

$$\mathbb{P}(\lim_{t\to\infty} A^i_{ot} / t = a) = 1,$$

it follows that

$$\mathbb{P}\,(\lim_{t\to\infty} X_{ot} / t = a) = 1$$

Hence $\gamma = a$,

and any convex combination

$$A_{st} = \sum_i p_i A^i_{st} \qquad (p_i \geqslant 0,\ \sum p_i = 1)$$

satisfies (1.4.2).

This example suggests that a subadditive process might be analysed in terms of the convex class of additive processes lying below it. This is not the case, since if that class determined the process uniquely it would necessarily do so through the formula

$$X_{st} = \sup\,\{A_{st}\ ;\ (A_{st})\ \text{additive},\ A_{st} \leqslant X_{st}\} \qquad (1.5.1)$$

and an example given in [11] shows that this identity can fail.

(ii) Throughout the argument, we have taken the parameter set to consist of all the intgers, positive and negative, although the theorems relate only to positive parameter values. This differs from the formulation in [15], but it has the advantage that the shift Θ (and therefore T and S) is invertible. In a a private communication, Professor C. G. Esseen pointed out that this fact is needed at one stage of the proof of the Theorem 1.6, and the present argument therefore corrects an error in my original argument.

It may be objected that the present definition is unduly restrictive, in that there might be families $(X_{st}\ ;\ s,\ t \geqslant 1)$ which satisfy (S_1), (S_2) and (S_3) on the positive integers but admit no extension to $\mathbb{Z}$. That no such loss of generality occurs may be seen by modifying in a slight and obvious way an argument of Doob ([3], page 456) to show that, because of (S_2), such an extension is always possible.

(iii) Theorem 1.7 starts from assumptions about the two-parameter family (X_{st}), and draws conclusions about the one-parameter sequence (X_{ot}). This suggests the question : given a random sequence $(Z_t ; t \geq 1)$, under what conditions does there exist a subadditive process with $Z_t = X_{ot}$? A similar question is : under what conditions are there subadditive processes (X^1_{st}) and (X^2_{st}) with

$$Z_t = X^1_{st} - X^2_{st} \text{ ?}$$

A necessary condition in each case is that

$$\lim_{t\to\infty} Z_t / t$$

should exist with probability one, but I know of no necessary and sufficient conditions.

(iv) It is sometimes useful to know what happens when (1.1.4) is violated, so that $\gamma = -\infty$. In such a case there is of course no L_1 ergodic theorem, but the probability one result remains true so long as (S_3) is replaced by the much weaker condition

$$(S_{3a}) \qquad \mathbb{E}\,(X^+_{01}) < \infty \quad , \tag{1.5.2}$$

where $x^+ = \max\,(x, 0)$.

Theorem 1.8

Suppose that (X_{st}) satisfies (S_1), (S_2) and (S_{3a}). Then the limit

$$\xi = \lim_{t\to\infty} X_{ot}/t \tag{1.5.3}$$

exists with probability one in $-\infty \leq \xi < \infty$, and

$$\mathbb{E}\,(\xi) = \lim_{t\to\infty} \mathbb{E}\,(X_{ot})/t. \tag{1.5.4}$$

Proof

From (S_1) and (S_2),

$$\mathbb{E}\,(X^+_{st}) \leq (t-s)\ \mathbb{E}\,(X^+_{01}) < \infty \quad ,$$

and it follows easily that

$$X_{st}^{(N)} = \max \{X_{st}, - N (t-s)\}$$

defines a subadditive process for each $N \geq 1$. Hence

$$\xi^{(N)} = \lim_{t\to\infty} X_{0t}^{(N)}/t = \lim_{t\to\infty} \max \{X_{ot}/t , -N\}$$

exists with probability one for all $N \geq 1$. This implies the existence of the limit (1.5.3) related to $\xi^{(N)}$ by

$$\xi^{(N)} = \max (\xi, - N).$$

The expectation

$$g_t = \mathbb{E} (X_{ot})$$

exists in $- \infty \leq g_t < \infty$, and satisfies $g_{m+n} \leq g_m + g_n$. Hence, as in the proof of Theorem 1.1,

$$\gamma = \lim_{t\to\infty} g_t/t$$

exists in $- \infty \leq \gamma < \infty$. If γ is finite, (S_3) is satisfied, and (1.5.4) has already been proved. Thus we have only to show that $\mathbb{E} (\xi) = - \infty$ when $\gamma = - \infty$. To do this, note that

$$\mathbb{E} (\xi) \leq \mathbb{E} (\xi^{(N)}) = \gamma^{(N)} = \inf_t t^{-1} \mathbb{E} (X_{ot}^{(N)}) \leq t^{-1} \mathbb{E} (X_{ot}^{(N)})$$

for any $t, N \geq 1$. Letting $N \to \infty$,

$$\mathbb{E} (\xi) \leq t^{-1} \mathbb{E} (X_{ot}) = g_t/t ,$$

and letting $t \to \infty$,

$$\mathbb{E} (\xi) \leq - \infty,$$

which is enough to complete the proof.

Note that there are two ways in which (S_{3a}) can be true if (S_3) is false. One is that $g_t = - \infty$ for some (and then for all larger) t , and the other is that g_t is finite for all t, but $g_t/t \to - \infty$ as $t \to \infty$.

(v) By Theorem 1.5, ξ is an invariant random variable, and therefore

$$\lim_{t\to\infty} X_{st}/t = \xi$$

with probability one, for every fixed s. For some applications, it is useful to extend this result to cover the case in which s and t both tend to infinity. The following theorem is such a generalisation, rather trivial for additive processes, but less so for subadditive processes.

Theorem 1.9

Let (X_{st}) be a subadditive process, and $c > 1$ a constant. Then

$$\frac{X_{st}}{t-s} \to \xi$$

with probability one as $s, t \to \infty$ subject to the condition $cs \leqslant t$.

Proof

Use 'Lim' to denote the limit described in the statement of the theorem. By (S_1),

$$X_{st} \geqslant X_{ot} - X_{os} = (\xi t + o(t)) - (\xi s + o(s))$$
$$= \xi\,(t-s) + o\,(t-s),$$

so that

$$\text{Lim inf } X_{st}/(t-s) \geqslant \xi \qquad (1.5.6)$$

To obtain an inequality in the reverse direction, fix $k \geqslant 1$ and let M and N be the smallest and largest integers respectively such that

$$s \leqslant Mk \leqslant Nk \leqslant t.$$

Then

$$X_{st} \leqslant X_{s,Mk} + \sum_{r=M+1}^{N} X_{(r-1)k,rk} + X_{Nk,t}\ ,$$

and arguing as in the proof of Theorem 1.4,

$$\operatorname{Lim\,sup} \frac{X_{st}}{t-s} \leqslant \operatorname{Lim\,sup} \frac{1}{(N-M)k} \sum_{r=M+1}^{N} X_{(r-1)k,rk}$$

with probability one. Since

$$\xi_k = \lim_{N\to\infty} \frac{1}{Nk} \sum_{r=1}^{N} X_{(r-1)k,rk}$$

exists, and since $N/M \sim t/s \geqslant c > 1$, it follows that

$$\operatorname{Lim\,sup} \frac{X_{st}}{t-s} \leqslant \xi_k$$

and thus

$$\mathbb{E}\left\{\operatorname{Lim\,sup} \frac{X_{st}}{t-s}\right\} \leqslant \frac{g_k}{k} .$$

Letting $k \to \infty$, and comparing the result with (1.5.6), completes the proof.

(vi) We shall encounter applications in which the random variables X_{st} are defined for real values of s and t, rather than just for integer values. Thus we have the concept of a <u>continuous-parameter subadditive process</u>, a collection

$$(X_{st} \; ; \; -\infty < s < t < \infty \quad) \tag{1.5.7}$$

satisfying (S_1), (S_2) generalised to arbitrary shifts $(X_{st}) \to (X_{s+\tau,t+\tau})$, and (S_3). Some new phenomena arise in dealing with such processes, and it is for example not always the case that

$$\mathbb{P}\left\{\lim_{t\to\infty} X_{ot}/t \text{ exists}\right\} = 1, \tag{1.5.8}$$

even under assumptions of separability.

These complications are explored in detail in Section 1.4 of [16], but they do not cause difficulties in the examples of interest, since these have the additional property that X_{ot} is monotone in t. With this extra information (1.5.8) follows from

$$\mathbb{P}\left\{\lim_{n\to\infty} X_{on}/n \text{ exists}\right\} = 1, \tag{1.5.9}$$

and (1.5.9) can be deduced form Theorem 1.7, by noting that the "discrete skeleton"

$$(X_{mn} \; ; \; m, n \in \mathbb{Z} \; , \; m < n)$$

is a subadditive process in the sense of our original definition.

CHAPTER 2

SOME APPLICATIONS

2.1 - Shortest paths

Hammersley and Welsh [8] were led to the axioms for subadditive processes by problems of the following type. Consider a connected Graph G, and suppose that with every edge e of G is associated a non-negative random variables u (e) with finite expectation. The u (e) for distinct edges are assumed to be independent and to have a common distribution. For any two vertices v and v' of G, define

$$U(v, v') = \inf \sum u(e), \tag{2.1.1}$$

where the infimum is taken over all paths from v to v', and the sum extends over all the edges in the path. It is shown in [8] that the properties of the random variable U (v, v') are of great importance in problems of percolation theory and related areas of applied probability.

In only the most trivial cases can the distribution of U (v, v') be determined analytically, and it is necessary to rely on approximate and limiting results. That these can be found rests on the fact that, for vertices v, v', v", we have

$$U(v, v'') \leqslant U(v, v') + U(v', v'') \tag{2.1.2}$$

because the right hand side is the restricted infinum when the path from v to v" is required to pass through v'. This inequality can be exploited when the graph G has some homogeneity of structure.

Suppose for example that the vertex set of G is the integer lattice $\mathbb{Z} \times \mathbb{Z}$ in $\mathbb{R}^2$, and that the edges join each point to its four nearest neighbours. Write

$$X_{st} = U\{(s, 0), (t, 0)\} \tag{2.1.3}$$

for the "shortest distance" from the point s on the axis to the point t.

Then (2.1.2) shows at once that (S_1) is satisfied. Moreover, the assumptions on u (e), and the homogeneity of G under the shift $(x, y) \longmapsto (x+1, y)$, show that (S_2) is satisfied, and (S_3) is satisfied because

$$0 \leq \mathbb{E}(X_{st}) \leq (t - s)\,\mathbb{E}(u)$$

Hence the theory of subadditive processes is applicable. It is easy to see (from the zero-one law for independent sequences, cf [15], Section 4) that $\underline{\underline{I}}$ is trivial, so that the theorems of Chapter 1 imply that there is a constant γ such that

$$\lim_{t\to\infty} U\big((0, 0), (t, 0)\big) / t = \gamma \tag{2.1.4}$$

with probability one (and in L_1). The constant γ depends only on the common distribution of the u (e), and its evaluation seems to be of great difficulty (for partial results, see [8]).

Note that, for any $v = (v_1, v_2) \in \mathbb{Z} \times \mathbb{Z}$, we can modify (2.1.3) by defining

$$X_{st} = U(sv, tv) \tag{2.1.5}$$

for $s, t \in \mathbb{Z}$, $s < t$. Once again this is a subadditive process for which $\underline{\underline{I}}$ is trivial (when $v \neq (0, 0)$), so that there is a constant $\gamma(v)$ with

$$\lim_{t\to\infty} U\big((0, 0), (tv_1, tv_2)\big) / t = \gamma(v) \tag{2.1.6}$$

with probability one.

Although the determination of $\gamma(v)$ is at present impossible, there are general properties which can be asserted. Clearly

$$0 \leq \gamma(v) \leq (|v_1| + |v_2|)\,\mathbb{E}(u) \tag{2.1.7}$$

and

$$\gamma(nv) = |n|\ \gamma(v) \tag{2.1.8}$$

for $n \in \mathbb{Z}$. Moreover, (2.1.2) shows, using the technique of the proof of Theorem 1.9, that

$$\gamma(v + v') \leq \gamma(v) + \gamma(v') \tag{2.1.9}$$

It follows easily from these facts that γ can be extended to the whole of $\mathbb{R}^2$ in such a way that it is a seminorm, i.e. that

$$\gamma(x+y) \leqslant \gamma(x) + \gamma(y) \ , \ \gamma(ax) = |a| \ \gamma(x) \qquad (2.1.10)$$

for $a \in \mathbb{R}$, $x, y \in \mathbb{R}^2$. Such a function γ is determined by the corresponding unit ball

$$\{x \in \mathbb{R}^2 \ ; \ \gamma(x) \leqslant 1\} \qquad (2.1.11)$$

and it would be interesting to know the shape of this convex set. Is it true that the shape (though not the size) does not depend on the distribution of u ? Under what conditions is γ a norm ?

Very closely related ideas have been applied by Richardson [20] to certain models of growth or contagion in $\mathbb{R}^2$ or $\mathbb{R}^3$. These lead to qualitative conclusions about the growth of an organism on the spread of some contagion, described by a set which increases its dimensions linearly with time and whose shape is asymptotically that of a convex set of the form (2.1.11). In Richardson's models there is numerical evidence that this set is a circle, so that

$$\gamma(x_1, x_2) = \gamma_0 (x_1^2 + x_2^2)^{1/2} \qquad (2.1.12)$$

for some constant γ_0, but in the Hammersley-Welsh problem it would be plausible to conjecture that

$$\gamma(x_1, x_2) = \gamma_0 (|x_1| + |x_2|) \qquad (2.1.13)$$

In view of the different possible applications, it seems useful to give a more general context into which the special cases fit. To do this, let V be a countable set, and let Φ be a bijection from V onto itself. For distincts elements v, v' of V, let u (v, v') denote a random variable with values in $[0, \infty]$, and define

$$U(v, v') = \inf \sum_{r=1}^{n} u(v_{r-1}, v_r) \qquad (2.1.14)$$

where the infimum is taken over all finite sequences

$$v = v_0, v_1, v_2, \ldots , v_n = v'$$

Then the inequality (2.1.2) is clearly valid.

For any fixed $v_0 \in V$, define v_n $(n \in \mathbb{Z})$ by

$$v_n = \Phi(v_{n-1})$$

Then (2.1.2) shows that

$$X_{st} = U\ (v_s,\ v_t) \tag{2.1.15}$$

satisfies (S_1). If the u have the property that the joint distribution of the two families

$$\{u\ (v,\ v')\ ;\ v,\ v' \in V\}\ ,\ \{u\ (\Phi v,\ \Phi v')\ ;\ v,\ v' \in V\}$$

are the same, then (S_2) is satisfied, and if

$$\mathbb{E}\ \{u\ (v_o,\ v_1)\} < \infty,$$

then (S_3) also holds. Hence in these circumstances (2.1.15) defines a subadditive process, and we may conclude that

$$\lim_{n\to\infty}\ U\ (v_o,\ v_n)\ /\ n$$

exists with a probability one.

2.2 - Products of non-commuting random elements

A non-homogeneous Markov chain (with a finite number N of states and a discrete time parameter) is described by its family of transition matrices P_n $(n \in \mathbb{Z})$, where the $(i,\ j)^{th}$ element of P_n is the probablility of moving from the ith state to the jth state between time (n-1) and time n. The transition probabilities between time s and time t $(s,\ t \in \mathbb{Z}\ ,\ s < t)$ are then given by the corresponding elements of the matrix product

$$P\ (s,\ t) = P_{s+1}\ P_{s+2}\ \ldots\ P_t, \tag{2.2.1}$$

and therefore considerable interest attaches to the properties of the matrix product P (s, t) [17].

In some applications, the lack of homogeneity expressed by the variation of P_n with n may be due to random fluctuations in the environment of the chain, so that it may be useful to think of (P_n) as a stationary random sequence of matrices.

This is one reason for studying products of stationary sequences of matrices, and there are others. More general Markov processes give rise to a similar formal structure, where the P_n are now linear operators on on infinite dimensional spaces. To illustrate the power of subadditive ergodic theory as applied to such problems, we prove a theorem first established for matrices by Furstenberg and Kesten [5].

Theorem 2.1

Let A be a semigroup, and $||.|| : A \to \mathbb{R}^+$ a function such that

$$||PQ|| \leqslant ||P||.||Q|| \tag{2.2.2.}$$

for P, Q in A. Let (P_n) be a stationary sequence of random elements of A, and suppose that

$$\mathbb{E}\,\{(\log ||P_1||)^+\} < \infty \tag{2.2.3.}$$

Then, if P(s, t) is defined by (2.2.1), the limit

$$\xi = \lim_{t\to\infty} t^{-1} \log ||P(0, t)|| \tag{2.2.4.}$$

exists, and

$$\mathbb{E}\,(\xi) = \lim_{t\to\infty} \mathbb{E}\,(t^{-1} \log ||P(0, t)||) \tag{2.2.5.}$$

Proof

It is only necessary to observe that

$$X_{st} = \log ||P\,(s, t)||$$

satisfies (S_1), (S_2) and (S_{3a}), and to use Theorem 1.8.
(In formulating this theorem and its proof, we have omitted obvious measure theoretic details. Obviously A must have a measurable structure compatible with multiplication, and $||.||$ must be measurable. Such omissions will be made without comment throughout these notes).

Note that, as in Theorem 1.5, ξ is measurable with respect to the invariant σ-field of (P_n). If this σ-field is trivial, as it will be for example if the P_n are independent, then ξ is equal to its expectation (2.2.5) with probability one.

We have used the emotive notation $||.||$ for obvious reasons, but there is no need for this function to be a norm even when A is an algebra of matrices. One important example occurs in the case when the P_n are stochastic matrices, as in the application which began this section. This concerns the so-called "convergence norm"

$$||P|| = \frac{1}{2} \max_{i.j} \sum_k |p_{ik} - p_{jk}| \qquad (2.2.6)$$

It is easy to check that this satisfies (2.2.2) on

$$A = \{P = (p_{ij} ; i, j = 1, 2, \ldots, N) ; p_{ij} \geqslant 0, \sum_j p_{ij} = 1\},$$

but it is not a norm in the usual sense because its kernel is non-trivial, consisting of all stochastic matrices with identical rows.

With this choice of $||.||$, (2.2.3) is trivially satisfied since $||P|| \leqslant 1$ for all P. Hence the limit (2.2.4) exists, and is non-positive. In particular, $||P(0, t)||$ converges to zero exponentially fast as $t \to \infty$ on the event $\{\xi < 0\}$. Since such convergence implies weak ergodicity [17] this is an important conclusion. If the invariant σ-field is trivial, it takes a stronger form ; the non-homogeneous Markov chain determined by (P_n) is weakly ergodic with probability one if $\mathbb{E}(\xi) < 0$. In view of (2.2.5) this is true if

$$\mathbb{P}(||P(0, t)|| < 1) > 0 \quad \text{for some } t \geqslant 1. \qquad (2.2.7)$$

a very weak "scrambling" condition.

Just as the function $P \to ||P||$ is submultiplicative, the function $P \to p_{11}$ (the $1, 1)^{th}$ element of P) is supermultiplicative for positive matrices, since the $(1, 1)^{th}$ element of the product PQ of two such matrices is

$$\sum_k p_{1k} q_{k1} \geqslant p_{11} q_{11} \qquad (2.2.8)$$

This leads to a simple proof of another result of Furstenberg and Kesten [5] which they proved by more elaborate methods. The assumption of strict positivity can be weakened at the cost of some complication.

Theorem 2.2

Let (P_n) be a stationary sequence of random $(N \times N)$ matrices, the elements of which are strictly positive random variables whose logarithms have finite expectations. Let $p_{ij}(s, t)$ denote the $(i, j)^{th}$ element of the matrix $P(s, t)$ defined by (2.2.1). Then the limit

$$\rho = \lim_{t} \{p_{ij}(0, t)\}^{1/t} \tag{2.2.9}$$

exists and does not depend on i and j, with probability one.

Proof

Applying (2.2.8) to the identity

$$P(s, u) = P(s, t)\, P(t, u) \qquad (s < t < u)$$

we have $p_{11}(s, u) \geq p_{11}(s, t)\, p_{11}(t, u)$

so that $X_{st} = -\log p_{11}(s, t)$

satisfies (S_1). Because (P_n) is stationary, (S_2) holds and by hypothesis X_{st} has finite expectation g_{t-s} (say). Using the matrix norm

$$||P|| = \max_i \sum_j |p_{ij}|$$

we have

$$\begin{aligned} -g_t &= \mathbb{E}\{\log p_{11}(0, t)\} \\ &\leq \mathbb{E}\{\log ||P_1 P_2 \cdots P_t||\} \\ &\leq \sum_{n=1}^{t} \mathbb{E}\{\log ||P_n||\} \\ &= t\,\mathbb{E}\{\log ||P_1||\} \end{aligned}$$

showing that (S_3) is also satisfied. Hence (X_{st}) is a subadditive process, and Theorem 1.7 establishes (2.2.9) when $i = j = 1$, with $\rho = e^{-\xi}$.

Because ξ is invariant, so is ρ, and thus

$$\lim_{t\to\infty} \{p_{11}(s, t)\}^{1/t} = \rho \tag{2.2.10}$$

for each fixed s. Since

$$p_{ij}(0, t) \geq p_{i1}(0,1)\, p_{1i}(1,t-1)\, p_{1j}(t-1,t),$$

we have

$$\liminf_{t\to\infty} \{p_{1j}(0,\ t)\}^{1/t} \geqslant \lim_{t\to\infty} \{p_{11}(0,\ 1)\}^{1/t} \lim_{t\to\infty} \{p_{11}(1,\ t-1)\}^{1/t} \lim_{t\to\infty} \{p_{1j}(t-1,\ t)\}^{1/t}$$

$$\geqslant \rho\ ,$$

since on the right hand side the first limit is clearly 1, the second is ρ by (2.2.10), and the third is 1 (by the simple argument used in the proof of Theorem 1.4 to show that $W_N / N \to 0$). Similarly, the inequality

$$p_{11}(-1,\ t+1) \geqslant p_{11}(-1,\ 0)\ p_{1j}(0,\ t)\ p_{j1}(t,\ t+1)$$

shows that

$$\limsup_{t\to\infty} \{p_{1j}(0,\ t)\}^{1/t} \leqslant \rho,$$

which suffices to establish (2.2.9).

Notice that the argument applies also to infinite matrices so long as we impose the condition

$$E\ \{\log ||P_1||\} < \infty \qquad (2.2.11)$$

a condition trivially satisfied if P_1 is a stochastic matrix. Note also that, for finite (but not for infinite) matrices, the limits (2.2.4) and (2.2.9) are necessarily related by

$$\rho = e^{\xi} \qquad (2.2.12)$$

2.3 - A problem of Ulam

Let π be any permutation of $\{1, 2, \ldots, n\}$, and define $l(\pi)$ to be the largest value of r for which there is a sequence

$$1 \leqslant k_1 < k_2 < k_3 < \ldots < k_r \leqslant n \tag{2.3.1}$$

with

$$\pi(k_1) < \pi(k_2) < \ldots < \pi(k_r). \tag{2.3.2}$$

Define $l^*(\pi)$ similarly, with the inequalities in (2.3.2) reversed. It is a well known fact that, for every π,

$$\max\{l(\pi), l^*(\pi)\} \geqslant n^{1/2} \tag{2.3.3}$$

Ulam has asked how much better than $n^{1/2}$ can be achieved if we require only that the inequality hold for most π. In other words, what is the value of $l(\pi)$ for a "typical" π, when n is large. A remarkable answer to this question has been given by Hammersley (whose paper [9] should be consulted for the history of the problem) ; the numerical values represent an improvement on those given in [9].

Theorem 2.3

There is an absolute constant c in the range

$$1.59 < c < 2.49 \tag{2.3.4}$$

with the following property. If $b > c$, the number of π with $l(\pi) \geqslant bn^{1/2}$ is $o(n!)$ as $n \to \infty$, while if $b < c$, the number of π with $l(\pi) \leqslant bn^{1/2}$ is $o(n!)$.

Proof

Construct a Poisson process Π in $\mathbb{R}^2$ of unit rate. Thus Π is to be a random countable subset of $\mathbb{R}^2$, whose intersection with any fixed Borel set of finite area a consists of n points with probability

$$e^{-a} a^n / n! \qquad (n = 0, 1, 2, \ldots),$$

and whose intersections with disjoint Borel sets are independent.

For $s < t$, we define a random variable L_{st} to be the largest integer for which there exists points (x_j, y_j) of Π satisfying

$$s < x_1 < x_2 < \dots < x_r < t, \qquad (2.3.5)$$

$$s < y_1 < y_2 < \dots < y_r < t.$$

Then it is clear that, with probability one,

$$L_{su} \geqslant L_{st} + L_{tu} \qquad (s < t < u) \qquad (2.3.6)$$

Hence, if we restrict s, t to $\mathbb{Z}$, then

$$X_{st} = - L_{st}$$

satisfies (S_1). It also satisfies (S_2) because Π is stationary, and (1.5.2) holds trivialy. Moreover, an easy zero-one argument for Π shows that the invariant σ-field is trivial, and Theorem 1.9 implies that there is a constant $\gamma \geqslant -\infty$ such that

$$\lim_{n\to\infty} X_{on} / n = \gamma$$

with probability one. Taking $c = -\gamma$ and noting that L_{ot} increases with t, it follows that there is a constant c in $0 \leqslant c \leqslant \infty$ such that

$$\lim_{t\to\infty} L_{ot} / t = c \qquad (2.3.7)$$

with probability one.

The lower bound $c \geqslant 0$ can easily be improved. To do this, we define recursively a sequence of points of Π as follows. First let (x_1, y_1) be the point of Π with the smallest value of $x + y$ subject to $x > 0$, $y > 0$. If (x_r, y_r) $(r = 1, 2, \dots, n)$ have been choosen, then (x_{n+1}, y_{n+1}) is the point of Π with the smallest value of $x + y$ subject to

$$x > x_n \quad , \quad y > y_n .$$

Then $$0 < x_1 < x_2 < \dots,$$

$$0 < y_1 < y_2 < \dots,$$

so that, if $z(n) = \max (x_n, y_n)$

$$L_{0,\, z(n)} \geqslant n.$$

Now it is clear that the differences $x_{n+1} - x_n$ are independent and identically distributed, with expectation

$$\int_0^\infty \int_0^\infty x\, e^{-\frac{1}{2}(x+y)^2}\, dx\; dy \;=\; (\pi/8)^{1/2}$$

Hence the strong law of large numbers implies that, with probability one,

$$\lim_{n\to\infty} x_n / n = (\pi / 8)^{1/2}$$

Similarly

$$\lim_{n\to\infty} y_n / n = (\pi / 8)^{1/2}$$

so that

$$\lim_{n\to\infty} z(n) / n = (\pi / 8)^{1/2}$$

with probability one. Therefore, from (2.3.7),

$$c = \lim_{n\to\infty} L_{o,z(n)} / z(n) \geqslant \lim_{n\to\infty} n / z(n) = (8/\pi)^{1/2} > 1.59,$$

proving the left hand inequality in (2.3.5).

(The original argument in [9] minimised $x^2 + y^2$ instead of $x + y$, and deduced the inequality $c \geqslant \frac{1}{2}\pi > 1.57$. It is easy to show that there is no way of improving the result further by ingenious choice of the minimising function).

To exploit these results for the Ulam problem, let $t(n)$ be the smallest value of t for which there are exactly n points of Π in the square $\{(x, y)\ ;\ 0 < x < t,\ 0 < y < t\}$. Then the strong law shows that

$$\lim_{n\to\infty} n\, \{t(n)\}^{-2} = 1,$$

and then (2.3.7) implies that

$$\lim_{n\to\infty} L_{o,t(n)} / n^{1/2} = c$$

But

$$L_{o,t(n)} = l\,(\pi_n),$$

where π_n is the permutation π of $\{1, 2, \ldots, n\}$ such that the n points of Π in the square may be labelled (X_r, Y_r) with

$$X_{\pi 1} < X_{\pi 2} < \ldots < X_{\pi n}\ ,\ Y_1 < Y_2 < \ldots < Y_n.$$

Moreover, the symetry properties of Π make it clear that the random permutation π_n is uniformly distributed over the n! permutations.

The assertion of the theorem now follows from the fact that

$$\lim_{n\to\infty} l(\pi_n) / n^{1/2} = c \qquad (2.3.8)$$

For if b < c, the number of π with $l(\pi) \leqslant bn^{1/2}$ is

$$n!\ \ \mathbb{P}\{l(\pi_n) / n^{1/2} \leqslant b\} = o(n!),$$

and if b < c (which is not possible of course if $c = \infty$), the number of π with $l(\pi) \geqslant bn^{1/2}$ is

$$n!\ \ \mathbb{P}\{l(\pi_n) / n^{1/2} \geqslant b\} = o(n!)$$

as $n \to \infty$. Hence to complete the proof, it remains only to prove the upper bound for c.

To do this, let π be a permutation of {1, 2, ..., n}, and let $\nu(r, \pi)$ be a number of sequences of length r satisfying (2.3.1) and (2.3.2). For any sequence (2.3.1), there are exactly n! / r! permutations π for which (2.3.2) holds, so that

$$\sum_{\pi} \nu(r, \pi) = \sum_{k_1<k_2<\ldots<k_r} \frac{n!}{r!} = \binom{n}{r}\frac{n!}{r!}$$

On the other hand, if $l(\pi) \geqslant r$, there is an ascending sequence of length $l(\pi)$, so that each subsequence is ascending, and so

$$\nu(r, \pi) \geqslant \binom{l(\pi)}{r}$$

Hence the random permutation π_n satisfies

$$\mathbb{E}\binom{l(\pi_n)}{r} \leqslant \binom{n}{r} / r!$$

Since $\binom{\cdot}{r}$ is non-decreasing, this means that

$$\mathbb{P}\{l(\pi_n) \geqslant m\} \leqslant \binom{n}{r} / r!\binom{m}{r}$$

for $r \leqslant m \leqslant n$. Now fix constants α and β in $0 < \alpha < \beta$, and let $r, m, n \to \infty$ in such a way that

$$r\,n^{-1/2} \to \alpha\ ,\quad m\,n^{-1/2} \to \beta.$$

Then, by Stirling's formula,

$$\log \{\binom{n}{r} / r! \binom{m}{r}\} = (n \log n - n) - (r \log r - r) - [(n-r) \log(n-r) - (n-r)]$$
$$- (m \log m - m) + [(m-r) \log (m-r) - (m-r)] + o(n^{1/2})$$
$$= - r \log (\frac{r}{n^{1/2}}) - (n-r) \log (\frac{n-r}{n}) - m \log (\frac{m}{n^{1/2}})$$
$$+ (m-r) \log (\frac{m-r}{n^{1/2}}) + r + o(n^{1/2})$$
$$= - r \log \alpha + (n-r) \frac{r}{n} - m \log \beta + (m-r) \log(\beta-\alpha) + r + o(n^{1/2})$$
$$= \{2\alpha - \alpha \log \alpha - \beta \log \beta + (\beta-\alpha) \log (\beta-\alpha) + o(1)\} n^{1/2}$$

Accordingly, if α and β satisfy

$$0 < \alpha < \beta, \; 2\alpha - \alpha \log \alpha - \beta \log \beta + (\beta-\alpha) \log (\beta-\alpha) < 0,$$

then

$$\mathbb{P} \{l(\pi_n) \geqslant \beta n^{1/2}\} \to 0$$

as $n \to \infty$, and therefore $c \leqslant \beta$. The original treatment in [9] used a disguised form of this argument with $r = m$, and concluded that $c \leqslant e$. The present argument does a little better ; when $\beta > 2$ is fixed the minimum of

$$2\alpha - \alpha \log \alpha - \beta \log \beta + (\beta-\alpha) \log (\beta-\alpha)$$

occurs when

$$\alpha > 1, \; \alpha (\beta-\alpha) = 1.$$

It follows from this that $c \leqslant \alpha + \alpha^{-1}$, where $\alpha > 1$ is the root of

$$2 \alpha^2 = (1+\alpha^2) \log (1+\alpha^2).$$

Numerical solution of this equation then yields $c < 2.49$, and the proof of the theorem is complete.

The theorem therfore shows that, when n is large, most permutations π of $\{1, 2, \ldots, n\}$ have values of $l(\pi)$ near $c\, n^{1/2}$. The next problem is clearly the evaluation of c, and this remains unsolved, althrough there is some evidence to support the conjecture that $c = 2$.

One aspect of the logical structure of the theorem and its proof deserves remark. The theorem is really about convergence in probability ; it asserts that if (π_n) is a sequence of random permutations, such that for each n, π_n is uniformly distributed over the $n!$ permutations of $\{1, 2, \ldots, n\}$, then

$$\lim_{n\to\infty} l(\pi_n) / n^{1/2} = c \qquad (2.3.9)$$

in the sense of convergence in probability (which can be strengthened to L_1 convergence using (2.3.8)). This is proved by constructing a particular such sequence, in which the π_n are all defined on the same probability space, and for which (2.3.9) holds with probability one. That this is always possible was proved by Skorokhod ([21], page 281), but it rises an interesting question. Is it true that (2.3.9) holds with probability one whenever the π_n are defined on a common probability space ? By the Borel-Cantelli lemmas this will occur if and only if

$$\sum_{n=1}^{\infty} P \{|n^{-1/2} l(\pi_n) - c| > \varepsilon\} < \infty \qquad (2.3.10)$$

for all $\varepsilon > 0$. I do not know whether this is true, althrough (2.3.8) does imply that

$$\limsup_{n\to\infty} l(\pi_n) / n^{1/2} < 2.49 \qquad (2.3.11)$$

whenever the π_n are so defined.

There is however a slightly weaker result which is known to be true. It follows from a theorem of Kesten to be described in Section 3.1 that (2.3.9) holds with probability one whenever the π_n are defined on a common probability space in such a way that $l(\pi_n)$ is non-decreasing in n.

2.4 - An application to potentiel theory

Spitzer [23] has pointed out that subadditive theory gives a neat approach to a result in the potential theory of Markov processes. Suppose for instance that Z_t is Brownian motion in $\mathbb{R}^3$, and let A be any compact subset of $\mathbb{R}^3$. Attach A to the Brownian particle, by considering the random set

$$Z_t + A = \{x \in \mathbb{R}^3 ; x - Z_t \in A\} \qquad (2.4.1)$$

and consider the volume V_t swept out by this random set in time t. Thus $V_t = X_{ot}$, where

$$X_{st} = \left| \bigcup_{\tau \in [s,t)} (Z_\tau + A) \right| \qquad (2.4.2)$$

where $|.|$ denotes volume.

It is now very easy to see that (X_{st}) is a subadditive process, and that the invariant σ-field is trivial. Hence Theorem 1.7, and the fact that V_t is non-decreasing in t, show that these is a finite constant γ such that

$$\lim_{t\to\infty} V_t / t = \gamma \qquad (2.4.3)$$

with probability one and in L_1.

This problem has the unusual feature that the constant γ can be computed, for

$$\gamma = \lim_{t\to\infty} t^{-1} \mathbb{E}(V_t)$$

$$= \lim_{t\to\infty} t^{-1} \int_{\mathbb{R}^3} \mathbb{P} \{x \in Z_\tau + A \text{ for some } \tau \leqslant t\} dx$$

$$= \lim_{t\to\infty} t^{-1} \int_{\mathbb{R}^3} \mathbb{P} \{Z_\tau \in A - x \text{ for some } \tau \leqslant t\} dx$$

A theorem of Spitzer [22] identifies this limit with the electrostatic capacity of A.

The result can be very widely generalized. It applies to any transient spatially homogeneous Markov process with values in $\mathbb{R}^k$, and the identification of γ with the corresponding generalized capacity is (under the usual conditions of Hunt potential theory for Markov processes) a consequence of Getoor's generalization [7] of Spitzer's theorem.

There are probably many other applications of the subadditive axioms still to be realised. It will be noted that the general theory usually takes only the first steps towards an understanding of the process, and tends to raise as many questions as it answers. In particular, the determination of the fundamental constant γ is typically a considerable challenge.

CHAPTER 3

INDEPENDENT SUBADDITIVE PROCESSES

3.1 - Kesten's Theorem

In several of the problems described in chapter 2, the subadditive processes constructed have a further property :

(S_4) *for any increasing sequences* $(t_1, t_2, \dots, t_n)$ *in* $\mathbb{Z}$ *, the variables* $X_{t_{r-1} t_r}$ *are independent*

For examples the processes of Section 2.2 satisfy (S_4) if the P_n are independent, and those of Sections 2.3 and 2.4 always do. It would be surprising if this additional structure could not be exploited to strengthen the results of chapter 1.

Theorem 3.1

Let (X_{st}) *satisfy* (S_1), (S_2), (S_4) *and* (1.5.2). *Then*

$$\lim_{t\to\infty} X_{ot} / t = \gamma \qquad (3.1.1.)$$

with probability one, where

$$\gamma = \lim_{t\to\infty} g_t / t \geqslant -\infty \qquad (3.1.2.)$$

Moreover, the convergence takes place in L_1 *norm if* $\gamma > -\infty$.

Proof

Follow the proof of Theorem 1.4 as far as (1.3.4), noting that (S_4) implies that the limit ξ_k is non-random. Thus

$$\xi = \limsup_{t\to\infty} X_{ot} / t \leqslant g_k / k \qquad (3.1.3.)$$

with probability one. Hence, letting $k \to \infty$, $\xi \leqslant \gamma$, and (1.3.3) shows that $\xi = \gamma$ with probability one if γ is finite. Thus Theorem 1.7 establishes the present result if γ is finite. On the other hand, if $\gamma = -\infty$, (3.1.3.) shows that

$$\limsup_{t\to\infty} X_{ot} / t = -\infty ,$$

and the proof is complete.

Hammersley [11] has pointed out that a rather more general, and in some respects simpler, formulation can be given if attention is concentrated on the distribution of X_{on} for each $n \geq 1$. We shall describe the theory in the most important case, that in which the random variables X_{st} are non-negative.

Theorem 3.2

Suppose that the non-negative random variables X_{st} satisfy (S_1), (S_2) and (S_4). Then the function

$$F_{t-s}(x) = \mathbb{P}(X_{st} \leqslant x) \tag{3.1.4}$$

depends only on (t-s), and satisfies

$$F_{m+n} \geqslant F_m * F_n \tag{3.1.5}$$

for $m, n \geqslant 1$.

In (3.1.5) the symbol $*$ denotes Stieltjes convolution

$$(F * G)(x) = \int_{[0,x]} F(x-y)\, d\, G(y) \tag{3.1.6}$$

The functions F_n, $F_m * F_n$ are distribution functions on $[0, \infty)$, that is, they are non-decreasing and right-continuous, with

$$F(0) \geqslant 0, \quad \lim_{n\to\infty} F(x) = 1.$$

Proof

That (3.1.4) depends only on (t-s) follows from (S_2). Because of (S_4), $F_m * F_n$ is the distribution function of

$$X_{om} + X_{m,\, m+n}\ ,$$

and (S_1) then shows that

$$F_{m+n}(x) = \mathbb{P}(X_{o,m+n} \leqslant x) \geqslant \mathbb{P}(X_{om} + X_{m,m+n} \leqslant x) = (F_m * F_n)(x).$$

It is a surprising fact that the converse is false, as the following example [11] shows. Define

$$F_1 = \tfrac{1}{2} + \tfrac{1}{2} H\ ,\ F_2 = \tfrac{1}{4} + \tfrac{3}{4} H\ ,\ F_3 = F_1 * F_2\ ,\quad F_n = 1\ (n \geqslant 4) \tag{3.1.7}$$

where $H(x) = 0$ if $x < 1$, and $= 1$ if $x \geqslant 1$. Then F_n is a distribution function on $[0, \infty]$, and (3.1.5) holds, the only non-trivial case being :

$$F_2 \geqslant \frac{1}{4} + \frac{1}{2} H + \frac{1}{4} H * H = F_1 * F_1.$$

But suppose if possible that there are random variables X_{st} satisfying (S_1), (S_2), (S_4) and (3.1.4). Then, with probability one,

$$X_{01} \leqslant 1 , \quad X_{02} \leqslant 1 , \quad X_{13} \leqslant 1 , \quad X_{23} \leqslant 1 .$$

and (S_1) implies that

$$\begin{aligned} \frac{3}{8} = \mathbb{P}\,(X_{03} = 2) &\leqslant \mathbb{P}\,(X_{01} = X_{13} = X_{02} = X_{23} = 1) \\ &\leqslant \mathbb{P}\,(X_{01} = X_{23} = 1) \\ &= \mathbb{P}\,(X_{01} = 1)\ \mathbb{P}\,(X_{23} = 1) = \frac{1}{4} \ ; \end{aligned}$$

a contradiction.

It is not known which sequences (F_n) can arise from a (non-negative) <u>independant subadditive process</u> (i.e. a family satisfying (S_1), (S_2), (S_3), and (S_4)), and therefore it is useful to try to establish results for all sequences (F_n) satisfying (3.1.5). For example, if (F_n) arises from an independent subadditive process, Theorem 3.1 shows that there is a constant γ with

$$\begin{aligned} \lim_{n\to\infty} F_n\,(nx) &= 1 \qquad (x > \gamma) \\ &= 0 \qquad (x < \gamma) \end{aligned} \tag{3.1.8}$$

Is this true for all sequences (F_n) satisfying (3.1.5) and the integrability condition

$$\int_0^\infty x \, d\, F_1\,(x) < \infty \ ? \tag{3.1.9}$$

This result is in fact an easy consequence of an interesting theorem of Kesten [14], which we now prove. (Kesten stated without proof a somewhat more general result, and his argument is reproduced in [11]).

Theorem 3.3

Let $(F_n ; n \geq 1)$ *be a sequence of distribution functions on* $[0, \infty)$ *which satisfy* (3.1.5) *and*

$$\int_0^\infty x^2 \, dF_1(x) < \infty \tag{3.1.10}$$

Then there is a constant γ *such that*

$$\lim_{n\to\infty} \int_0^\infty \left(\frac{x}{n} - \gamma\right)^2 dF_n(x) = 0 \tag{3.1.11}$$

and, whenever $\alpha < \gamma < \beta$ *and* $s(k) = m\,2^k$ $(m \geq 1)$,

$$\sum_{k=0}^{\infty} \{1 - F_{s(k)}(\beta\, s(k)) + F_{s(k)}(\alpha\, s(k))\} < \infty \tag{3.1.12}$$

Proof

Write

$$g_n = \int_0^\infty x \, dF_n(x) \ , \quad G_n = \{\int_0^\infty x^2 \, dF_n(x)\}^{1/2}$$

By Schwarz's inequality $0 \leq g_n \leq G_n \leq \infty$, and by (3.1.10), $g_1 \leq G_1 < \infty$. Integrating by parts and using (3.1.5),

$$\begin{aligned} G^2_{m+n} &= \int_0^\infty 2x\left[1 - F_{m+n}(x)\right] dx \\ &\leq \int_0^\infty 2x\left[1 - (F_m * F_n)(x)\right] dx \\ &= \int_0^\infty x^2 \, d(F_m * F_n)(x) \\ &= \int_0^\infty \int_0^\infty (x+y)^2 \, dF_n(x)\, dF_n(y) \end{aligned}$$

so that

$$G^2_{m+n} \leq G^2_m + G^2_n + 2\, g_m g_n .$$

It follows in particular that $G_n < \infty$ and so $g_n < \infty$ for all n.

An exactly similar argument now yields

$$g_{m+n} \leq g_m + g_n$$

so that

$$\gamma = \lim_{n\to\infty} g_n / n \tag{3.1.14}$$

exists in $0 \leq \gamma \leq g_1$. From (3.1.13) and the fact that $g_n \leq G_n$, we have

$$G_{m+n} \leq G_m + G_n$$

so that

$$\Gamma = \lim_{n\to\infty} G_n / n \tag{3.1.15}$$

exists in $\gamma \leqslant \Gamma \leqslant G_1 < \infty$

Now write

$$V_n = G_n^2 - g_n^2 = \int_0^\infty (x - g_n)^2 \, dF_n(x)$$

and use (3.1.13) with m = n to give

$$\frac{V_{2n}}{(2n)^2} - \frac{1}{2}\frac{V_n}{n^2} \leqslant (\frac{g_n}{n})^2 - (\frac{g_{2n}}{2n})^2.$$

Write $n = s(k)$ and sum from $k = 0$ to $k = K - 1$, so that

$$\sum_{k=1}^{K} \frac{V_{s(k)}}{s(k)^2} - \frac{1}{2}\sum_{k=0}^{K-1} \frac{V_{s(k)}}{s(k)^2} \leqslant (\frac{g_m}{m})^2$$

Hence

$$\sum_{k=0}^{K} \frac{V_{s(k)}}{s(k)^2} \leqslant 2\,\frac{V_m}{m^2} + (\frac{g_m}{m})^2,$$

and therefore

$$\sum_{k=0}^{\infty} V_{s(k)} \Big/ s(k)^2 < \infty. \tag{3.1.16}$$

From (3.1.14) et (3.1.15),

$$\lim_{n\to\infty} V_n / n^2 = \Gamma^2 - \gamma^2$$

If this limit is non-zero, the convergence (3.1.16) is impossible. Thus $\Gamma^2 - \gamma^2 = 0$, and (3.1.11) is proved.

Finally, if $\alpha < \gamma < \beta$, there exist $N \geqslant 1$ and $\varepsilon > 0$ such that

$$\alpha + \varepsilon < g_n / n < \beta - \varepsilon \qquad (n \geqslant N)$$

Then Tchebychev's inequality gives

$$1 - F_n(n\beta) + F_n(n\alpha) \leqslant V_n / n^2 \varepsilon^2,$$

and (3.1.12) follows from (3.1.16). Hence the proof is complete.

Kesten pointed out that this theorem is strong enough to settle the question of almost sure convergence left open in our discussion of Ulam's problem. Use the notation of the proof of theorem 2.3. and fix ε in $0 < \varepsilon < 1$. Then

$$\mathbb{P}\{l(\pi_n) \leqslant (1-\varepsilon)\, cn^{1/2}\} = \mathbb{P}\{L_{o,z(n)} \leqslant (1-\varepsilon)\, cn^{1/2}\}$$

$$\leqslant \mathbb{P}\{L_{o,z(n)} \leqslant (1-\varepsilon)\, cn^{1/2},\ |z(n) - n^{1/2}| \leqslant \varepsilon^2 n^{1/2}\}$$

$$+ \mathbb{P}\{|z(n) - n^{1/2}| > \varepsilon^2 n^{1/2}\}$$

$$\leqslant \mathbb{P}\{L_{o,(1-\varepsilon^2)n^{\frac{1}{2}}} \leqslant (1-\varepsilon)\, cn^{1/2}\} + \varepsilon^{-4} n^{-1}\, \mathbb{E}\, (z(n) - n^{1/2})^2\}$$

using the fact that L_{ot} increases with t. It is easy to check that

$$\mathbb{E}\{(z(n) - n^{1/2})^2\} \leqslant 1.$$

Applying theorem 3.3. where F_n is the distribution of

$$(2\, cn - L_{o,(1-\varepsilon^2)n})^+,$$

we see that

$$\sum_{k=0}^{\infty} \mathbb{P}\{l(\pi_{m\, 2^{2k}}) \leqslant (1-\varepsilon)\, cm^{1/2}\, 2^k\} < \infty.$$

A similar argument operates in the reverse direction, and thus

$$\sum_{k=0}^{\infty} \mathbb{P}\left\{ \left| \frac{l(\pi_{m\, 2^{2k}})}{m^{1/2}\, 2^k} - c \right| \geqslant \varepsilon \right\} < \infty. \qquad (3.1.17)$$

In (3.1.17), π_n is the particular random permutation constructed from the Poisson process Π. But since each probability refers only to one π_n, the result remains true if (π_n) is any random sequence such that, for each n, π_n is uniformly distributed over the n! permutations of $\{1, 2, \ldots, n\}$. In particular, if the π_n are all defined on the same probability space, the Borel-Cantelli lemma shows that, with probability one,

$$\lim_{k\to\infty} l(\pi_{m\, 2^{2k}}) / m^{1/2}\, 2^k = c \qquad (3.1.18)$$

for all $m \geqslant 1$.

Now suppose that the π_n are so defined that $l(\pi_n)$ is non-decreasing in n. Fix $p \geq 1$ and remark that, for each $n \geq 2^{2p}$, there are integers $k \geq 1$ and m in the finit set $2^{2p} \leq m < 2^{2p+2}$, so that

$$m\, 2^{2k} \leq n < (m+1)\, 2^{2k}$$

Hence

$$l(\pi_{m\,2^{2k}}) \leq l(\pi_n) \leq l(\pi_{(m+1)2^{2k}}),$$

and from (3.1.18) and the fact that

$$\frac{m+1}{m} \leq 1 + 2^{-2p},$$

we have

$$(1 + 2^{-2p})^{-1/2}\, c \leq \liminf \frac{l(\pi_n)}{n^{1/2}} \leq \limsup \frac{l(\pi_n)}{n^{1/2}} \leq (1 + 2^{-2p})^{1/2}\, c.$$

Letting $p \to \infty$, it follows that

$$\lim_{n\to\infty} l(\pi_n) / n^{1/2} = c \qquad (3.1.19)$$

with probability one.

Kesten's theorem applies, of course, to any independent non-negative subadditive process (X_{st}) such that $\mathbb{E}(X_{01}^2) < \infty$. There is however an alternative approach which yields (3.1.12) without this condition.

Suppose that $(s(k)\ ;\ k \geq 1)$ is any increasing sequence of integers for which

$$s(k) \geq \alpha S(k) = \alpha \sum_{j=1}^{k-1} s(j) \qquad (3.1.20)$$

for some $\alpha > 0$ and all k. If (X_{st}) is an independent subadditive process, Theorems 3.1 and 1.9 (with $c = 1 + \alpha$) imply that

$$\mathbb{P}\{\lim_{k\to\infty} X_{S(k),S(k+1)} / s(k) = \gamma\} = 1.$$

But the variables $X_{S(k),S(k+1)}$ are independant, and the Borel-Cantelli lemma implies that

$$\sum_{k=1}^{\infty} \mathbb{P} \left\{ \left| \frac{X_{S(k),S(k+1)}}{s(k)} - \gamma \right| > \varepsilon \right\} < \infty$$

for all $\varepsilon > 0$.

By (S_2) this is equivalent to

$$\sum_{k=1}^{\infty} \mathbb{P} \left\{ \left| \frac{X_{o,s(k)}}{s(k)} - \gamma \right| > \varepsilon \right\} < \infty \tag{3.1.21}$$

When $s(k) = m\, 2^k$, we recover (3.1.12), but only of course in the case when the F_n arise from a subadditive process.

3.2 - A generalisation of a theorem of Chernoff

We have seen that, when $x < \gamma$, $F_n(nx) \to 0$ as $n \to \infty$, and it is natural to ask how rapidly this convergence takes place. In the additive case, the answer is given by a famous theorem of Chernoff [2], and Hammersley [11] has shown how this may be extended to independent subadditive processes, or more generally to families of distributions (not necessarily confined to $[0, \infty)$) satisfying (3.1.5).

Theorem 3.4

Suppose that, for each $n \geq 1$, F_n is a distribution function on $\mathbb{R}$ for which

$$\phi_n(\theta) = \int_{-\infty}^{\infty} e^{-\theta x}\, dF_n(x) \tag{3.2.1.}$$

is finite in $0 \leq \theta \leq \Theta$ for some $\Theta > 0$ and suppose that

$$F_{m+n} \geq F_m * F_n \tag{3.2.2}$$

where $*$ denotes Stieljes convolution. Then the limits

$$\psi(x) = \lim_{n\to\infty} n^{-1} \log F_n(nx) \tag{3.2.3}$$

and

$$K(\theta) = \lim_{n\to\infty} n^{-1} \log \phi_n(\theta) \tag{3.2.4}$$

exist for all x and for all $\theta \geq 0$, and are connected by the relations

$$\psi(x) = \operatorname*{Inf}_{\theta} \{K(\theta) + \theta x\} \tag{3.2.5}$$

$$K(\theta) = \operatorname*{Sup}_{x} \{\psi(x) - \theta x\} \tag{3.2.6}$$

for x in the interior of $\{y ; \psi(y) > -\infty\}$ and θ in the interior of $\{\theta' ; K(\theta') < \infty\}$.

Proof

From (3.2.2)

$$F_{m+n}(x+y) \geq \int_{\mathbb{R}} F_m(x+y-z)\, dF_n(z) \geq \int_{(-\infty,y]} F_m(x+y-z)\, dF_n(z)$$

$$\geq \int_{(-\infty,y]} F_m(x)\, dF_n(z)$$

so that

$$F_{m+n}(x+y) \geq F_m(x)\, F_n(y) \tag{3.2.7}$$

In this inequality replace x by mx, y by nx, to show that for fixed x the sequence

$$-\log F_n(nx)$$

is subadditive. Hence (cf. Theorem 1.1), the limit (3.2.3) exists, and moreover,

$$F_n(nx) \leq e^{n\psi(x)} \tag{3.2.8}$$

(Note incidentally that $\psi(x) = -\infty$ only if $F_n(nx) = 0$ for all n).

Now replace x by mx and y by ny in (3.2.7), and let $m, n \to \infty$ in such a way that $m/(m+n) \to \theta \in (0,1)$. This gives

$$\psi\left[\theta x + (1-\theta) y\right] \geq \theta\psi(x) + (1-\theta)\psi(y) \quad ;$$

ψ is a concave function.

Integrating by parts and using (3.2.2),

$$\phi_{m+n}(\theta) = \int_{-\infty}^{\infty} \theta e^{-\theta x} F_{m+n}(x)$$

$$\geq \int_{-\infty}^{\infty} \theta e^{-\theta x} (F_m * F_n)(x)$$

$$= \int_{-\infty}^{\infty} e^{-\theta x}\, d(F_m * F_n)(x)$$

$$= \phi_m(\theta)\ \phi_n(\theta).$$

which shows that the limit (3.2.4.) exists, and that

$$\phi_n(\theta) \leqslant e^{nK(\theta)} \tag{3.2.9}$$

Since

$$\phi_n(\theta) \geqslant \int_{(-\infty,nx]} e^{-\theta y}\, dF_n(y) \geqslant e^{-\theta nx} F_n(nx)$$

this gives

$$F_n(nx) \leqslant e^{nK(\theta)+\theta nx}$$

so that

$$\psi(x) \leqslant K(\theta) + \theta x$$

for all $\theta \geqslant 0$. Hence

$$\psi(x) \leqslant \inf_{\theta} \{K(\theta) + \theta x\} \tag{3.2.10}$$

Now fix $\theta > 0$ so that

$$M = \sup_x \{\psi(x) - \theta x\} < \infty,$$

and let α, β, a, b, satisfy $0 < \beta < \theta < \alpha$ and $a < b$. Then (3.2.8) and (3.2.10) imply that

$$F_n(nx) \leqslant \exp\{n \min [K(\alpha) + \alpha x,\ M + \theta x,\ K(\beta) + \beta x]\},$$

so that

$$\phi_n(\theta) = \int_{-\infty}^{\infty} \theta\, e^{-\theta y} F_n(y)\, dy = n\theta \int_{-\infty}^{\infty} e^{-\theta nx} F_n(nx)\, dx$$

$$\leqslant n\theta \left\{\int_{-\infty}^{a} e^{n[K(\alpha)+(\alpha-\theta)x]}\, dx + \int_a^b e^{nM}\, dx + \int_b^{\infty} e^{n[K(\beta)-(\theta-\beta)x]}\, dx\right\}$$

$$= \frac{\theta}{\alpha - \theta}\, e^{n[K(\alpha)+(\alpha-\theta)a]} + n\theta\,(b-a)\, e^{nM} + \frac{\theta}{\theta - \beta}\, e^{n[K(\beta)-(\theta-\beta)b]}$$

Taking logarithms and letting $n \to \infty$,

$$K(\theta) \leqslant \max \{K(\alpha) + (\alpha-\theta)\, a,\ M,\ K(\beta) - (\theta-\beta)\, b\},$$

and letting $a \to -\infty$, $b \to \infty$,

$$K(\theta) \leqslant M.$$

This inequality is trivial if $M = \infty$, so that

$$K(\theta) \leqslant \sup_x \{\psi(x) - \theta x\}$$

for all $\theta > 0$. On the other hand, (3.2.10) shows that

$$\psi(x) \leqslant K(\theta) + \theta x$$

holds for all x and all $\theta > 0$, so that

$$K(\theta) \geqslant \psi(x) - \theta x,$$

so
$$K(\theta) \geqslant \sup_x \{\psi(x) - \theta x\}.$$

Hence (3.2.6) is proved.

Since ψ is concave and non-decreasing, there exists, for every x in the interior of $\{\psi > -\infty\}$, a value of $\theta \geqslant 0$ for which

$$\psi(y) \leqslant \psi(x) + \theta(y-x)$$

for all y. Then (3.2.6.) gives

$$K(\theta) \leqslant \sup_y \{\psi(x) + \theta(y-x) - \theta y\} = \psi(x) - \theta x.$$

Thus
$$\psi(x) \geqslant K(\theta) + \theta x$$

for this particular θ, which combined with (3.2.10) establishes (3.2.5) and completes the proof.

The function ψ is non-positive, non-decreasing and concave, but may take the value $-\infty$. The function K is convex, with $K(0) = 0$; it may take the value $+\infty$, but not $-\infty$. It is not difficult to construct examples of independent subadditive processes with any given ψ , or any given K, satisfying these conditions.

To link theorem 3.4 with theorem 3.1, note that $F_n(nx) \to 0$ exponentially fast if and only if $\psi(x) < 0$. This occurs if and only if $K(\theta) + \theta x < 0$ for some $\theta \geqslant 0$, and recalling that K is convex and that $K(0) = 0$, this occurs if and only if $K'(0) + x < 0$. It is not difficult to show that $K'(0) = \gamma$, so that we can conclude that $F_n(nx) \to 0$ exponentially fast for all $x < \gamma$ if $K(\theta)$ is finite for small positives values of θ.

3.3 - The first birth problem for branching processes

Theorem 3.4 was first proved with a view to its application to an interesting problem in the theory of branching processes. Consider a process in which an initial ancestor is born at time $t = 0$, after which he produces children in a random way. Thus, if $Z_1(t)$ denotes the number of children born before time t, then Z_1 is assumed to be a random process with values in the non-negative integers, which is non-decreasing and right-continuous. Each children, from its birth, behaves just like its father, producing children according to a random process with the same stochastic structure as Z_1, and this process is assumed independent of those related to the father and his other children. Thus the child's children, grand-children of the ancestor, themselves produce children, and so on.

Let B_n be the instant at which the first birth in the nth generation takes place ; B_n is well-defined on the event S that there are descendants in every generation. Then Hammersley [10], [11] shows that there is a constant γ such that

$$\lim_{n\to\infty} B_n / n = \gamma \tag{3.3.1}$$

almost everywhere on S. Indeed, this is an easy consequence of Kesten's theorem if we impose the additional conditions

$$\mathbb{P} \; \{Z_1(t) \geqslant 1 \;\; \text{for some } t\} = 1 \tag{3.3.2}$$

and

$$\mathbb{E} \; (B_1^2) < \infty \tag{3.3.3}$$

To see this, note that (3.3.2) means that $\mathbb{P}(S) = 1$, and write F_n for the distribution function of B_n. Consider the individual born a B_m in the mth generation, and let $B'_{m+n} \geqslant B_{m+n}$ be the instant at which the first of his descendants is born in the $(m+n)^{th}$ generation. Then it is clear that $B'_{m+n} - B_m$ is independant of B_m and has distribution function F_n. It follows that (F_n) satisfies (3. 1.5) and (3.3.3) allows us to apply

Theorem 3.3. This establishes (3.3.1) if n is restricted to any sequence of the form $(m\,2^k\,;\,k \geq 0)$. The "filling-in" procedure used at the end of Section 3.1, together with the fact that $B_n \leq B_{n+1}$, then proves (3.3.1) in general.

To remove the conditions (3.3.2) and (3.3.3) requires a technique of truncation, which is described in detail in [18]. Essentially one forces (3.3.2) by concentrating attention of individuals with descendants of all generations, and (3.3.3) by forced sterilisation at a large fixed age.

The most interesting feature of this problem is however that it is possible to evaluate γ explicitly. Thus suppose that

$$\phi\,(\theta) = \mathbb{E} \int e^{-\theta t}\, dZ_1(t) \tag{3.3.4}$$

is finite for large θ. Note that $\phi\,(0)$ is the mean number of children in a family, so that we must asume $\phi\,(0) > 1$ to ensure that $\mathbb{P}(S) > 0$. Hammersley shows (for the slightly less general process studied in [10]) that

$$\gamma = \text{Sup}\,\{a\,;\,\mu\,(a) < 1\} \tag{3.3.5}$$

where

$$\mu\,(a) = \text{Inf}\,\{\phi(\theta)\, e^{\theta a}\,;\,\theta \geq 0\} \tag{3.3.6}$$

This remarkable result is proved in [11] by an elaborate argument from Theorem 3.4, but an alternative approach [18], also informed by subadditive theory, seems more natural and more applicable to more complicated problems.

This approach depends on the remark that, if θ is so large that $\phi\,(\theta) < \infty$, then

$$W_n\,(\theta) = \phi\,(\theta)^{-n} \sum_r e^{-\theta B_{nr}} \tag{3.3.7}$$

defines a martingale. Here

$$B_{n1} \leq B_{n2} \leq \dots$$

are the birth times of the individuals in the nth generation. When $\theta = 0$, this martingale is a familiar tool of branching process theory [13], and for general θ it can be regarded as obtained from $W_n(0)$ by a certain modification of the process.

The fact that $\mathbb{E}\{W_n(\theta)\} = \mathbb{E}\{W_1(\theta)\} = 1$ already has a usefull consequence, for it implies that

$$\phi(\theta)^n = \mathbb{E}\{\sum_r e^{-\theta B_{nr}}\} \geq \mathbb{E}\{e^{-\theta B_n}; S\}$$

$$\geq e^{-\theta na}\,\mathbb{P}\{B_n \leq na, S\}$$

for any a. Hence

$$\mathbb{P}\{B_n \leq na, S\} \leq \{\phi(\theta)\, e^{\theta a}\}^n$$

and the Borel-Cantelli lemma shows that

$$\liminf_{n\to\infty} B_n / n \geq a \tag{3.3.8}$$

almost surely on S, whenever $\phi(\theta)\, e^{\theta a} < 1$. Since θ is at our disposal, this shows that (3.3.8) holds whenever $\mu(a) < 1$.

To get an asymptotic upper bound for B_n / n is rather more difficult. It turns out that, under (3.3.2) and (3.3.3), we can parody the proof of Theorem 1.4 to show that the limit (3.3.1) holds in L_1 norm.

Now use the martingale convergence theorem to establish the existence of the limit

$$W(\theta) = \lim_{n\to\infty} W_n(\theta) \tag{3.3.9}$$

almost surely. By Fatou's lemma, $\mathbb{E}\{W(\theta)\} \leq 1$; suppose if possible that

$$\mathbb{E}\{W(\theta)\} = 1, \tag{3.3.10}$$

in which case the limit (3.3.9) holds in L_1 norm. Differentiate the identity for $\phi(\theta)^n$ to give

$$-n\,\phi(\theta)^{n-1}\,\phi'(\theta) = \mathbb{E}\{\sum_r B_{nr}\, e^{-\theta B_{nr}}\}$$

$$\geq \mathbb{E}\{B_n \sum_r e^{-\theta B_{nr}}\}$$

so that

$$-\phi'(\theta) / \phi(\theta) \geq \mathbb{E}\{n^{-1} B_n W_n(\theta)\}$$

$$= \mathbb{E}\{n^{-1} B_n W(\theta)\},$$

using the martingale property. Letting $n \to \infty$,

$$- \phi'(\theta) / \phi(\theta) \geq \mathbb{E} \{\gamma W(\theta)\} = \gamma$$

Hence

$$\gamma \leq - \phi'(\theta) / \phi(\theta) \qquad (3.3.11)$$

for all θ such that (3.3.10) holds.

Thus the analysis turns on the question, which is also of interest in its own right, of deciding for what values of θ (3.3.10) is valid. The usual way of proving such a fact is to estimate $\mathbb{E} \{W_n(\theta)^2\}$, assuming that

$$\mathbb{E} \{W_1(\theta)^2\} < \infty \qquad (3.3.12)$$

and this turns out to establish (3.3.10) so olong as

$$\phi (2\theta) < \phi (\theta)^2$$

Unfortunately, this is not enough, and a more delicate argument is needed, based on the estimation of $\mathbb{E} \{W_n(\theta)^{\alpha}\}$ for $1 < \alpha < 2$. This shows that (3.3.10) holds if

$$\phi(\alpha\theta) < \phi(\theta)^{\alpha}$$

for some such α, and this is true if

$$\theta \phi'(\theta) \leq \phi(\theta) \log \phi(\theta).$$

An examination to the values of θ for which this inequality holds shows that they enable (3.3.11) to assert that $\gamma < a$ whenever $\mu(a) > 1$, and this suffices to identify γ according to (3.3.5) and (3.3.6). The proof is then completed by using truncation methods to remove the redundant conditions (3.3.2), (3.3.3) and (3.3.12).

The details of this argument may be found in [18]. More recently, in a yet unpublished work, J.D. Biggins has extended it to cover multi-type branching processes, and has also used a time-reversal argument to prove a corresponding result for the last birth in the nth generation (when this has meaning).

BIBLIOGRAPHIE

[1] D.L. BURKHOLDER, Contribution to the discussion of [16].

[2] H. CHERNOFF, A measure of asymptotic efficiency for tests of a hypothesis on the sum of observations, Ann. Math. Statist. 23 (1952) 493-507.

[3] J.L. DOOB, Stochastic processes, Wiley, New-York (1953).

[4] N. DUNFORD and J.T. SCHWARZ, Linear Operators (Part I), Interscience, New-York (1958)

[5] H. FURSTENBERG and H. KESTEN, Products of random matrices, Ann. Math. Statist. 31 (1960) 457-469

[6] A.M. GARSIA, A simple proof of E. Hopf's maximal ergodic theorem, J. Math. Mech. 14 (1965) 381-382.

[7] R.K. GETOOR, Some assymptotic formulas involving capacity, Z. Wahrscheinlichkeitsth 4 (1964) 248-252

[8] J.M. HAMMERSLEY and J.A.D. WELSH, First-passage percolation, subadditive processes, stochastic networks and generalised reneval theory, Bernoulli-Bayes-Laplace Anniversary Volume, Springer, Berlin (1965).

[9] J.M. HAMMERSLEY, A few seedlings of research, Proc. Sixth Berkeley Symp. University of California Press, 1 (1972) 345-394.

[10] J.M. HAMMERSLEY, Contribution to the discussion of [16].

[11] J.M. HAMMERSLEY, Postulates for subadditive processes, Ann. Prob. 2 (1974) 652-680.

[12] J.M. HAMMERSLEY, Poking about for the vital juices of mathematical research, Bull. Inst. Math. Appl. 10 (1974) 235-247.

[13] T.E. HARRIS, The theory of branching processes, Springer, Berlin (1963)

[14] H. KESTEN, Contribution to the discussion of [16].

[15] J.F.C. KINGMAN, The ergodic theory of subadditive stochastic processes, J. Roy. Statist. Soc. B 30 (1968) 499-510.

[16] J.F.C. KINGMAN, Subadditive ergodic theory, Ann. Prob. 1 (1973) 883-909.

[17] J.F.C. KINGMAN, Geometrical aspects of the theory of non-homogeneous Markov chains, Math. Proc. Camb. Phil. Soc. 77 (1975) 171-183.

[18] J.F.C. KINGMAN, The first birth problem for an age-dependent branching process, Ann. Prob. 3 (1975).

[19] J. KOMLOS, A generalisation of a problem of Steinhaus, Acta Math. Acad. Sci. Hungar 18 (1967) 217-229.

[20] D. RICHARDSON, Random growth in a tessellation, Proc. Camb. Phil. Soc. 74 (1973) 515-528.

[21] A.V. SKOROKHOD, Limit theorems for stochastic processes, Theor. Prob. Appl. 1 (1956) 261-290.

[22] F. SPITZER, Electrostatic capacity, heat flow and brownian motion, Z. Wahrscheinlichkeitsch 3 (1964) 110-121.

[23] F. SPITZER, Contribution to the discussion of [16].

[24] K. YOSIDA and E. HEWITT, Finitely additive measures, Trans. Amer. Math. Soc. 72 (1952) 46-66.

THE LAW OF THE ITERATED LOGARITHM AND RELATED STRONG CONVERGENCE THEOREMS FOR BANACH SPACE VALUED RANDOM VARIABLES

PAR J. KUELBS

1. <u>Introduction.</u> These lectures contain a survey of recent results on the law of the iterated logarithm (LIL) for Banach space valued random variables, and related topics. Many of the results can be found in the papers included in the bibliography, but the contents of Sections five and six, and Theorem 3.2 and its corollaries are new. Furthermore, a number of the results have been improved as is evident in the case of Theorem 3.1 and Corollary 3.1 when contrasted with similar results in [12]. Theorem 4.1 is also improved over its counterpart in [12], and this is due largely to [3].

The law of the iterated logarithm is, without doubt, one of the crowning achievements of probability theory. It has caught the interest of a great number of mathematicians, and in order to motivate related limit theorems for Banach space valued random variables we first provide a brief outline of some of the history leading up to what we call the classical law of the iterated logarithm. There are many gaps in this historical sketch, but hopefully it will provide the motivation intended.

For the moment assume $X_1, X_2, \ldots$ are independent random variables such that $P(X_k = \pm 1) = 1/2$ for $k = 1, 2, \ldots,$ and set $S_n = X_1 + \ldots + X_n$ for $n \geq 1$. Under these conditions Hausdorff (1913) proved that $P(\lim_n \frac{S_n}{n^\alpha} = 0) = 1$ for each $\alpha > 1/2$. If $\alpha \geq 1$ this result was already known from the law of large numbers, and if $\alpha \leq 1/2$ then it is not hard to prove (combining the central limit theorem and that $\{\overline{\lim}_n \frac{S_n}{n^{1/2}} \leq M\}$ is a tail event for the sequence $\{X_k : k \geq 1\}$) that

$$P\left(\overline{\lim_n}\ \frac{S_n}{n^{1/2}} = +\infty\right) = P\left(\underline{\lim_n}\ \frac{S_n}{n^{1/2}} = -\infty\right) = 1 \quad .$$

In 1914 Hardy and Littlewood improved Hausdorff's result by proving

$$P\left(\lim_n \frac{S_n}{\sqrt{n \log n}} = 0\right) = 1 \quad ,$$

and in 1924 Khintchine proved

$$P\left(\overline{\lim_n}\ \frac{S_n}{\sqrt{2n \log \log n}} = 1\right) = P\left(\underline{\lim_n}\ \frac{S_n}{\sqrt{2n \log \log n}} = -1\right) = 1 \quad .$$

All of the previous results were for Bernoulli random variables, but in 1941 Hartman and Wintner proved that if $X_1, X_2, \ldots$ are arbitrary independent identically distributed random variables such that $E(X_k) = 0$ and $E(X_k^2) = \sigma^2$, then

$$P\left(\overline{\lim_n}\ \frac{S_n}{\sqrt{2n \log \log n}} = +\sigma\right) = P\left(\underline{\lim_n}\ \frac{S_n}{\sqrt{2n \log \log n}} = -\sigma\right) = 1 \quad .$$

In 1964 Strassen proved (among many interesting things) that under the conditions used by Hartman and Wintner we have

$$P\left(\lim_n d\left(\frac{S_n}{\sqrt{2n \log \log n}}, [-\sigma, \sigma]\right) = 0\right) = 1$$

and

$$P\left(C\left(\left\{\frac{S_n}{\sqrt{2n \log \log n}} : n \geq 3\right\}\right) = [-\sigma, \sigma]\right) = 1$$

where $d(x, A) = \inf_{y \in A} |x - y|$ and $C(\{a_n\})$ denotes all possible limit points of the sequence $\{a_n\}$.

Strassen's version of the law of the iterated logarithm is what we generalize to the Banach space setting.

Henceforth we use Lx to denote $\log x$ for $x \geq e$ and 1 otherwise.

Now assume that B is a real separable Banach space with norm $\|\cdot\|$, and that $X_1, X_2, \ldots$ are i.i.d. B-valued random variables on $(\Omega, \mathfrak{F}, P)$ such that $E(X_k) = 0$ and $E\|X_k\|^2 < \infty$. Again let $S_n = X_1 + \ldots + X_n$ for $n \geq 1$. The optimist's version of the law of the iterated logarithm for Banach space valued random variables would be that there exists a bounded symmetric set $K \subseteq B$ such that

$$(1.1) \qquad P\left\{\omega : \lim_n d\left(\frac{S_n(\omega)}{\sqrt{2n\, LLn}}, K\right) = 0\right\} = 1 ,$$

and

$$(1.2) \qquad P\left\{\omega : C\left(\left\{\frac{S_n(\omega)}{\sqrt{2n\, LLn}} : n \geq 1\right\}\right) = K\right\} = 1$$

where

$$d(x, A) = \inf_{y \in A} \|x - y\|$$

and

$$C(\{a_n\}) = \text{all limit points of } \{a_n\} \text{ in } B \quad .$$

This result, however, is simply too optimistic as can easily be seen from some interesting examples due to R. M. Dudley. Dudley has examples involving random variables in $C[0, 1]$ [5] and also in $L^p[0, 1]$ $(p < 2)$. The example in the $C[0, 1]$ setting was constructed in connection with the central limit theorem for $C[0, 1]$ valued random variables, but applies to the law of the iterated logarithm as well. The examples in L^p $(p < 2)$ are similar, but are interesting since both the central limit theorem and the law of the iterated logarithm are true in L^p, $2 \leq p < \infty$. Of course, $L^p[0, 1]$ is not a Banach space if $0 < p < 1$, but we can formulate the LIL in linear metric spaces as well, and the cases $0 < p < 1$ serve as counterexamples in this setting. Some positive results for locally convex Frechet spaces will be given as an application to Theorem 3.2, and others appear in [15] and [16].

The examples of Dudley in $L^p[0, 1]$ $(0 < p < 2)$ are the following. Fix $0 < p < 2$ and choose α such that $p < \alpha < 2$. Let Y be a symmetric stable law of index α defined on the probability space $[0, 1]$ with probability given by the Lebesgue measure λ. Let $\{\theta_k : k \geq 1\}$ be independent identically distributed (i. i. d.) random variables uniformly distributed on $[0, 1]$ defined on the probability space $(\Omega, \mathfrak{F}, P)$. Define $X_k(\omega)(t) = Y((\theta_k(\omega) + t) \bmod 1)$ for $0 \leq t \leq 1$, $\omega \in \Omega$, $k \geq 1$. Then $EX_k = 0$, and since $p < \alpha$

$$\|X_k(\cdot)\|_p = \left(\int_0^1 |X_k(\cdot)(t)|^p \, dt \right)^{1/p}$$

$$= \left(\int_0^1 |Y(s)|^p \, ds \right)^{1/p} = c < \infty$$

independent of $\omega \in \Omega$. Hence X_k has all possible moments. Furthermore, for each $t \in [0, 1]$ we have $\mathcal{L}(X_k(\cdot)(t)) = \mathcal{L}(Y)$, and by standard properties of the symmetric stable laws we also have

$$\mathcal{L}\left(\frac{S_n(\cdot)(t)}{n^{1/\alpha}}\right) = \mathcal{L}(Y) \qquad (0 \leq t \leq 1)$$

where, of course, $S_n = X_1 + \ldots + X_n$ $(n \geq 1)$. Thus, independent of t, we have for each $M > 0$

$$\lim_{n\to\infty} P\left\{\omega : \frac{|S_n(\omega)(t)|}{n^{1/2}} > M\right\} = \lim_{n\to\infty} \lambda\left\{|Yn^{1/\alpha - 1/2}| > M\right\} = 1 \quad .$$

Now fix $M > 0$ and choose $\varepsilon > 0$. Let $0 < \delta < 1/\alpha - 1/2$ and define

$$A_n = \left\{(\omega, t) : \frac{|S_n(\omega)(t)|}{n^{1/2+\delta}} > M\right\} .$$

Then, $\exists n_0$ such that $n \geq n_0$ implies

$$P\{\omega : (\omega, t) \in A_n\} > 1 - \varepsilon \qquad (0 \leq t \leq 1) \quad .$$

Therefore, for $n \geq n_0$

$$(\lambda \times P)(A_n) > 1 - \varepsilon \quad ,$$

and hence

$$P\left\{\omega : \lambda\left\{t : \frac{|S_n(\omega)(t)|}{n^{1/2+\delta}} > M\right\} > 1 - \sqrt{\varepsilon}\right\} > 1 - \sqrt{\varepsilon} \quad .$$

That is, if $P\{\omega : \lambda\{t : \frac{|S_n(\omega)(t)|}{n^{1/2+\delta}} > M\} \leq 1 - \sqrt{\varepsilon}\} \geq \sqrt{\varepsilon}$, then by Fubini's theorem for $n \geq n_0$

$$1 - \varepsilon < \int_\Omega \int_0^1 1_{A_n}(\omega, t)\, dt\, dP(\omega)$$

$$\leq (1 - \sqrt{\varepsilon})\sqrt{\varepsilon} + 1(1 - \sqrt{\varepsilon}) = 1 - \varepsilon$$

which is a contradiction. Now $n \geq n_0$ implying

$$P\left\{\omega : \lambda\left\{t : \frac{|S_n(\omega)(t)|}{n^{1/2+\delta}} > M\right\} > 1 - \sqrt{\varepsilon}\right\} > 1 - \sqrt{\varepsilon},$$

implies that

$$P\left\{\omega : \frac{\|S_n(\omega)\|_p}{n^{1/2+\delta}} > \frac{M}{1 - \sqrt{\varepsilon}}\right\} > 1 - \sqrt{\varepsilon} \quad ,$$

and since M and ε are arbitrary we actually have $\frac{\|S_n\|_p}{n^{1/2+\delta}}$ converging to infinity in probability. Hence $\frac{S_n}{\sqrt{2n\, LLn}}$ does not converge in probability to any bounded set K in $L^p[0, 1]$ and thus (1.1) is impossible for such sets K.

In the previous examples $\|X_k\|_p$ is uniformly bounded with probability one, so any result valid for i.i.d. sequences in all separable Banach spaces must involve something more than moment conditions. What is more important, however, is that these examples suggest a number of directions of possible research.

They are:

(1) For which infinite dimensional Banach spaces, if any, does the LIL always hold for i.i.d. random variables under the classical moment conditions $E(X_k) = 0$ and $E\|X_k\|^2 < \infty$?

(2) If one considers special sequences of i.i.d. random variables with values in spaces like $C[0,1]$ can one then prove a LIL for these sequences?

(3) Is there a "general result" holding for all real separable Banach spaces B and all i.i.d. B-valued random variables satisfying the classical moment assumptions?

We turn to the answer of (3) first. Its proof will follow easily from the first theorem we give.

<u>Corollary 3.1.</u> (N.A.S.C. for the LIL in the Banach space setting). Let $X_1, X_2, \ldots$ be i.i.d. B-valued such that $E(X_k) = 0$ and $E\|X_k\|^2 < \infty$. Then:

I. There exists a compact, symmetric, convex set $K \subseteq B$ such that

$$P\left\{\omega : C\left(\left\{\frac{S_n(\omega)}{\sqrt{2n\,LLn}} : n \geq 1\right\}\right) \not\subseteq K\right\} = 0 \quad . \tag{1.3}$$

II. In addition, there exists a compact, symmetric, convex set K satisfying (1.3) such that

$$P\left\{\omega : \lim_n d\left(\frac{S_n(\omega)}{\sqrt{2n\,LLn}}, K\right) = 0\right\} = 1 \tag{1.4}$$

and

$$(1.5) \qquad P\left\{\omega : C\left(\left\{\frac{S_n(\omega)}{\sqrt{2n\,LLn}} : n \geq 1\right\}\right) = K\right\} = 1 \quad ,$$

iff

$$(1.6) \qquad P\left\{\omega : \left\{\frac{S_n(\omega)}{\sqrt{2n\,LLn}} : n \geq 1\right\} \text{ is conditionally compact in } B\right\} = 1 \quad .$$

The event in (1.6) is a tail event for the sequence $X_1, X_2, \ldots$ so it has probability zero or one, and hence the LIL holds with limit set K or not at all.

We now turn to the details required to get at the limit set K.

2. <u>Construction of the limit set K.</u> The limit set K in our limit theorems depend on the covariance function of the random variables involved, and is intimately related to the mean-zero Gaussian measure on B with the given covariance function provided this measure exists.

A measure μ on B is called a mean-zero Gaussian measure if every $f \in B^*$ has a mean-zero Gaussian distribution with variance $\int_B [f(x)]^2 \, d\mu(x)$.

If μ is a measure on B (not necessarily Gaussian) such that $\int_B x \, d\mu(x) = 0$ and $\int_B \|x\|^2 \, d\mu(x) < \infty$, then the bilinear function T defined on $B^* \times B^*$ by

$$T(f, g) = \int_B f(x)\, g(x)\, d\mu(x) \qquad (f, g \in B^*)$$

is called the covariance function of μ.

If μ is a mean-zero Gaussian measure then it is well known that $\int_B \|x\|^2 \, d\mu(x) < \infty$, and that μ is uniquely determined by its covariance function. However, a mean-zero Gaussian measure μ is also determined by a unique subspace H_μ of B which has a Hilbert space structure. We describe this relationship by saying μ is generated by H_μ, and mention that the pair (B, H_μ) is an abstract Wiener space in the sense of [9].

One method of finding this Hilbert space is given in the next lemma which applies to non-Gaussian measures as well. It also provides a construction of the limit set K used in our results, and the relationship to Gaussian measures is given in part (vi) of the lemma. Finally, I emphasize that most of Lemma 2.1 is known in one form or another, but to avoid sending the reader to various references the crucial facts regarding K are collected here.

Lemma 2.1. Let μ denote a Borel probability measure on B (not necessarily Gaussian) such that $\int_B \|x\|^2 \, d\mu(x) < \infty$ and $\int_B x \, d\mu(x) = 0$. Let S denote the linear operator from B^* to B defined by the Bochner integral

$$(2.1) \qquad Sf = \int_B xf(x) \, d\mu(x) \qquad (f \in B^*) \quad .$$

Let H_μ denote the completion of the range of S with respect to the norm obtained from the inner product

$$(2.2) \qquad (Sf, Sg)_\mu = \int_B f(x) \, g(x) \, d\mu(x) \quad .$$

Then: (i) H_μ can be realized as a subset of B and the identity map $i : H_\mu \to B$ is continuous. In fact, for $x \in H_\mu$

$$(2.3) \qquad \|x\| \le \left(\int_B \|y\|^2 \, d\mu(y) \right)^{1/2} \|x\|_\mu \, .$$

(ii) If $e : B^* \to H_\mu^*$ is the linear map obtained restricting an element in B^* to the subspace H_μ of B and if we identify H_μ^* and H_μ in the usual way then

$$e = S \quad .$$

(iii) Let $\{f_k : k \ge 1\}$ be a weak-star dense subset of the unit ball of B^*. Let $\{\alpha_k : k \ge 1\}$ be an orthonormal sequence obtained from the sequence $\{f_k\}$ by the usual Gram-Schmidt orthogonalization method with respect to the inner product given by the right-side of (2.2). Then each $\alpha_k \in B^*$, and $\{S\alpha_k : k \ge 1\}$ is a C.O.N.S. in $H_\mu \subseteq B$. Further, the linear operators

$$(2.4) \qquad \Pi_N(x) = \sum_{k=1}^{N} \alpha_k(x)\, S\alpha_k \quad \text{and} \quad Q_N(x) = x - \Pi_N(x) \qquad (N \geq 1)$$

are continuous from B into B where by $\alpha_k(x)$ we mean the linear functional α_k applied to x.

(iv) If K is the unit ball of H_μ, then K is a compact symmetric convex set in B. Further, for each $f \in B^*$ we have

$$(2.5) \qquad \sup_{x \in K} f(x) = \left\{ \int_B [f(y)]^2 \, d\mu(y) \right\}^{1/2} .$$

(v) If μ and ν are two measures on B satisfying the basic hypothesis of the lemma and having common covariance function, then $H_\mu = H_\nu$.

(vi) If μ is a mean-zero Gaussian measure on B, then $\int_B \|x\|^2 \, d\mu(x) < \infty$ and H_μ is the generating Hilbert space for μ.

Proof. Take $f \in B^*$. Then $\int_B \|y\|^2 \, d\mu(y) < \infty$ implies the Bochner integral defining $Sf = \int_B y f(y) \, d\mu(y)$ exists and $Sf \in B$. Further,

$$(2.6) \qquad \|Sf\| \leq \left(\int_B \|y\|^2 \, d\mu(y) \right)^{1/2} \|Sf\|_\mu ,$$

and hence the map $i : S(B^*) \to B$ is continuous. Now (2.6) also implies the completion of $S(B^*)$ with respect to the norm given by the inner product in (2.2) can be realized as a subspace of B, and that the map $i : H_\mu \to B$ is continuous as indicated. Further, (2.3) follows from (2.6) since $S(B^*)$ is dense in H_μ with respect to the norm $\| \cdot \|_\mu$. Hence (i) holds.

Let $e : B^* \to H_\mu^* \equiv H_\mu$ as in (ii). Take $f \in B^*$. Then for $g \in B^*$ we have

$$f(Sg) = \int_B f(x)\, g(x)\, d\mu(x) = (Sf, Sg)_\mu$$

and hence $e(f) = Sf$ when acting on the elements in SB^*. Since SB^* is dense in H_μ we have $e(f) = Sf$ provided we identify H_μ and H_μ^* in the canonical way.

The assertions of (iii) are obvious since each α_k is a finite linear combination of the f_j's. To see that $\{S\alpha_k : k \geq 1\}$ is complete in H_μ simply observe that the f_j's separate points of B (and hence in H_μ). That is, if $\alpha_k(y) = 0$ for every k and some $y \in H_\mu$, then by undoing the Gram-Schmidt proceedure we thus have $f_j(y) = 0$ for every j. Since the f_j's separate points we have $y = 0$ as required. Perhaps it should be pointed out that when we undo the Gram-Schmidt proceedure we omit all f_j's which are linear combinations of previous f_i $(i < j)$ and those such that $\int_B [f_j(x)]^2\, d\mu(x) = 0$. However, if f_j is a finite linear combination of f_i $(i < j)$ and $f_i(y) = 0$ for $i < j$ then $f_j(y) = 0$ as asserted. On the other hand, if $\int_B [f_j(x)]^2\, d\mu(x) = 0$, then $S(f_j) \equiv e(f_j) = 0$ and hence $f_j(y) = 0$ again.

To verify (2.5) note that

$$\sup_{x \in K} f(x) = \sup_{Sg \in K} f(Sg) = \sup_{Sg \in K} \int_B f(x)\, g(x)\, d\mu(x)$$

$$\leq \left(\int_B f^2(x) \cdot d\mu(x) \right)^{1/2}$$

since $Sg \in K$ implies $\left(\int_B g^2(x)\, d\mu(x) \right)^{1/2} \leq 1$. Now set

$$g = \frac{f}{\left(\int_B f^2(x)\, d\mu(x) \right)^{1/2}}$$ and (2.5) holds.

To finish the proof of (iv) we show K is compact in B by first showing K is closed in B and then verifying that every subsequence $\{y_n\} \subseteq K$ has a convergent subsequence in B.

Take $\{y_n\} \subseteq B$ and assume $\|y_n - y\| \to 0$ for $y \in B$. Since K is compact in the weak topology induced by H_μ^* we have a subsequence $\{y_{n_j}\}$ such that $y_{n_j} \xrightarrow{\text{weakly}} z$ and $z \in K$. Thus $\{y_{n_j}\}$ converges weakly to z in the weak topology on B induced by B^* as $i : H_\mu \to B$ is continuous by (i). Since B^* separates points of B we have $y = z$ so $y \in K$ and K is closed.

Since $SB^* \cap K$ is dense in K it now suffices to prove that if $\{y_n\} \subseteq SB^* \cap K$ then $\{y_n\}$ has a convergent subsequence.

Let U denote the unit ball of B^* with the weak-star topology. Since B is separable we have that U is a compact metric space in the weak-star topology. For $x \in B$, $f \in B^*$ let $\theta x(f) = f(x)$. Then $\theta : B \to C(U)$ is an isometry from B into the Banach space $C(U)$ with the supremum norm. Thus to show $\{y_n\}$ has a B-convergent subsequence we need only show that $\{\theta y_n\}$ is an equicontinuous and uniformly bounded sequence in $C(U)$ (apply Ascoli's Theorem).

Let $f, g \in U$. Then since $\{y_n\} \subseteq K \cap SB^*$ we have $y_n = Sr_n$ for $r_n \in B^*$ and such that $\int_B r_n^2(x)\, d\mu(x) \leq 1$. Hence

$$|\theta y_n(f) - \theta y_n(g)| = |(f - g)(Sr_n)|$$

$$= \left| \int_B (f - g)(x)\, r_n(x)\, d\mu(x) \right| \tag{2.7}$$

$$\leq \left\{ \int_B [(f - g)(x)]^2\, d\mu(x) \right\}^{1/2} .$$

Now

$$\int_B [(f - g)(x)]^2 \, d\mu(x) \le \|f - g\|^2_{B^*} \int_B \|x\|^2 \, d\mu(x)$$

so setting $g = 0$ we have from (2.7) that

$$\sup_{f \in U} |\theta y_n(f)| \le \left(\int_B \|x\|^2 \, d\mu(x) \right)^{1/2} .$$

Thus $\{\theta y_n : n \ge 1\}$ is uniformly bounded on U and it remains to prove $\{\theta y_n : n \ge 1\}$ is equicontinuous on U.

Recall that the weak star topology on U is equivalent to that given by the metric

$$d(f, g) = \sum_{j=1}^{\infty} \frac{1}{2^j} \frac{|f(x_j) - g(x_j)|}{1 + |f(x_j) - g(x_j)|}$$

where $\{x_1, x_2, \ldots\}$ is dense in B.

Fix ε such that $0 < \varepsilon \le 1$. In view of (2.7) to establish equicontinuity of $\{\theta y_n : n \ge 1\}$, we need only show that there exists a $\delta > 0$ such that $d(f, g) < \delta$ implies

$$\int_B [(f - g)(x)]^2 \, d\mu(x) < \varepsilon .$$

Our first step is to choose a compact set C in B such that

$$\int_{B-C} \|x\|^2 \, d\mu(x) < \varepsilon/2 .$$

Then we observe that since weak-star convergence of elements in U is equivalent to uniform convergence on compact subsets of B we have a $\delta > 0$ such that $d(f, g) < \delta$ implies

$$\int_C [(f - g)(x)]^2 \, d\mu(x) < \varepsilon/2 \quad .$$

Combining these two inequalities we have for $f, g \in U$ and $d(f, g) < \delta$ that

$$\int_B [(f - g)(x)]^2 \, d\mu(x) < \varepsilon/2 + \varepsilon/2 = \varepsilon \quad .$$

Thus K is compact as asserted.

If μ and ν have the same covariance function then for every $f \in B^*$ we have

$$\int_B xf(x) \, d\mu(x) = \int_B xf(x) \, d\nu(x) \quad .$$

This follows since applying $g \in B^*$ to both sides we get $T(f, g)$, the common covariance function of μ and ν. Since such elements are dense in $H_\mu (H_\nu)$ and the norms induced by μ and ν are identical on these elements $H_\mu = H_\nu$ as asserted.

The verification of (vi) follows from well known results on Gaussian measures so the details are omitted.

3. <u>A basic convergence result and some corollaries.</u> We first give a general result which will have corollaries dealing with sums of independent identically distributed B-valued random variables as well as with other stochastic processes. In the applications of Theorem 3.1 which we have in mind the Y_n's should be viewed as approximately Gaussian with approximately a fixed covariance structure, and the ϕ_n's are positive constants taken to provide the necessary convergence.

<u>Theorem 3.1.</u> Let K denote the unit ball of the Hilbert space $H_\mu \subseteq B$ where μ is a mean-zero measure on B such that $\int_B \|x\|^2 \, d\mu(x) < \infty$. Let $\{Y_n : n \geq 1\}$ be a sequence of B-valued random variables such that for some sequence of positive constants $\{\phi_n\}$ we have

$$(3.1) \qquad P\left\{\omega : \overline{\lim_n}\, f\left(\frac{Y_n(\omega)}{\phi_n}\right) \leq \sup_{x \in K} f(x)\right\} = 1 \qquad (f \in B^*) .$$

Then:

I. We have

$$(3.2) \qquad P\left\{\omega : C\left(\left\{\frac{Y_n(\omega)}{\phi_n}\right\}\right) \not\subseteq K\right\} = 0 ,$$

and hence $P\left\{\omega : \left\{\frac{Y_n(\omega)}{\phi_n} : n \geq 1\right\} \text{ is conditionally compact in } B\right\} = 1$ iff

$$(3.3) \qquad P\left\{\omega : \lim_n d\left(\frac{Y_n(\omega)}{\phi_n}, K\right) = 0\right\} = 1 .$$

Here $d(x, K) = \inf_{y \in K} \|x - y\|$.

II. If $P\left\{\omega : \overline{\lim_n} f\left(\frac{Y_n(\omega)}{\phi_n}\right) = \sup_{x \in K} f(x)\right\} = 1$ for f in B^* and if

$$P\left\{\omega : \left\{\frac{Y_n(\omega)}{\phi_n} : n \geq 1\right\} \text{ is conditionally compact in } B\right\} = 1 \quad , \tag{3.4}$$

then H_μ infinite dimensional implies

$$P\left\{\omega : C\left(\left\{\frac{Y_n(\omega)}{\phi_n} : n \geq 1\right\}\right) = K\right\} = 1 \quad . \tag{3.5}$$

Proof. Let $K(\omega) = C\left(\left\{\frac{Y_n(\omega)}{\phi_n} : n \geq 1\right\}\right)$ for $\omega \in \Omega$. If $K(\omega) = \phi$, then, of course, $\phi = K(\omega) \subseteq K$. Now $B - K$ is open and B is separable so

$$B - K = \bigcup_{r=1}^{\infty} N_r \quad ,$$

where each N_r is a closed sphere in B. Then

$$\{\omega : K(\omega) \not\subseteq K\} = \bigcup_{r=1}^{\infty} \{\omega : K(\omega) \cap N_r \neq \phi\} \quad ,$$

and hence if P^* denotes the outer measure induced by P

$$P^*(\omega : K(\omega) \not\subseteq K) \leq \sum_{r=1}^{\infty} P^*(\omega : K(\omega) \cap N_r \neq \phi) \quad .$$

If $P^*(\omega : K(\omega) \not\subseteq K) > 0$, then $P^*(\omega : K(\omega) \cap N_r \neq \phi) > 0$ for some r and this will produce a contradiction.

To verify this last assertion choose $g \in B^*$ such that

(3.6) $$\sup_{x \in K} g(x) = \gamma_1 < \gamma_2 = \inf_{x \in N_r} g(x) .$$

Then

$$\{\omega : K(\omega) \cap N_r \neq \phi\} \subseteq \left\{\omega : \overline{\lim_n}\, g\left(\frac{Y_n(\omega)}{\phi_n}\right) \geq \gamma_2\right\} ,$$

so $P^*(\omega : K(\omega) \cap N_r \neq \phi) > 0$ implies

$$P\left(\omega : \overline{\lim_n}\, g\left(\frac{Y_n(\omega)}{\phi_n}\right) \geq \gamma_2\right) > 0 .$$

This contradicts (3.1) since (3.6) holds for g. Thus we have

$$P^*(\omega : K(\omega) \not\subseteq K) = 0 ,$$

and since we assume our probability space to be complete this gives (3.2).

If (3.4) holds, then (3.2) implies (3.3), and the proof of (I) is complete.

Now we establish (II). To do so we need the linear operators Π_N and Q_N defined in (2.4) with $\{S\alpha_k : k \geq 1\}$ a C.O.N.S. in H_μ such that each $\alpha_k \in B^*$.

Fix $\varepsilon > 0$. First we shown there exists N_0 such that $N \geq N_0$ implies

(3.7) $$Q_N K \subseteq \{x \in B : \|x\| < \varepsilon\} .$$

If (3.7) does not hold, then we have a sequence $\{x_j\}$ such that

$$x_j \in Q_j K \quad \text{and} \quad \|x_j\| \geq \varepsilon \quad (j = 1, 2, \ldots) \quad .$$

Now $Q_j K \subseteq K$ for all $j \geq 1$ and K compact implies there exists a subsequence j' such that

$$\lim_{j' \to \infty} x_{j'} = z$$

in B. Thus $\|z\| \geq \varepsilon$ and since $\{x_j : j \geq N\} \subseteq Q_N K$ for $N = 1, 2, \ldots$ $(Q_1 K \supseteq Q_2 K \supseteq \ldots)$ with each $Q_N K$ compact we have $z \in \bigcap_{N \geq 1} Q_N K$. This is impossible since $\bigcap_{N \geq 1} Q_N K = \{0\}$ and $\|z\| \geq \varepsilon > 0$. Hence (3.7) holds as indicated.

Therefore for $N \geq N_0$ we have

$$\left\{\omega : \overline{\lim_k}\ d\left(Q_N \frac{Y_k(\omega)}{\phi_k}, Q_N K\right) \leq \varepsilon\right\} \tag{3.8}$$
$$\subseteq \left\{\omega : \overline{\lim_k}\ \left\|Q_N \frac{Y_k(\omega)}{\phi_k}\right\| \leq 2\varepsilon\right\} \quad .$$

Since (3.4) holds we have, as mentioned previously, that (3.3) holds. Since Q_N maps B into B continuously we have

$$P\left\{\omega : \overline{\lim_k}\ d\left(Q_N \frac{Y_k(\omega)}{\phi_k}, Q_N K\right) = 0\right\} = 1 \quad , \tag{3.9}$$

and hence for $N \geq N_0$ (3.8) implies

$$P\left\{\omega : \overline{\lim_k}\ \left\|Q_N \frac{Y_k(\omega)}{\phi_k}\right\| \leq 2\varepsilon\right\} = 1 \quad . \tag{3.10}$$

Choose $h \in K$ and take $N \geq N_0$ such that $\|Q_N h\| \leq \varepsilon$. Then for an ω-set of probability one we have

$$(3.11) \qquad \left\|\frac{Y_k(\omega)}{\phi_k} - h\right\| \leq \left\|\Pi_N\left(\frac{Y_k(\omega)}{\phi_k} - h\right)\right\| + \left\|Q_N \frac{Y_k(\omega)}{\phi_k}\right\| + \|Q_N h\|$$

$$\leq \left\|\Pi_N\left(\frac{Y_k(\omega)}{\phi_k} - h\right)\right\| + 3\varepsilon$$

for all k sufficiently large (the largeness of k depends, of course, on ω).

Since K is separable (3.5) follows from (3.11) if

$$(3.12) \qquad P\left\{\omega : \left\|\Pi_N\left(\frac{Y_k(\omega)}{\phi_k} - h\right)\right\| < \varepsilon \text{ for infinitely many } k\right\} = 1$$

for any $\varepsilon > 0$.

Now $\Pi_N B = \Pi_N H_\mu$ and all norms on a finite dimensional space are equivalent so (3.12) holds if

$$(3.13) \qquad P\left\{\omega : \left\|\Pi_N\left(\frac{Y_k(\omega)}{\phi_k} - h\right)\right\|_\mu \leq \varepsilon \text{ i.o. in } k\right\} = 1$$

for each $\varepsilon > 0$.

To show (3.13) we first prove that for every $g \in \Pi_{N+1}K$ such that $\|g\|_\mu = 1$ we have

$$(3.14) \qquad P\left(\omega : \left\|\Pi_{N+1}\left(\frac{Y_k(\omega)}{\phi_k} - g\right)\right\|_\mu \leq \varepsilon \text{ i.o. in } k\right) = 1$$

for each $\varepsilon > 0$. Then (3.13) follows from (3.14) by taking $g = \Pi_N h + c\alpha_{N+1}$ where c is such that $\|g\|_\mu = 1$. That is,

$$\left\|\Pi_{N+1}\left(\frac{Y_k(\omega)}{\phi_k}\right) - g\right\|_\mu^2 = \left\|\Pi_N \frac{Y_k(\omega)}{\phi_k} - \Pi_N h\right\|_\mu^2 + \left|\alpha_{N+1}\left(\frac{Y_k(\omega)}{\phi_k}\right) - c\right|^2$$

so the event in (3.13) contains the event in (3.14).

Therefore (3.14) is to be established to complete the proof. Take $g \in \Pi_{N+1}K$ such that $\|g\|_\mu = 1$. Then $g = \sum_{k=1}^{N+1} \alpha_k(g)\, S\alpha_k$ where $\sum_{k=1}^{N+1} \alpha_k^2(y) = 1$. Furthermore, $g = Sf_0$ where $f_0 = \sum_{k=1}^{N+1} \alpha_k(g)\, \alpha_k$ is in B^*. Thus if (3.14) does not hold there exists a $\delta > 0$ such that

$$P\left\{\omega : \overline{\lim_k}\; f_0\left(\frac{Y_k(\omega)}{\phi_k}\right) \le 1 - \delta\right\} > 0 \quad . \tag{3.15}$$

That is,

$$f_0\left(\frac{Y_k(\omega)}{\phi_k}\right) = \sum_{j=1}^{N+1} \alpha_j(g)\, \alpha_j\left(\frac{Y_k(\omega)}{\phi_k}\right) = \left(\Pi_{N+1}\left(\frac{Y_k(\omega)}{\phi_k}\right), g\right)_\mu , \tag{3.16}$$

and hence $f_0\left(\frac{Y_k(\omega)}{\phi_k}\right)$ denotes the length of $\Pi_{N+1}\left(\frac{Y_k(\omega)}{\phi_k}\right)$ in the direction g (computed in H_μ). Letting

$$A = \left\{\omega : \lim_k d\left(\frac{Y_k(\omega)}{\phi_k}, K\right) = 0 \text{ and } \underline{\lim_{k\to\infty}}\; \left\|\Pi_N \frac{Y_k(\omega)}{\phi_k} - g\right\| \ge \varepsilon\right\}$$

we have that $P(A) > 0$ if (3.14) fails. Therefore for each $\omega \in A$ there exists a $\delta > 0$ (depending only on N and ε) such that $\overline{\lim_k}\; f_0\left(\frac{Y_k(\omega)}{\phi_k}\right) \le (g, g)_\mu - \delta = 1 - \delta$. Thus $P(A) > 0$ implies (3.15). Now (3.15) contradicts the condition

$$P\left\{\omega : \overline{\lim_k}\; f_0\left(\frac{Y_k(\omega)}{\phi_k}\right) = \sup_{x\in K} f_0(x)\right\} = 1$$

since $\sup_{x \in K} f_0(x) = \sup_{x \in K} (x, g)_\mu = 1$. Thus (3.14) holds and the proof is complete.

Remark. If H_μ is infinite dimensional, then Corollary 3.1 of section one is an immediate corollary of Theorem 3.1.

To see this recall that K is compact so (1.4) implies (1.6). For the remainder let $Y_n = \frac{S_n}{\sqrt{n}}$ and $\phi_n = \sqrt{2\, LLn}$ in Theorem 3.1. Then by the Hartman-Wintner result applied to the i.i.d. real valued random variables

$$f(X_1), f(X_2), \ldots$$

we have

$$P\left\{\omega : \overline{\lim_n}\, f\left(\frac{Y_n(\omega)}{\phi_n}\right) = \left(\int_B [f(y)]^2 \, d\mu(y)\right)^{1/2}\right\} = 1$$

for each $f \in B^*$ where $\mu = \mathcal{L}(X_1)$. By Lemma 2.1 (iv) we have

$$\left\{\int_B [f(y)]^2 \, d\mu(y)\right\}^{1/2} = \sup_{x \in K} f(x)$$

and hence the conditions of Theorem 3.1 hold proving the corollary.

If $\dim H_\mu < \infty$ then Corollary 3.1 follows from a result of H. Finkelstein (Ann. Math. Stat., 42, 1971, 607-615). That is, if $\dim H_\mu < \infty$, then $P(S_n \in H_\mu) = 1$ for every n and we can work with the H_μ norm instead of the B-norm on H_μ because all locally convex Hausdorff topologies compatible with the vector space structure are equivalent on finite dimensional vector spaces.

For an example where the normalizing constants ϕ_n appearing in Theorem 3.1 are something other than $\sqrt{2\, LLn}$ we turn to a generalization of some of the recent work of T. L. Lai [20].

First, however, we need the definition of the Prohorov metric for probability measures.

Let μ_1 and μ_2 be probability measures on the Borel subsets of the metric space (M, d) which is complete and separable. Let $\mathcal{C}$ denote the closed sets of (M, d) and define for each $\varepsilon > 0$ and subset A of M the set $A^\varepsilon = \{y \in M : d(y, A) < \varepsilon\}$ where, of course, $d(y, A) = \inf_{x \in A} d(y, x)$. Let

$$\varepsilon_{12} = \inf\{\varepsilon > 0 : \mu_1(F) \le \mu_2(F^\varepsilon) + \varepsilon \ \forall F \in \mathcal{C}\}$$

$$\varepsilon_{21} = \inf\{\varepsilon > 0 : \mu_2(F) \le \mu_1(F^\varepsilon) + \varepsilon \ \forall F \in \mathcal{C}\}$$

and define

$$L(\mu_1, \mu_2) = \max(\varepsilon_{12}, \varepsilon_{21}) \quad .$$

Then L is the Prohorov metric on the class of all Borel probability measures on M and weak convergence for these measures is equivalent to L-convergence.

Corollary 3.2. Let $\{Y_k : k \ge 1\}$ be a sequence of B-valued random variables and assume μ is a mean-zero Gaussian measure on B. Let K denote the unit ball of H_μ. If

$$L(\mathcal{L}(Y_k), \mu) = b_k \qquad (k \ge 1) \tag{3.17}$$

where $\sum_k b_k < \infty$, and L is the Prohorov metric for probability measures on $(B, \|\cdot\|)$, then

$$(3.18) \qquad P\left\{\omega : \lim_n d\left(\frac{Y_n(\omega)}{\sqrt{2\,Ln}}, K\right) = 0\right\} = 1 \quad .$$

If the Y_k's are independent as well, then (3.17) also implies

$$(3.19) \qquad P\left\{\omega : C\left(\left\{\frac{Y_n(\omega)}{\sqrt{2\,Ln}} : n \geq 1\right\}\right) = K\right\} = 1 \quad .$$

Remark. If $\mathcal{L}(Y_k) = \mu$ for every k then $L(\mathcal{L}(Y_k), \mu) = 0$ so we have (3.17) with $b_k = 0$, and hence (3.18) holds. If, in the case $\mathcal{L}(Y_k) = \mu$ for $k \geq 1$, we also have

$$(3.20) \qquad \lim_{\substack{m \to \infty \\ k - m \to \infty}} E\left(\left\{E\left(f(Y_k) \mid \mathcal{F}_m\right)\right\}^2\right) = 0$$

for every $f \in B^*$ where $\mathcal{F}_m = \mathcal{F}(Y_k : k \leq m)$, then (3.19) holds.

Corollary 3.2 and the remark following it are not immediate consequences of Theorem 3.1. Nevertheless, we ask the reader to examine Theorem 4.3 and Corollary 4.2 of [12] for the details of their proofs, so that we may continue our development of the LIL.

Before returning to the LIL, however, perhaps it is worthwhile to point out that Strassen's functional law of the iterated logartihm [22] can be obtained from Corollary 3.2. This observation was first made by T. L. Lai in [20]. Let $\{W(t) : 0 \leq t < \infty\}$ be standard Brownian motion on $(\Omega, \mathcal{F}, P)$ and define

$$(3.21) \qquad Z_n(t, \omega) = \frac{W(nt, \omega)}{\sqrt{n}} \qquad (0 \leq t \leq 1, n \geq 1, \omega \in \Omega) \quad .$$

Then each $\{Z_n(t) : 0 \leq t \leq 1\}$ is Brownian motion on $[0, 1]$ and it induces Wiener measure μ on $C[0, 1]$ which, of course, is a Gaussian measure. Further, it is well known that

$$H_\mu = \left\{ f \in C[0,1] : f(t) = \int_0^t g(s)\, ds \text{ where } \int_0^1 g^2(s)\, ds < \infty \right\}$$

with inner product

$$(f_1, f_2)_\mu = \int_0^1 f_1'(s) f_2'(s)\, ds \quad ,$$

and hence

$$K = \left\{ f \in C[0,1] : f(t) = \int_0^t g(s)\, ds \text{ where } \int_0^1 g^2(s)\, ds \leq 1 \right\} \quad .$$

Now let $\beta > 1$ and define n_k = greatest integer in β^k. Let

(3.22) $$Y_k = Z_{n_k} \qquad (k \geq 1) \quad .$$

Then, by (3.18) we have

(3.23) $$P\left\{ \omega : \lim_k d\left(\frac{W(n_k(\cdot), (\omega)}{\sqrt{2n_k \, Lk}}, K \right) = 0 \right\} = 1$$

where $d(x, K) = \inf_{y \in K} \sup_{0 \leq t \leq 1} |x(t) - y(t)|$. Given $\varepsilon > 0$ we define

$$A_k = \left\{\omega : \max_{n_k \le j \le n_{k+1}} \sup_{0 \le t \le 1} \left| \frac{W(n_k t, \omega)}{\sqrt{2n_k\, LLn_k}} - \frac{W(jt, \omega)}{\sqrt{2j\, LLj}} \right| > \varepsilon/2 \right\} .$$

Then there exists $\beta > 1$ such that

$$P(A_k \text{ i.o.}) = 0 \quad , \tag{3.24}$$

and since $LLn_k \sim Lk$ as $k \to \infty$ we have by combining (3.23) and (3.24) that

$$P\left\{\omega : \frac{W(n(\cdot), \omega)}{\sqrt{2n\, LLn}} \in K^{\varepsilon} \text{ for all } n \text{ sufficiently large}\right\} = 1 \quad .$$

Thus

$$P\left\{\omega : \lim_{n \to \infty} d\left(\frac{W(n(\cdot), \omega)}{\sqrt{2n\, LLn}}, K\right) = 0\right\} = 1 \quad . \tag{3.25}$$

In view of (3.25) to show

$$P\left\{\omega : C\left(\left\{\frac{W(n(\cdot), \omega)}{\sqrt{2n\, LLn}} : n \ge 1\right\}\right) = K\right\} = 1 \tag{3.26}$$

it suffices to show

$$P\left\{\omega : C\left(\left\{\frac{W(n_k(\cdot), \omega)}{\sqrt{2n_k\, LLn_k}} : k \ge 1\right\}\right) = K\right\} = 1 \quad . \tag{3.27}$$

Now (3.27) follows by using (3.22), $LLn_k \sim Lk$, and (3.19) if we can verify (3.20). If $f \in C^*[0, 1]$ with $f(x) = \int_0^1 x(s)\, dF(x)$ for $x \in C[0, 1]$, then

$$E(f(Y_k) \mid \mathfrak{F}_m) = E\left(\int_0^1 \frac{W(n_k t)}{\sqrt{n_k}} \, dF(t) \mid \mathfrak{F}_m \right)$$

$$= E\left(\int_0^{\frac{n_m}{n_k}} \frac{W(n_k t)}{\sqrt{n_k}} \, dF(t) + \int_{\frac{n_m}{n_k}}^1 \frac{W(n_k t)}{\sqrt{n_k}} \, dF(t) \mid \mathfrak{F}_m \right)$$

$$= \int_0^{a_{m,k}} \frac{W(n_k t)}{\sqrt{n_k}} \, dF(t) + \int_{a_{m,k}}^1 \frac{W(n_m)}{\sqrt{n_k}} \, dF(t)$$

where $a_{m,k} = \frac{n_m}{n_k}$. Hence

$$E(E(f(Y_k) \mid \mathfrak{F}_m)^2) = \frac{1}{n_k} \Bigg\{ \int_0^{a_{m,k}} \int_0^{a_{m,k}} \min(n_k s, n_k t) \, dF(s) \, dF(t)$$

$$+ 2 \int_0^{a_{m,k}} n_k s \, dF(s) \int_{a_{m,k}}^1 dF(t)$$

$$+ \int_{a_{m,k}}^1 \int_{a_{m,k}}^1 n_m \, dF(s) \, dF(t) \Bigg\} ,$$

and (3.20) holds since $\lim_{\substack{m \to \infty \\ k - m \to \infty}} a_{m,k} = \lim_{\substack{m \to \infty \\ k - m \to \infty}} \frac{n_m}{n_k} = 0.$

The functional law of the iterated logarithm for Brownian motion in B given in [18] also follows exactly as above.

With the many important topologies on linear topological spaces other than norm topologies, it is reasonable to examine the LIL for random variables when the convergence to the limit set K, and clustering throughout K, is computed with respect to a topology other than one given by a norm. Our next theorem and its corollaries deal with this situation in some special cases. However, perhaps the main point of these results is that they leave little doubt about the "correctness of K" for the limiting set in the LIL.

If τ is a topology on some set M and $\{a_n\} \subseteq M$, then $C_\tau(\{a_n\})$ denotes the set of all limit points of the sequence $\{a_n\}$ in the τ-topology.

Theorem 3.2. Let B denote a real separable Banach space with norm $\|\cdot\|$, and assume $X_1, X_2, \ldots$ is a sequence of B valued random variables such that

$$(3.28) \qquad E(X_k) = 0 \text{ and } E\|X_k\|^2 < \infty \ .$$

Let τ be a locally convex Haudorff topology on B which is weaker then the norm topology. If K is the unit ball of $H_{\mathcal{L}(H_1)}$, then

$$(3.29) \qquad P\left\{\omega : \left\{\frac{S_n(\omega)}{\sqrt{2n\, LLn}} : n \geq 1\right\} \text{ is eventually in } V \text{ for every } \tau\text{-open set } V \supseteq K\right\} = 1$$

and

$$(3.30) \qquad P\left\{\omega : C_\tau\left(\left\{\frac{S_n(\omega)}{\sqrt{2n\, LLn}} : n \geq 1\right\}\right) = K\right\} = 1$$

iff

$$(3.31) \qquad P\left\{\omega : \left\{\frac{S_n(\omega)}{\sqrt{2n\, LLn}} : n \geq 1\right\} \text{ is } \tau\text{-conditionally compact in } B\right\} = 1 \ .$$

Remark. The events in (3.29) and (3.31) are assumed to be completion measurable events. This will always be the case in the corollaries we indicate, but it seems that it must be assumed for the general result. In fact, if the event in (3.29) is completion measurable with probability one, then we show the event in (3.31) is completion measurable with probability one. Similarly, if the event in (3.31) is completion measurable with probability one, then we show the events in (3.29) and (3.30) are completion measurable with probability one.

Proof. If (3.29) holds, then (3.31) holds. That is, if

$$A = \left\{ \omega : \left\{ \frac{S_n(\omega)}{\sqrt{2n\, LLn}} : n \geq 1 \right\} \text{ is eventually in } V \text{ for every } \tau\text{-open set } V \supseteq K \right\}$$

and $\omega \in A$, then $\left\{ \frac{S_n(\omega)}{\sqrt{2n\, LLn}} : n \geq 1 \right\} \cup K$ is τ-compact in B. To see this last assertion let $\{U_t : t \in T\}$ be a τ-open cover of $\{x_n : n \geq 1\} \cup K$ where $x_n = \frac{S_n(\omega)}{\sqrt{2n\, LLn}}$ for $n \geq 1$. Recall K is compact in B and hence also τ-compact. Since K is compact there exists $t_1, \ldots, t_n \in T$ such that $W \equiv \bigcup_{j=1}^{n} U_{t_j} \supseteq K$. Now let $\{x_{n'}\}$ denote the subset of $\{x_n\}$ not in W. Then $\{x_{n'}\}$ is finite since $\omega \in A$ implies $\{x_n\}$ is eventually in W. Hence the arbitrary open cover $\{U_t : t \in T\}$ of $K \cup \{x_n : n \geq 1\}$ has a finite subcover and $K \cup \{x_n : n \geq 1\}$ is compact as asserted. Hence $\{x_n : n \geq 1\}$ is τ-conditionally compact, and since $P(A) = 1$ we have (3.31) holding. In fact, we have shown that A is a subset of the event appearing in (3.31) and since A is completion measurable with $P(A) = 1$ we thus have the event in (3.31) completion measurable and of probability one.

Now assume (3.31) holds. We first show that

$$\text{(3.32)} \qquad P\left\{ \omega : C_\tau\left(\left\{ \frac{S_n(\omega)}{\sqrt{2n\, LLn}} : n \geq 1 \right\} \right) \not\subseteq K \right\} = 0 \quad .$$

In fact, (3.32) holds without using (3.31) and this is something we wish to make note of.

Since B is separable in the norm topology we can write $B - K = \bigcup_{r=1}^{\infty} N_r$ where each N_r is a τ-closed convex set having non-empty τ-interior. That is, cover each point $x \in B - K$ with a τ-open convex set U_x whose closure lies within $B - K$. This is possible since x fixed, $x \notin K$, τ-Hausdorff, and K compact implies there are open sets U_x and V_x such that $K \subseteq V_x$, $x \in U_x$, U_x is convex and $U_x \cap V_x = \phi$. Since τ is weaker than the norm topology and B is separable in the norm topology, we can reduce any open cover to a countable open cover, and hence

$$B - K = \bigcup_{j=1}^{\infty} U_{x_j} \quad .$$

Let N_r denote the τ-closure of U_{x_r} and then N_r is as indicated.

Let $K(\omega) = C_\tau\left(\left\{\frac{S_n(\omega)}{\sqrt{2n\, LLn}} : n \geq 1\right\}\right)$ for $\omega \in \Omega$. If $K(\omega) = \phi$, then, of course, $K(\omega) \subseteq K$. Thus

$$\{\omega : K(\omega) \not\subseteq K\} = \bigcup_{r=1}^{\infty} \{\omega : K(\omega) \cap N_r \neq \phi\} \quad ,$$

and hence if P^* denotes the outer measure induced by P, then

$$P^*\{\omega : K(\omega) \not\subseteq K\} \leq \sum_{r=1}^{\infty} P^*\{\omega : K(\omega) \cap N_r \neq \phi\} \quad .$$

If $P^*\{\omega : K(\omega) \cap N_r \neq 0\} > 0$ we produce a contradiction. Choose $f \in B^*$ such that f is τ-continuous, and

$$\sup_{x \in K} f(x) = \gamma_1 < \gamma_2 = \inf_{x \in N_r} f(x) \quad .$$

Then f τ-continuous implies

$$\left\{\omega : K(\omega) \cap N_r \neq \phi\right\} \subseteq \left\{\omega : \overline{\lim_n}\, f\left(\frac{S_n(\omega)}{\sqrt{2n\, LLn}}\right) \geq \gamma_2\right\} .$$

Hence $P^*\{\omega : K(\omega) \cap N_r \neq \phi\} > 0$ implies

$$P\left\{\omega : \overline{\lim_n}\, f\left(\frac{S_n(\omega)}{\sqrt{2n\, LLn}}\right) \geq \gamma_2\right\} > 0 \quad \text{(and hence } = 1) .$$

This is a contradiction to the LIL applied to the real valued random variables $f(X_1), f(X_2), \ldots$ which implies by (2.5) that

$$P\left\{\omega : \overline{\lim_n}\, f\left(\frac{S_n(\omega)}{\sqrt{2n\, LLn}}\right) = \left(\int_B f^2(x)\, d\mu(x)\right)^{1/2}\right.$$

$$\left. = \sup_{x \in K} f(x) = \gamma_1 < \gamma_2\right\} = 1$$

where $\mu = \mathcal{L}(X_1)$. Therefore $P^*\{\omega : K(\omega) \not\subseteq K\} = 0$, and since our probability space is always assumed complete we have established (3.32).

Next we show (3.29). Let

$$D = \left\{\begin{array}{l} \omega : \left\{\dfrac{S_n(\omega)}{\sqrt{2n\, LLn}} : n \geq 1\right\} \text{ is } \tau\text{-conditionally compact in} \\ B \text{ and } C_\tau\left(\left\{\dfrac{S_n(\omega)}{\sqrt{2n\, LLn}} : n \geq 1\right\}\right) \subseteq K \end{array}\right\} . \tag{3.33}$$

Since we are assuming (3.31) holds and (3.32) holds we have $P(D) = 1$, and if $\omega_0 \in D$ then

$$\omega_0 \in \left\{ \omega : \left\{ \frac{S_n(\omega)}{\sqrt{2n\,LLn}} : n \geq 1 \right\} \text{ is eventually in } V \text{ for every } \tau\text{-open set } V \supseteq K \right\}.$$

That is, let V be τ-open and assume $V \supseteq K$. Denote the points of $\left\{ \frac{S_n(\omega_0)}{\sqrt{2n\,LLn}} : n \geq 1 \right\}$ outside of V by N. Then N is a finite set or, since $\omega_0 \in D$, the infinite set would be τ-conditionally compact and hence have a limit point which is outside of V. This, of course contradicts the fact that $C_\tau \left(\left\{ \frac{S_n(\omega_0)}{\sqrt{2n\,LLn}} : n \geq 1 \right\} \right) \subseteq K$. Hence N is finite and (3.29) holds as $V \supseteq K$ was arbitrary.

We finish the proof by establishing (3.30). Now K is separable in the τ-topology and since (3.32) holds we have (3.30) if we show

$$P\left\{ \omega : h \in C_\tau \left(\left\{ \frac{S_n(\omega)}{\sqrt{2n\,LLn}} : n \geq 1 \right\} \right) \right\} = 1 \tag{3.34}$$

for any $h \in K$.

The next step of the proof of (3.30) is to choose a sequence $\{\alpha_k : k \geq 1\}$ from the τ-continuous linear functionals on B such that $\{S\alpha_k : k \geq 1\}$ is a C.O.N.S. in H_μ. We then define the operators

$$\Pi_N(x) = \sum_{k=1}^{N} \alpha_k(x)\, S\alpha_k \quad \text{and} \quad Q_N(x) = x - \Pi_N(x) \qquad (N \geq 1)$$

which are τ-continuous from B into B and of use in the proof of (3.30). If H_μ is only finite dimensional, then the related remark following the proof of Theorem 3.1 applies so we assume without loss of generality that H_μ is infinite dimensional.

To find the sequence $\{\alpha_k : k \geq 1\}$ let $M = \{f \in B^* : f$ is τ-continuous on $B\}$. Then M is a linear subspace of $L^2(B, \mu)$ and since $L^2(B, \mu)$ is separable we have that M has a countable dense subset $\{f_1, f_2, \ldots\}$ in the $L^2(B, \mu)$ topology. Using the Gram-Schmidt proceedure on $\{f_1, f_2, \ldots\}$ we obtain an orthonormal sequence $\{\alpha_1, \alpha_2, \ldots\}$ in $L^2(B, \mu)$ such that $\{\alpha_k : k \geq 1\} \subseteq M$ and the

$$\text{span}\{f_k : k \geq 1\} = \text{span}\{\alpha_k : k \geq 1\} \quad .$$

Then $\{S\alpha_k : k \geq 1\}$ is a C.O.N.S. in $H_\mu \subseteq B$ as indicated. To see this, note that if $h \in H_\mu$, $\|h\|_\mu = 1$, and $(h, S\alpha_k)_\mu = 0$ for each k, then by (ii) of Section two we have $\alpha_k(h) = 0$ for every k. Undoing the Gram-Schmidt proceedure we then have $f_k(h) = 0 \, \forall k$. Since M separates points of B, and hence of H_μ, there exists $f \in M$ such that $f(h) = 1$. Then from (2.5) we have

$$\int_B [(f - f_k)(y)]^2 \, d\mu(y) = \sup_{x \in K} |(f - f_k)(x)|^2 \geq 1 \quad .$$

This contradicts the fact that $\{f_1, f_2, \ldots\}$ is dense in M. Thus $\{S\alpha_k : k \geq 1\}$ is a C.O.N.S. in H_μ and, of course, $\{\alpha_k\} \subseteq M$. This makes the operators Π_N and Q_N $(N \geq 1)$ τ-continuous as asserted.

Let $\mathfrak{n}$ denote the collection of all τ-open neighborhoods of zero. Let $U \in \mathfrak{n}$ and choose $V \in \mathfrak{n}$ so that

$$V + V \subseteq U \quad .$$

Next, since the τ-topology is weaker than the norm topology on B, we take $\varepsilon > 0$ sufficiently small so that $\{x : \|x\| < \varepsilon\} \subseteq V$. Since $Q_1 K \supseteq Q_2 K \supseteq \cdots$ and $\bigcap_N Q_N K = \{0\}$ with K compact in B we can choose N_V such that $N \geq N_V$ implies

$$Q_N K \subseteq \{x : \|x\| < \tfrac{\varepsilon}{2}\} \subseteq V \quad ,$$

and hence

$$(3.35) \qquad Q_N K \cap Q_N B = Q_N K \subseteq V \cap Q_N B \quad .$$

Note that $V \cap Q_N B$ is an open set in $Q_N B$ in the τ-induced topology on $Q_N B$.

Since $Q_N B = \{y : \Pi_N y = 0\}$, and Π_N is continuous in both the norm and τ-topologies we have $Q_N B$ closed in both of these topologies. Applying the argument used to establish (3.32) to the random variables

$$Q_N X_1, Q_N X_2, \ldots$$

taking values in the Banach space $Q_N B$ (which is also τ-closed in B) we have

$$(3.36) \qquad P\left\{\omega : C_\tau\left(\left\{Q_N\left(\frac{S_n(\omega)}{\sqrt{2n\, LLn}}\right) : n \geq 1\right\}\right) \subseteq Q_N K\right\} = 1 \quad .$$

Furthermore, from (3.31) and that Q_N is τ-continuous, we have for each N that

$$(3.37)\qquad P\left\{\omega : \left\{Q_N\left(\frac{S_n(\omega)}{\sqrt{2n\,LLn}}\right) : n \geq 1\right\} \text{ is } \tau\text{-conditionally compact in } Q_N B\right\} = 1 .$$

Let

$$(3.38)\qquad \Omega_0 = \left\{\begin{array}{l}\omega : \left\{\dfrac{S_n(\omega)}{\sqrt{2n\,LLn}} : n \geq 1\right\} \text{ is } \tau\text{-conditionally compact in } B \text{ and} \\ C_\tau\left(\left\{Q_N\left(\dfrac{S_n(\omega)}{\sqrt{2n\,LLn}}\right) : n \geq 1\right\}\right) \subseteq Q_N K \text{ for } N = 1, 2, \ldots\end{array}\right\}$$

From (3.36) and (3.37) we have $P(\Omega_0) = 1$. Further, by the argument used to establish (3.29) and using (3.38) we have for $N \supseteq N_V$ (recalling (3.35)) that

$$(3.39)\qquad \Omega_0 \subseteq \left\{\omega : Q_N\left(\frac{S_n(\omega)}{\sqrt{2n\,LLn}}\right) \in V \cap Q_N B \supseteq Q_N K \text{ for all sufficiently large } n\right\}.$$

Note that Ω_0 is independent of N and of the τ-open set $V \in \mathfrak{h}$. Further, if $h \in K$, and

$$(3.40)\qquad \Omega_1 = \Omega_0 \cap \bigcap_{N=1}^{\infty} \bigcap_{k=1}^{\infty} \left\{\omega : \left\| \pi_N \frac{S_n(\omega)}{\sqrt{2n\,LLn}} - \pi_N h\right\| < 1/k \text{ i.o. in } n\right\} ,$$

then by the LIL in finite dimensional spaces (Theorem 4.1) $P(\Omega_1) = 1$.

Now for $h \in K$

$$(3.41)\qquad \frac{S_n(\omega)}{\sqrt{2n\,LLn}} - h = \Pi_N\left(\frac{S_n(\omega)}{\sqrt{2n\,LLn}} - h\right) + Q_N h + Q_N\left(\frac{S_n(\omega)}{\sqrt{2n\,LLn}}\right) ,$$

and hence for $\omega_0 \in \Omega_1$ we have

$$(3.42) \qquad \omega_0 \in \bigcap_{U \in \mathfrak{h}} \left\{ \omega : \frac{S_n(\omega)}{\sqrt{2n\, LLn}} - h \in U \text{ i.o. in } n \right\}$$

$$= \left\{ \omega : h \in C_\tau\left(\left\{ \frac{S_n(\omega)}{\sqrt{2n\, LLn}} : n \geq 1 \right\} \right) \right\} .$$

That is, if $\omega_0 \in \Omega_1$ and $U \in \mathfrak{h}$ is given choose V and N_V as above. Then, for fixed $N \geq N_V$ and infinitely many n,

$$(3.43) \qquad \frac{S_n(\omega_0)}{\sqrt{2n\, LLn}} - h = \pi_N\left(\frac{S_n(\omega_0)}{\sqrt{2n\, LLn}} - h \right) + Q_N h + Q_N\left(\frac{S_n(\omega_0)}{\sqrt{2n\, LLn}} \right) \in U$$

since for fixed $N \geq N_V$ we have

(a) $\left\| \pi_N\left(\frac{S_n(\omega_0)}{\sqrt{2n\, LLn}} - h \right) \right\| < \varepsilon/2$ i.o. in n,

(b) $\|Q_N h\| < \varepsilon/2$,

(c) $Q_N\left(\frac{S_n(\omega_0)}{\sqrt{2n\, LLn}} \right) \in V \cap Q_N B \subseteq V$ for all sufficiently large n,

(d) $V + V \subseteq U$, and $\{x : \|x\| < \varepsilon\} \subseteq V$.

Since $P(\Omega_1) = 1$ and (3.42) holds we have (3.34) for any $h \in K$. This completes the proof.

<u>Corollary 3.3</u>. Let B denote a real separable Banach space with norm $\|\cdot\|$, and assume $X_1, X_2, \ldots$ is a sequence of i.i.d. B-valued random variables such that

$$(3.44) \qquad E(X_k) = 0 \quad \text{and} \quad E\|X_k\|^2 < \infty \quad .$$

Furthermore, assume B is the dual of the Banach space E and let $\tau = \sigma(B, E)$ denote the weak-star topology on B. If K is the unit ball of $H_{\mathcal{L}(X_1)}$, then (3.29) and (3.30) hold iff

$$P\left\{\omega : \sup_n \left\|\frac{S_n(\omega)}{\sqrt{2n\,LLn}}\right\| < \infty\right\} = 1 \quad . \tag{3.45}$$

Proof. The condition defining the event in (3.45) is equivalent to being τ-conditionally compact in B. Hence (3.45) is equivalent to (3.31) and the corollary follows immediately from Theorem 3.2.

Remark. If B is as in Corollary 3.3, then (3.45) implies convergence to K and clustering throughout K in the weak-star topology. However, I am unaware of a situation where (3.45) holds, and we do not, in fact, have

$$P\left\{\omega : \left\{\frac{S_n(\omega)}{\sqrt{2n\,LLn}} : n \geq 1\right\} \text{ is norm conditionally compact in } B\right\} = 1 \quad .$$

Corollary 3.4. Let E denote a real separable Frechet space, and assume $X_1, X_2, \ldots$ is a sequence of i.i.d. E-valued random variables such that

$$E(f(X_k)) = 0 \text{ and } E\|X_k\|_j^2 < \infty \qquad (j \geq 1,\ f \in E^*) \tag{3.46}$$

where $\{\|\cdot\|_j : j \geq 1\}$ is an increasing sequence of semi-norms on E generating the metric topology τ of E. Then, there is a compact convex symmetric set K of E such that

$$(3.47)\qquad P\left\{\omega : \left\{\frac{S_n(\omega)}{\sqrt{2n\,LLn}} : n \geq 1\right\} \text{ is eventually in } V \text{ for every } \tau\text{-open set } V \supseteq K\right\} = 1$$

and

$$(3.48)\qquad P\left\{\omega : C_\tau\left\{\frac{S_n(\omega)}{\sqrt{2n\,LLn}} : n \geq 1\right\} = K\right\} = 1 ,$$

iff

$$(3.49)\qquad P\left\{\omega : \left\{\frac{S_n(\omega)}{\sqrt{2n\,LLn}} : n \geq 1\right\} \text{ is } \tau\text{-conditionally compact in } E\right\} = 1 .$$

Proof. First of all, let $B \subseteq E$ denote the Banach space determined as in Corollary 1 of [15], and choose the sequence $\{b_j\}$ such that the sequence $\{a_j\}$ involved in the definition of the norm $\|\cdot\|_0$ [15, p. 30] satisfies

$$(3.50)\qquad \sum_j a_j (E\|X_k\|_j^2)^{1/2} < \infty .$$

Then

$$(3.51)\qquad E\|X_k\|_0^2 < \infty ,$$

and since $E(f(X_k)) = 0$ for every $f \in E^*$ we also have $E(X_k) \equiv \int_B x\, d\mu(x) = 0$ where $\mu = \mathcal{L}(X_k)$ on B. That is, the Bochner integral $\int_B x\mu(dx)$ is perfectly well defined, and for each $f \in E^* \subseteq B^*$ (recall B maps continuously into E under the identity) we have

$$f\left(\int_B x\, d\mu(x)\right) = \int_B f(x)\, d\mu(x) = E(f(X_k)) = 0 .$$

Hence $\int_B x \, d\mu(x) = 0$ since E^* separates points of B.

Recall, also, that $\mu(B) = 1$ so the random variables $X_1, X_2, \ldots$ can also be viewed as B valued random variables.

Let K be the unit ball of $H_{\mathcal{L}(X_1)}$ where $H_{\mathcal{L}(X_1)}$ is constructed from the operator $S : B^* \to B$ defined by $Sf = \int_B xf(x) \, d\mu(x)$. The construction used, of course, is that indicated in Section two. Then, since E^* is dense in B^* with respect to the topology induced by $L^2(\mu)$, we see that $H_{\mathcal{L}(X_1)}$ and, of course, K are independent of the choice of B provided B is as in Corollary 1 of [15] and (3.51) holds.

To see E^* is $L^2(\mu)$-dense in B^* suppose the $L^2(\mu)$ closure of E^* is H_1 and the $L^2(\mu)$ closure of B^* is H_2. Then $H_1 \subseteq H_2$, and assume $h \in H_2$ is orthogonal to H_1. Now H_2 is isometric to $H_{\mathcal{L}(X_1)}$ under the extension to all of H_2 of the densely defined linear map S which we will denote by $\tilde{S}$. Then $\tilde{S}h$ is orthogonal to $\tilde{S}H_1 \supseteq SE^*$. Using property (ii) of Section two we have $f(\tilde{S}h) = 0$ for ever $f \in E^*$, and hence since E^* separates points of E (and hence H_μ) we have $\tilde{S}h = 0$. Thus $H_1 = H_2$ and E^* is $L^2(\mu)$ dense in B^* as asserted.

To see $H_{\mathcal{L}(X_1)}$, and hence K, are independent of the choice of B under the conditions stated note that for $f \in E^*$ we have

$$\int_{B_1} xf(x) \, d\mu(x) = \int_{B_2} xf(x) \, d\mu(x) \tag{3.52}$$

whenever B_1 and B_2 are possible choices of B. That is, the Bochner integrals are equal since $g \in E^*$ implies

$$(3.53) \qquad g\left(\int_{B_1} x f(x)\, d\mu(x)\right) = \int_{B_1} g(x)\, f(x)\, d\mu(x) = \int_{B_2} g(x)\, f(x)\, d\mu(x)$$

$$= g\left(\int_{B_2} x f(x)\, d\mu(x)\right)$$

where the middle equality holds since $\mu(B_1) = \mu(B_2) = 1$. Then, since E^* separates points of E (3.53) implies (3.52). Now (3.52) and E^* being $L^2(\mu)$ dense in B^* makes $H_{\mathcal{L}(X_1)}$ independent of the possible choices of B. Hence K is independent of the possible choices of B, and since K is compact in any such B we have K compact in E as asserted.

To finish the proof fix B as indicated earlier in the proof. Then $P\{\omega : S_n(\omega) \in B \text{ for all } n\} = 1$, and since E is a metric space (3.47) implies that

$$(3.54) \qquad P\left\{\omega : \lim \rho\left(\frac{S_n(\omega)}{\sqrt{2n\, LLn}}, K\right) = 0\right\} = 1$$

where $\rho(x, A) = \inf_{y \in A} \rho(x, y)$ and ρ is a metric giving the topology τ on E. Now K compact in E and (3.54) imply (3.49).

On the other hand, if (3.49) holds then the arguments used in the proof of Theorem 3.2 imply (3.47). That is, we establish (3.32) for E independent of (3.49) just as in Theorem 3.2. Combining (3.32) and (3.49) we then have (3.47) by the same argument used in Theorem 3.2 to prove (3.29) from (3.31). Now (3.47) and $P\{\omega : S_n(\omega) \in B \text{ for all } n\} = 1$ implies (3.31) (see the proof of Theorem 3.2). Hence (3.30) holds by Theorem 3.2 and this gives (3.48), so the proof is complete.

Remark. Situations where (3.49) can be demonstrated are considered in [16].

4. Further results for the LIL. Some situations where Corollary 3.1 applies are examined here. The first application provides sufficient conditions on B so that the classical assumptions for the LIL actually imply the LIL. The second group of results, namely Theorem 4.2 and Theorem 4.3, deals with special conditions on the random variables which suffice for the LIL.

Let B denote a real separable Banach space with norm $\|\cdot\|$. The norm $\|\cdot\|$ on B is twice directionally differentiable on $B - \{0\}$ if for $x, y \in B$, $x + ty \neq 0$, we have

$$\frac{d}{dt}\|x + ty\| = D(x + ty)(y) \tag{4.1}$$

where $D : B - \{0\} \to B^*$ is Lip(1) on the surface of the unit sphere of B (and hence Lip(1) away from zero), and

$$\frac{d^2}{dt^2}\|x + ty\| = D^2_{x+ty}(y, y) \tag{4.2}$$

where D^2_x is a bounded symmetric bilinear form on $B \times B$. We call D^2_x the second directional derivative of the norm, and, of course, if the norm is actually twice Frechet differentiable on B with second derivative at x given by Λ_x, then $\Lambda_x = D^2_x$.

If $D^2_x(y, y)$ is continuous in x $(x \neq 0)$ and for all $r > 0$ and $x, h \in B$ such that $\|x\| \geq r$, $\|h\| \leq r/2$ we have

$$|D^2_{x+h}(h, h) - D^2_x(h, h)| \leq C_r\|h\|^{2+\alpha} \tag{4.3}$$

for some fixed $\alpha > 0$ and some constant C_r then we say the second directional derivative is Lip(α) away from zero.

Remark. If $B = L^p(\Omega, \mathfrak{F}, m)$ where m is a σ-finite measure on Ω then the second directional derivative of the usual norm on L^p is Lip α with $\alpha = p - 2$ for $2 < p \leq 3$ and Lip(1) for $p = 2$, $3 \leq p < \infty$. For the proof of this assertion we ask the reader to consider Theorem 4.1 of [11].

The norms in these examples are, in fact, Frechet differentiable, but we will not use that information here. For motivational purposes, however, we will examine the differentiability of the norm when B is a real separable Hilbert space or, equivalently, in the above notation when $B = L^2(\Omega, \mathfrak{F}, m)$.

Example. Let B denote a real separable Hilbert space with norm $\|x\| = (x, x)^{1/2}$ as usual. Then for $x + ty \neq 0$

$$\frac{d}{dt}\|x + ty\| = \frac{d}{dt}(x + ty, x + ty)^{1/2} = \frac{(x + ty, y)}{\|x + ty\|}, \tag{4.4}$$

and hence $D(x) = \frac{x}{\|x\|}$ for $x \neq 0$. That is, $D(x)$ is the linear functional (continuous) on B generated by the element $\frac{x}{\|x\|}$ in the usual way. Furthermore, the mapping D is Lip(1) on the surface of the unit sphere of B since $\|x\| = \|y\| = 1$ implies

$$\|D(x) - D(y)\| = \left\|\frac{x}{\|x\|} - \frac{y}{\|y\|}\right\| = \|x - y\| .$$

Differentiating (4.4) we obtain

$$\frac{d^2}{dt^2}\|x + ty\| = \frac{\|x + ty\|^2 (y, y) - (x + ty, y)^2}{\|x + ty\|^3}, \tag{4.5}$$

and hence for $x \neq 0$, $u, v \in B$ we have

(4.6)
$$D_x^2(u, v) = \frac{\|x\|^2(u, v) - (x, u)(x, v)}{\|x\|^3} .$$

From (4.6) we see that for $\|x\| \geq r > 0$ and $\|h\| \leq r/2$ we have

(4.7)
$$|D_{x+h}^2(h, h) - D_x^2(h, h)| \leq C_r \|h\|^3$$

where $C_r = C(1/r^3 + 1/r^2)$ and C is some absolute constant. We also have for the bilinear form D_x^2 that

(4.8)
$$\sup_{\|x\| = 1} \|D_x^2\| = 2 ,$$

and this is a condition which appears in Theorem 4.1 below.

Theorem 4.1. Let B denote a real separable Banach space with norm $\|\cdot\|$. Let $\|\cdot\|$ be twice directionally differentiable on $B - \{0\}$ with the second directional derivative D_x^2 being Lip(1) away from zero, and such that

$$\sup_{\|x\| = 1} \|D_x^2\| < \infty .$$

Let $X_1, X_2, \ldots$ be i.i.d. B-valued such that

$$E(X_1) = 0 , \quad E\|X_1\|^2 < \infty ,$$

and define $S_n = X_1 + \ldots + X_n$. Then, if K is the unit ball of $H_{\mathcal{L}(X_1)}$ we have

$$(4.9) \qquad P\left(\omega : \lim_n d\left(\frac{S_n(\omega)}{\sqrt{2n\, LLn}}, K\right) = 0\right) = 1 ,$$

and

$$(4.10) \qquad P\left(\omega : C\left(\left\{\frac{S_n(\omega)}{\sqrt{2n\, LLn}} : n \geq 1\right\}\right) = K\right) = 1 .$$

Remark. The proof of Theorem 4.1 will follow several lemmas.

Lemma 4.1. If the norm on B is twice directionally differentiable on $B - \{0\}$ with first derivative $D(x)$ and second derivative D_x^2, then

(a) $D(\lambda x) = (\operatorname{sgn} \lambda)\, D(x)$ for all real $\lambda \neq 0$, $x \neq 0$.

(b) $\|D(x)\| = 1$ for $x \neq 0$.

(c) $D(x)(x) = \|x\|$ for $x \neq 0$.

(d) D is Lip(1) away from zero.

(e) If $\lambda \neq 0$, $x \neq 0$, then $D_{\lambda x}^2 = \frac{1}{|\lambda|} D_x^2$.

(f) $D_x^2(h, h) \geq 0$ for all $x \neq 0$, $h \in B$.

(g) If $X_1, X_2, \ldots, X_n$ are independent B-valued random variables such that $EX_j = 0$ and $E\|X_j\|^2 < \infty$ $(j = 1, \ldots, n)$ then there exists a constant A independent of n and the random variables such that

$$(4.11) \qquad E\|X_1 + \ldots + X_n\|^2 \leq A \sum_{j=1}^{n} E\|X_j\|^2 .$$

Proof. For all real $\lambda \neq 0$ and $x \neq 0$

$$D(\lambda x)(y) = \frac{d}{dt} \|\lambda x + ty\| \Big|_{t=0} = |\lambda| \frac{d}{dt} \|x + ty/\lambda\| \Big|_{t=0}$$

$$= |\lambda| \; D(x)(y/\lambda)$$

$$= (\operatorname{sgn} \lambda) \; D(x)(y) \quad .$$

Since $y \in B$ was arbitrary the linear functionals $D(\lambda x)$ and $(\operatorname{sgn} \lambda) D(x)$ are equal and (a) holds.

Fix x, $x \neq 0$, and take $y \in B$. Then

$$D(x)(y) = \frac{d}{dt} \|x + ty\| \Big|_{t=0} = \lim_{t \to 0} \frac{\|x + ty\| - \|x\|}{t} \quad , \tag{4.12}$$

and hence $|D(x)(y)| \leq \|y\|$. If $y = x$ in (4.12), then we see $D(x)(x) = \|x\|$. Hence (b) and (c) hold.

Take $\|x\| \geq r > 0$, $\|x + h\| \geq r$. Then

$$\|D(x + h) - D(x)\|_{B^*} = \|D(\frac{x + h}{\|x + h\|}) - D(\frac{x}{\|x\|})\|_{B^*}$$

(4.13) $$\leq C \| \frac{x + h}{\|x + h\|} - \frac{x}{\|x\|} \|$$ since D is Lip(1) on the surface of the unit ball of B

$$\leq \frac{2C\|h\|}{\|x\|} \leq \frac{2C}{r} \|h\| \quad ,$$

and hence D is Lip(1) away from zero.

If $\lambda \neq 0$, $x \neq 0$, and $h \in B$, then

$$D^2_{\lambda x}(h, h) = \lim_{t \to 0} \frac{D(\lambda x + th)(h) - D(\lambda x)(h)}{t}$$

$$= \lim_{t \to 0} \frac{\lambda}{t}[D(\lambda(x + th/\lambda))(h/\lambda) - D(\lambda x)(h/\lambda)]$$

$$= \lim_{t \to 0} \frac{\lambda \operatorname{sgn} \lambda}{t}[D(x + th/\lambda)(h/\lambda) - D(x)(h/\lambda)]$$

$$= |\lambda| \, D^2_x(h/\lambda, h/\lambda) = \frac{1}{|\lambda|} D^2_x(h, h) \quad .$$

Hence $D^2_{\lambda x} = \frac{1}{|\lambda|} D^2_x$ since a symmetric bilinear form is determined by its values on the diagonal of $B \times B$.

The non-negativity of D^2_x follows because the existence of the second derivative of $\|x + th\|$ implies

$$D^2_x(h, h) = \lim_{t \to 0} \frac{\|x + th\| + \|x - th\| - 2\,\|x\|}{t} \quad ,$$

and since $\|2x\| \le \|x + th\| + \|x - th\|$, we easily see that $D^2_x(h, h) \ge 0$.

To prove property (g) let $G(x) = \|x\| \, D(\frac{x}{\|x\|})$ for $x \ne 0$ and define $G(0) = 0$. Then $G(x)(x) = \|x\|^2$, $\|G(x)\|_{B^*} = \|x\|$, and $\|G(x) - G(y)\|_{B^*} \le A\|x - y\|$ for all $x, y \in B$. Here $A = 2C + 1$ where C is any constant such that $\|D(x) - D(y)\|_{B^*} \le C\|x - y\|$ for $\|x\| = \|y\| = 1$ (such a $C > 0$ exists since we assume D is Lip(1) on the surface of the unit ball of B).

With these properties of G we can easily prove (g). That is,

$$E\|X_1 + \ldots + X_n\|^2 = E\{G(X_1 + \ldots + X_n)(X_1 + \ldots + X_n)\}$$

$$= \sum_{j=1}^{n} E(G(T_j + X_j)(X_j))$$

$$\text{where } T_j = \sum_{\substack{i=1 \\ i \neq j}}^{n} X_i$$

$$= \sum_{j=1}^{n} E\{G(T_j)(X_j) + [G(T_j + X_j) - G(T_j)](X_j)\} \tag{4.14}$$

$$= \sum_{j=1}^{n} E\{[G(T_j + X_j) - G(T_j)](X_j)\}$$

since $E(G(T_j)(X_j)) = EG(T_j)(E(X_j)) = 0$ by independence and that $EX_j = 0$

$$\leq A \sum_{j=1}^{n} E\|X_j\|^2 \quad ,$$

and hence (g) holds.

Before stating our next lemma perhaps it should be mentioned that property (g) of Lemma 4.1 was first proved in [7], and in a more general set-up in [23]. The details were included here since (g) provides us with a first step in linking the derivatives of the norm and probability.

Remark. The proof of property (g) in Lemma 4.1 only requires that $D(x)$ $(x \neq 0)$ is Lip(1) on the surface of the unit ball of B. No second derivative is required.

Lemma 4.2.* Let B denote a real separable Banach space with norm $\|\cdot\|$, and assume $\|\cdot\|$ has derivative D on B - {0} such that D is Lip(1) on the surface of the unit ball of B. If $X_1, X_2, \ldots$ are independent identically distributed B-valued random variables such that

$$(4.15) \qquad E(X_1) = 0, \; E\|X_1\|^2 < \infty, \; T(f,g) = E(f(X_1)\,g(X_1)) \qquad (f, g \in B^*) \; ,$$

then there exists a unique mean zero Gaussian measure μ on B such that

$$(4.16) \qquad T(f,g) = \int_B f(x)\,g(x)\,d\mu(x) \qquad (f, g \in B^*)$$

and

$$(4.17) \qquad \mathcal{L}\left(\frac{X_1 + \ldots + X_n}{\sqrt{n}}\right) \xrightarrow{\text{weakly}} \mu \; .$$

Proof. If $\mathcal{L}\left(\frac{S_n}{\sqrt{n}}\right)$ converges weakly to any probability measure μ on B, then μ must be a mean zero Gaussian measure with covariance as in (4.16). This is so because any probability measure on B is uniquely determined by its finite-dimensional distributions (recall that B is separable), and, furthermore, because all finite-dimensional distributions of $\mathcal{L}\left(\frac{S_n}{\sqrt{n}}\right)$ converge weakly to finite-dimensional Gaussian distributions which are determined by the covariance in (4.15).

Hence all that remains to be proved is that the sequence of probability measures $\left\{\mathcal{L}\left(\frac{S_n}{\sqrt{n}}\right) : n \geq 1\right\}$ is weakly conditionally compact on B. This is done through the use of property (g) of Lemma 4.1 (see the remark following Lemma 4.1).

---

* A more general version of Lemma 4.2 was first proved by J. Hoffman-Jorgensen and G. Pisier using a different method.

Now the sequence of probability measures $\left\{\mathcal{L}\left(\frac{S_n}{\sqrt{n}}\right) : n \geq 1\right\}$ is conditionally compact in B if for every $\delta > 0$ there exists a compact set $C \subseteq B$ such that $\sup_n P\left\{\frac{S_n}{\sqrt{n}} \notin C^{2\delta}\right\} \leq \delta$. Here, of course, $E^{\varepsilon} = \{y : \inf_{x \in E} \|y - x\| < \varepsilon\}$ for $E \subseteq B$, $\varepsilon > 0$.

Fix $\delta > 0$ and define $\tau_\delta(x) = \sum_{j=1}^{r} x_j 1_{A_j}(x)$ such that $E(\tau_\delta(X_1)) = 0$ and $E\|X_1 - \tau_\delta(X_1)\|^2 \leq \frac{\delta^3}{2(A+1)}$ where A is as in (4.11). Then, there exists a compact set C such that $C \subseteq \text{span}\{x_1, \ldots, x_r\}$ and by the central limit theorem in finite dimensions

$$\sup_n P\left(\sum_{j=1}^{n} \frac{\tau_\delta(X_j)}{\sqrt{n}} \notin C^{\delta}\right) < \delta/2 .$$

Then,

$$\begin{aligned} P\left(\frac{S_n}{\sqrt{n}} \notin C^{2\delta}\right) &\leq P\left(\sum_{j=1}^{n} \frac{\tau_\delta(X_j)}{\sqrt{n}} \notin C^{\delta}\right) \\ &\quad + P\left(\left\|\sum_{j=1}^{n} \frac{(\tau_\delta(X_j) - X_j)}{\sqrt{n}}\right\| \geq \delta\right) \\ &\leq \delta/2 + \frac{A\delta^3}{2(A+1)\delta^2} \leq \delta \end{aligned}$$

where the second inequality follows by applying Chebyshev's estimate and property (g). Thus the lemma is proved.

Remark. As was mentioned previously, property (g) of Lemma 4.1 provides us with the first link between derivatives of the norm and probability. If we assume more about the derivatives of the norm we can obtain the following estimate which is crucial in our proof of Theorem 4.1.

Lemma 4.3. Let B denote a real separable Banach space with norm $\|\cdot\|$. Let $\|\cdot\|$ be twice directionally differentiable on B with the second directional derivative D_x^2 being Lip(α) away from zero for some $\alpha > 0$, and such that

$$\sup_{\|x\| = 1} \|D_x^2\| < \infty \quad . \tag{4.18}$$

Let $X_1, X_2, \ldots$ be independent B-valued r.v.'s such that for some $\delta > 0$

$$\begin{gathered} \sup E\|X_k\|^{2+\delta} < \infty \\ E(X_k) = 0 \qquad (k = 1, 2, \ldots) \end{gathered} \tag{4.19}$$

and having common covariance function

$$T(f, g) = E(f(X_k)\, g(X_k)) \qquad (f, g \in B^*) \quad . \tag{4.20}$$

Let μ denote the mean zero Gaussian measure on B determined by T. Then, it follows for $t \geq 0$ and any $\rho > 0$ that

$$P\left\{\frac{\|X_1 + \ldots + X_n\|}{\sqrt{n}} \geq t\right\} \leq \mu(x : \|x\| \geq t - \rho) + C \sup E\|X_k\|^{2+\delta} n^{-\frac{\min(\alpha, \delta)}{2}} \tag{4.21}$$

where C is an absolute constant uniform in t for $t \geq 2\rho$.

Proof. If $0 \leq t \leq \rho$ then (4.21) is obvious, so fix $t > \rho$ and define a a function $f : (-\infty, \infty) \to [0, 1]$ such that f is monotone increasing, $f(u) = 0$ for $u \leq t - \rho$, $f(u) = 1$ for $u \geq t$, and $f''(u)$ is Lipschitz continuous (and hence in this case bounded) on $(-\infty, \infty)$. Let $g(x) = f(\|x\|)$, $W_n = (X_1 + \ldots + X_n)/\sqrt{n}$, and assume $Y_1, Y_2, \ldots$ are independent random variables each with Gaussian distribution μ. To be specific, we assume the sequences $\{X_k\}$ and $\{Y_k\}$ are defined on the probability space $(\Omega, \mathfrak{F}, P)$. We also assume the Y_k's are independent of the X_k's and that $Z_n = (Y_1 + \ldots + Y_n)/\sqrt{n}$. Then the distribution Z_n induces on B is μ and

$$P\{\|W_n\| \geq t\} = \mu(x : \|x\| \geq t) + \{P(\|W_n\| \geq t) - \mu(x : \|x\| \geq t)\} \tag{4.22}$$

$$\leq \mu(x : \|x\| \geq t - \rho) + E\{g(W_n) - g(Z_n)\} \quad .$$

Now

$$g(W_n) - g(Z_n) = \sum_{k=1}^{n} V_k$$

where

$$V_k = g(U_k + X_k/\sqrt{n}) - g(U_k + Y_k/\sqrt{n}) \tag{4.23}$$

and

$$U_k = (X_1 + \ldots + X_{k-1} + Y_{k+1} + \ldots + Y_n)/\sqrt{n} \quad . \tag{4.24}$$

Let $h(\lambda) = g(U_k + \lambda X_k/\sqrt{n})$ for $-\infty < \lambda < \infty$. Since $g(x) = f(\|x\|)$ and f vanishes in a neighborhood of zero we have $h(\lambda)$ twice continuously differentiable on $(-\infty, \infty)$. Hence by Taylor's formula

$$g(U_k + X_k/\sqrt{n}) = h(0) + h'(0) + \frac{h''(0)}{2} + \left[\frac{h''(\tau) - h''(0)}{2}\right]$$

$$\begin{aligned} &= g(U_k) + f'(\|U_k\|)\, D(U_k)(X_k/\sqrt{n}) + \\ &+ \tfrac{1}{2} f''(\|U_k\|)\{D(U_k)(X_k/\sqrt{n})\}^2 \\ &+ \tfrac{1}{2} f'(\|U_k\|)\, D^2_{U_k}(X_k/\sqrt{n}, X_k/\sqrt{n}) + J_n(U_k, X_k) \end{aligned} \tag{4.25}$$

$$(0 < \tau < 1) \quad ,$$

where

$$\begin{aligned} 2J_n(U_k, X_k) &= f''(\|U_k + \tau X_k/\sqrt{n}\|)\ [D(U_k + \tau X_k/\sqrt{n})(X_k/\sqrt{n})]^2 \\ &+ f'(\|U_k + \tau X_k/\sqrt{n}\|)\, D^2_{U_k+\tau X_k/\sqrt{n}}(X_k/\sqrt{n}, X_k/\sqrt{n}) \\ &- f''(\|U_k\|)\,[D(U_k)(X_k/\sqrt{n})]^2 \\ &- f'(\|U_k\|)\, D^2_{U_k}(X_k/\sqrt{n}, X_k/\sqrt{n}) \end{aligned} \tag{4.26}$$

and τ is a non-negative random quantity bounded by one.

A similar expression holds for $g(U_k + X_k/\sqrt{n})$ except Y_k replaces X_k and τ is replaced by a random quantity τ^* which is also non-negative and bounded by one.

We will show below that

$$E(f'(\|U_k\|)\, D(U_k)(X_k)) = E(f'(\|U_k\|)\, D(U_k)(Y_k)) = 0 \quad ,$$

$$(4.27) \qquad E(f''(\|U_k\|)(D(U_k)(X_k))^2) = E(f''(\|U_k\|)(D(U_k)(Y_k))^2) \quad ,$$

$$E(f'(\|U_k\|)\, D^2_{U_k}(X_k, X_k)) = E(f'(\|U_k\|)\, D^2_{U_k}(Y_k, Y_k)) \quad ,$$

and hence by (4. 23), (4. 24), and (4. 25) we have

$$(4.28) \qquad |E(V_k)| \le E|J_n(U_k, X_k)| + E|J_n(U_k, Y_k)| \quad .$$

Further, by showing both $E|J_n(U_k, X_k)|$ and $E|J_n(U_k, Y_k)|$ are dominated by $C_{t,\rho} \sup_j E\|X_j\|^{2+\delta} n^{-\frac{(1+\min(\alpha,\delta))}{2}}$ where $C_{t,\rho}$ is uniformly bounded in t for $t \ge 2\rho$ we see from (4. 23) and (4. 28) that

$$(4.29) \qquad |E(g(W_n) - g(Z_n))| \le C_{t,\rho} \operatorname{Sup}_j E\|X_j\|^{2+\delta} n^{-\frac{(1+\min(\alpha,\delta))}{2}} \quad .$$

We first establish the equalities in (4. 27). Since U_k and X_k are independent, $\|D(x)\| = 1$, and f' vanishes in a neighborhood of zero we have $E(f'(\|U_k\|)\, D(U_k)(X_k)) = E(f'(\|U_k\|)\, D(U_k)(EX_k)) = 0$. Replacing X_k by Y_k we thus have the first equality in (4. 27).

The second equality in (4. 27) follows exactly in the same way except we use the fact that X_k and Y_k have common covariance functions.

The third part of (4. 27) is a bit more complicated and we turn to this now.

Since X_k and U_k are independent, Y_k and U_k are independent, and f' vanishes in a neighborhood of zero the third equality in (4.27) holds provided

$$E(D_x^2(X_k, X_k')) = E(D_x^2(Y_k, Y_k)) \tag{4.30}$$

for all $x \in B$, $x \neq 0$.

Fix $x \neq 0$, $x \in B$. Then D_x^2 is a non-negative, symmetric, bounded, bilinear form on $B \times B$. Hence there exists a non-negative bounded linear operator $A : B \to B^*$ such that

$$D_x^2(y, z) = \langle A(y), z \rangle = A(y)(z) \qquad (y, z \in B) \quad . \tag{4.31}$$

Hence A restricted to H_μ maps H_μ to B^*. Letting Γ denote the linear ma obtained by restricting an element in B^* to H_μ we have $\Gamma: B^* \to H_\mu^*$. Now let ϕ denote the usual linear isometry identifying H_μ^* and H_μ Then, as in Lemma 2. 1 (ii), we have $S = \phi \circ \Gamma$ so $\phi \circ \Gamma \circ A : H_\mu \to H_\mu$. Thus the bilinear form D_x^2 restricted to $H_\mu \times H_\mu$ satisfies

$$\begin{aligned} D_x^2(y, z) &= (\phi \circ \Gamma \circ A(y), z)_\mu \qquad (y, z \in H_\mu) \\ &= (S \circ A(y), z)_\mu \end{aligned} \tag{4.32}$$

and since D_x^2 is symmetric we have $S \circ A$ is symmetric on H_μ to H_μ. Since K (the unit ball of H_μ) is compact in B we have $S \circ A$ a compact, non-negative, symmetric operator on H_μ to H_μ.

Thus the spectral theorem for compact, symmetric, non-negative operators on H_μ implies that for $z \in H_\mu \subseteq B$

(4.33) $$S \circ A(z) = \sum_j \lambda_j (z, e_j)_\mu e_j \quad ,$$

where $\{e_j : j \geq 1\}$ are orthonormal eigenvectors of $S \circ A$ corresponding to the eigenvalues $\{\lambda_i : \lambda \geq 1\}$ all of which are non-negative. Note that $(S \circ A)(e_j) = \lambda_j e_j$ implies that $e_j = Sf_j$ for some $f_j \in B^*$. That is, letting $Ae_j = \lambda_j f_j$ (which is in B^* as $A : B \to B^*$) we get $Sf_j = e_j$ as asserted. Thus

$$(S \circ A(z), z)_\mu = \sum_j \lambda_j (z, e_j)_\mu^2$$

(4.34) $$= \sum_j \lambda_j (z, Sf_j)_\mu^2$$

$$= \sum_j \lambda_j [f_j(z)]^2$$

for every $z \in H_\mu$. Let

(4.35) $$I(z) = \left\{ \sum_j \lambda_j [f_j(z)]^2 \right\}^{1/2} \qquad (z \in B) \quad .$$

Then, if M equals the closure of H_μ in B an easy application of Fatou's lemma implies that $I(z)$ is finite for each z in M. That is, if $z \in M$ and $\{z_n\} \subseteq H_\mu$ is such that $z_n \to z$ in B then by Fatou's lemma, (4.32), (4.34) and (4.35) we have

(4.36) $$D_X^2(z, z) = \lim_n I^2(z_n) \geq \sum_j \lambda_j [f_j(z)]^2 \quad .$$

Since (4.36) holds $I(z)$ is a continuous finite semi-norm on M, and since $I^2(z) = D_X^2(z, z)$ for $z \in H_\mu$ we have

(4.37) $$I^2(z) = D_X^2(z, z) \qquad (z \in M) \quad .$$

Now each X_k and Y_k have covariance function T as defined in (4.20). Thus for $f \in B^*$ such that $f = 0$ on H_μ (and hence M) we have $T(f,f) = 0$, and hence it follows easily from the Hahn-Banach theorem that the support of the measure induced by each $X_k(Y_k)$ is a subset of M.

Thus with probability one we have for $k \geq 1$ that

$$D_x^2(X_k, X_k) = \sum_j \lambda_j [f_j(X_k)]^2 \tag{4.38}$$

$$D_x^2(Y_k, Y_k) = \sum_j \lambda_j [f_j(Y_k)]^2 \quad .$$

Since each $\lambda_j \geq 0$ and $E([f_j(X_k)]^2) = E([f_j(Y_k)]^2)$ $(j \geq 1)$ we have (4.30) immediately from (4.38).

Having established (4.27) we need only establish the estimate required for $E|J_n(U_k, X_k)|$ since a similar estimate applies for $E|J_n(U_k, Y_k)|$.

Set $\gamma = \min(\frac{t - \rho}{4}, \rho)$, and let $E = \{x : \|x\| \leq \rho\}$ and $E' = \{x : \|x\| > \rho\}$ throughout the remainder of the proof. Let $C = \sup_{-\infty < u < \infty} \{|f(u)| + |f'(u)| + |f''(u)|\}$.

First note that C can be taken uniform in t for $t \geq 2\rho$ since $\rho > 0$ is fixed.

Then from (4.26)

$$1_{E'}(\|X_k\|/\sqrt{n})\, |2J_n(U_k, X_k)| \leq \tag{4.39}$$

$$1_{E'}(\|X_k\|/\sqrt{n}) \left[2C \frac{\|X_k\|^2}{n} + 2C \sup_{\|x\| \geq \gamma} \|D_x\|^2 \|X_k\|^2/n \right]$$

since f' and f'' vanish on $(-\infty, t - \rho]$.

If $\|X_k\|/\sqrt{n} \leq \gamma$ we have two cases to consider. They are

(a) $$\|U_k + \tau X_k/\sqrt{n}\| \leq 3(t - \rho)/4 \quad ,$$

(b) $$\|U_k + \tau X_k/\sqrt{n}\| > 3(t - \rho)/4 \quad .$$

Now case (a) is simple since $\dfrac{\|X_k\|}{\sqrt{n}} \leq \gamma \leq \dfrac{t - \rho}{4}$ and (a) implies $J_n(U_k, X_k) = 0$ since f' and f'' vanish on $(-\infty, t - \rho]$.

Now $\dfrac{\|X_k\|}{\sqrt{n}} \leq \gamma$ and (b) implies $\|U_k\| \geq \dfrac{t - \rho}{2}$. For $x, y \in B$ and $0 < \tau < 1$ we have

$$
\begin{aligned}
2J_n(x, \sqrt{n}\, y) = {} & [f''(\|x + \tau y\|) - f''(\|x\|)]\,[D(x + \tau y)(y)]^2 \\
& + f''(\|x\|)\,[[D(x + \tau y)(y)]^2 - [D(x)(y)]^2] \\
& + [f'(\|x + \tau y\|) - f'(\|x\|)]\, D^2_{x+\tau y}(y, y) \\
& + f'(\|x\|)\,[D^2_{x+\tau y}(y, y) - D^2_x(y, y)] \quad .
\end{aligned}
\tag{4.40}
$$

We now estimate the right hand side of (4.40) under the assumption $\|y\| \leq \gamma \leq \dfrac{t - \rho}{4}$ and $\|x\| \geq \dfrac{t - \rho}{2}$.

Let C' denote a positive constant which dominates the Lipschitz constants of both f' and f'', and recall C from above. Note that C' can be made uniform in t since $\rho > 0$ is fixed.

Then we have

$$1_E(\|y\|)\,|f''(\|x+\tau y\|) - f''(\|x\|)|\ |D(x+\tau y)(y)|^2$$

$$\leq 1_E(\|y\|)\min(2C, C'\|y\|)\ \|y\|^2$$

(4.41)
$$1_E(\|y\|)\,|f'(\|x+\tau y\|) - f'(\|x\|)|\ D^2_{x+\tau y}(y,y)$$

$$\leq 1_E(\|y\|)\min(2C, C'\|y\|)\sup_{\|z\|\geq 3(t-\rho)/4}\|D^2_z\|\cdot\|y\|^2$$

Further, since $\|D(x)\|_{B^*} = 1$ for $x \neq 0$ we have

$$1_E(\|y\|)\,|f''(\|x\|)|\ |[D(x+\tau y)(y)]^2 - [D(x)(y)]^2|$$

(4.42)
$$\leq 1_E(\|y\|)\,2C\|y\|\ |D(x+\tau y)(y) - D(x)(y)|$$

$$= 1_E(\|y\|)\,2C\|y\|\int_0^\tau \frac{d^2}{dt^2}\|x+ty\|\Big|_{t=s}\,ds$$

$$\leq 1_E(\|y\|)\,2C\|y\|\sup_{\|z\|\geq\frac{t-\rho}{4}}\|D^2_z\|\ \|y\|^2$$

because $\|x\| \geq \frac{t-\rho}{2}$, $\|y\| \leq \gamma \leq \frac{t-\rho}{4}$, $0 < \tau < 1$. Finally, since D^2_x is Lip(α) away from zero we have for $\|x\| \geq \frac{t-\rho}{2}$, $\|y\| \leq \frac{t-\rho}{4}$ that

(4.43)
$$1_E(\|y\|)\,|f'(\|x\|)|\ |D^2_{x+\tau y}(y,y) - D^2_x(y,y)| \leq$$

$$\leq 1_E(\|y\|)\cdot C\cdot C_{\frac{t-\rho}{2}}\ \|y\|^{2+\alpha}$$

where C_r is defined in (4.3).

Combining (4.39), (4.40), (4.41) and (4.43) we have a constant $C_{t,\rho}$ which is uniform in t for $t \geq 2\rho$ such that

$$(4.44) \qquad E|J_n(U_k, X_k)| \leq C_{t,\rho} \sup_j E\|X_j\|^{2+\delta}/n^{1+\frac{\min(\alpha,\delta)}{2}} .$$

Now (4.44) and a similar estimate for $E|J_n(U_k, Y_k)|$ completes the proof of Lemma 4.3.

Proof of Theorem 4.1. In view of Corollary 3.1 we need only show that

$$(4.45) \qquad P\left\{\omega : \left\{\frac{S_n(\omega)}{\sqrt{2n\, LLn}} : n \geq 1\right\} \text{ conditionally compact in } B\right\} = 1 .$$

Now (4.45) holds if and only if for every $\varepsilon > 0$

$$(4.46) \qquad P\left\{\omega : \left\{\frac{S_n(\omega)}{\sqrt{2n\, LLn}} : n \geq 1\right\} \text{ is covered by finitely many } \varepsilon\text{-balls}\right\} = 1 .$$

Let Π_N and Q_N be defined as in Lemma 2.1 (iii). Then, by the Hartman-Wintner result of Section one, we have for each integer N that

$$(4.47) \qquad P\left\{\omega : \left\{\Pi_N \frac{S_n(\omega)}{\sqrt{2n\, LLn}} : n \geq 1\right\} \text{ is conditionally compact in } \Pi_N B \subseteq B\right\} = 1 .$$

That is, if (4.47) is false for some N, then since $\Pi_N B$ is of finite dimension we have

$$(4.48) \qquad P\left\{\omega : \left\{\Pi_N \frac{S_n(\omega)}{\sqrt{2n\, LLn}} : n \geq 1\right\} \text{ is unbounded in } \Pi_N B\right\} > 0 \quad (= 1) .$$

Now (4.48) easily contradicts the Hartman-Wintner result and hence (4.47) holds.

Now (4.47) implies (4.46) if for every $\varepsilon > 0$ there exists an N (depending on ε) such that

$$(4.49) \qquad P\left\{\omega : \overline{\lim_n} \left\|Q_N \frac{S_n(\omega)}{\sqrt{2n\, LLn}}\right\| > \varepsilon\right\} = 0 \quad .$$

Hence we fix $\varepsilon > 0$. Let $A = \left\{\overline{\lim_n} \left\|Q_N \frac{S_n}{\sqrt{2n\, LLn}}\right\| > \varepsilon\right\}$ and define

$$(4.50) \qquad B_r = \left\{\omega : \sup_{n_r \le n \le n_{r+1}} \left\|Q_N \frac{S_n(\omega)}{\sqrt{2n\, LLn}}\right\| > \varepsilon\right\}$$

for $n_r = 2^r$ and $r = 1, 2, \ldots$. Then $A \subseteq \{B_r \text{ i.o.}\}$ and for large values of r

$$(4.51) \qquad \begin{aligned} P(B_r) &\le P\left\{\sup_{n_r \le n \le n_{r+1}} \|Q_N S_n\| > \varepsilon \sqrt{2n_r\, LLn_r}\right\} \\ &\le 2P\left\{\|Q_N S_{n_{r+1}}\| > (\varepsilon/2)\sqrt{2n_r\, LLn_r}\right\} \\ &\le 2P\left\{\frac{\|Q_N S_{n_{r+1}}\|}{\sqrt{n_{r+1}}} > \varepsilon \sqrt{\frac{n_r}{n_{r+1}}}\, \sqrt{2\, LLn_r}\right\} \end{aligned}$$

where the second inequality follows in a standard way provided

$$(4.52) \qquad \sup_{n_r \le n \le n_{r+1}} P\left\{\|Q_N S_{n_{r+1}} - Q_N S_n\| > (\varepsilon/2)\sqrt{2n_r\, LLn_r}\right\} \le 1/2 \quad .$$

Now by property (g) of Lemma 4.1 we have (4.52) for large r by an easy application of Chebyshev's inequality.

Now $A \subseteq \{B_r \text{ i.o.}\}$ implies $P(A) = 0$ if $\sum_r P(B_r) < \infty$, and since $n_r = 2^r$ we have $\sum_r P(B_r) < \infty$ whenever

$$\sum_r P\left\{\frac{\|Q_N S_{n_r}\|}{\sqrt{n_r}} > (\varepsilon/4)\sqrt{2\, LLn_r}\right\} < \infty \quad . \tag{4.53}$$

Next we will choose N so that (4.53) and hence (4.49) holds. Recall $\varepsilon > 0$ is fixed and define for each $j, n \geq 1$ the random variables

$$X_{j,n} = \begin{cases} X_j & \text{if } \|X_j\| \leq \sqrt{n} \\ 0 & \text{otherwise} \end{cases} . \tag{4.54}$$

The $\{X_{j,n} : j \geq 1\}$ are independent and identically distributed and have common covariance function

$$T_n(f, g) = E(f(X_{1,n} - m_n)\, g(X_{1,n} - m_n)) \qquad (f, g \in B^*) \tag{4.55}$$

where

$$m_n = E(X_{1,n}) = \int_{\{\|x\| \leq \sqrt{n}\}} x\, v(dx)$$

$n = 1, 2, \ldots$ and $v = \mathcal{L}(X_1)$.

Let $\mu_n(\mu)$ denote the mean zero Gaussian measure on B determined by the covariance function $T_n(T)$ where

$$T(f, g) = E(f(X_1)\, g(X_1)) \qquad (f, g \in B^*) \quad .$$

Recall Lemma 4.2 for the existence of μ and μ_n and notice that

$$T_n(f, f) \le T(f, f) \qquad (n \ge 1,\ f \in B^*) \quad . \tag{4.56}$$

Then by T. W. Anderson's inequality [2] for every convex symmetric Borel set $C \subseteq B$ we have

$$\mu(C) \le \mu_n(C) \qquad (n \ge 1) \quad . \tag{4.57}$$

That is, Anderson's inequality applies directly to convex symmetric Borel cylinder sets C and yields (4.57) for such sets. The extension of (4.57) to arbitrary convex symmetric Borel sets follows in a fairly standard way since it suffices to show (4.57) when the C are compact, convex, and symmetric in B. That (4.57) follows for such C results from the fact that C compact implies

$$\mu(C) = \inf_{C \subseteq I} \mu(I) \tag{4.58}$$

where I is an open cylinder set. Now one can insert a convex symmetric open cylinder U between C and any open cylinder set $I \supseteq C$ so we have

$$\mu(C) = \inf_{U \in \mathfrak{U}} \mu(U)$$

where $\mathfrak{U}$ denotes all open, convex, symmetric cylinder sets containing C. Similarly, $\mu_n(C) = \inf_{U \in \mathfrak{U}} \mu_n(U)$ and hence (4.57) follows from its validity on $\mathfrak{U}$.

Let $\Gamma(x)$ denote a continuous semi-norm on B. If $0 < s$ is such that $\mu(x : \Gamma(x) \leq s) > 1/2$ then by a lemma of Fernique [6] we have for each $t \geq 0$

$$\mu(x : \Gamma(x) \geq t) \leq \exp\left\{\frac{-t^2}{24s^2} \log\left[\frac{\mu(x : \Gamma(x) \leq s)}{\mu(x : \Gamma(x) > s)}\right]\right\} . \tag{4.59}$$

Next observe that by a well known result we have

$$\lim_{N \to \infty} \|Q_N x\| = 0 \qquad \text{a.e. } (\mu)$$

so fix s such that $0 < s = \tau(\varepsilon) \equiv \sqrt{\frac{\varepsilon^2 \log 3}{16.24}}$. Now choose N sufficiently large so that

$$\mu(x \in B : \|Q_N x\| \leq s) \geq 3/4 . \tag{4.60}$$

Then by (4.59) and inequality (4.57) we have

$$\begin{aligned} \mu_n(x : \|Q_N x\| \geq t) \leq \mu(x : \|Q_N x\| \leq t) &\leq \exp\left\{\frac{-t^2}{24s^2} \log 3\right\} \\ &\leq \exp\left\{\frac{-t^2 \log 3}{24\tau^2(\varepsilon)}\right\} . \end{aligned} \tag{4.61}$$

Assuming N fixed so that (4.60) and (4.61) holds we now apply Lemma 4.3.

Let $S'_n = X_{1,n} + \ldots + X_{n,n}$ for $n \geq 1$. Then

$$P\left\{\|Q_N\frac{S_n}{\sqrt{n}}\| \geq t\right\} \leq P\left\{\|Q_N\frac{S'_n}{\sqrt{n}}\| \geq t\right\} + nP\left\{\|X_1\| > \sqrt{n}\right\}$$

$$\text{(4.62)} \qquad \leq P\left\{\|Q_N\frac{(S'_n - n m_n)}{\sqrt{n}}\| \geq t - \sqrt{n}\,\|Q_N m_n\|\right\} + nP\left\{\|X_1\| > \sqrt{n}\right\}$$

$$\leq P\left\{\|Q_N\frac{(S'_n - n m_n)}{\sqrt{n}}\| \geq 3t/4\right\} + nP\left\{\|X_1\| > \sqrt{n}\right\}$$

uniformly in $t \geq 1$ as $\lim_n \sqrt{n}\, m_n = 0$ and Q_N is continuous. To see that $\lim_n \sqrt{n}\, m_n = 0$ notice that $\int_B x\, dv(x) = 0$ implies

$$\|\sqrt{n}\, m_n\| = \sqrt{n}\, \|\int_{\{\|x\| > \sqrt{n}} x\, dv(x)\| \leq \int_{\{\|x\| > \sqrt{n}} \|x\|^2\, dv(x) \xrightarrow[n\to\infty]{} 0$$

Now apply Lemma 4.3 to the random variables $Q_N(X_{j,n} - m_n)$ for $1 \leq j \leq n$. Therefore

$$\text{(4.63)} \qquad P\left\{\|Q_N\frac{S_n}{\sqrt{n}}\| \geq t\right\} \leq \mu_n(x \in B : \|Q_N x\| \geq t/2) + \frac{C_n}{n^{1/2}} + nP\left\{\|X_1\| > \sqrt{n}\right\}$$

uniformly in $t \geq 1$ with $\rho = 1/4$ (so $3/4\, t - \rho \geq t/2$), and with an absolute constant C such that

$$C_n = CE\|Q_N(X_{1,n} - m_n)\|^3$$

$$\text{(4.64)} \qquad \leq 4C\{E\|Q_N X_{1,n}\|^3 + \|Q_N m_n\|^3\}$$

$$\leq 4C'\{E\|X_{1,n}\|^3 + \|m_n\|^3\}$$

as N is fixed and Q_N is continuous and linear.

Combining (4.63), (4.64), and (4.61) we have

$$(4.65)\qquad P\left\{\|Q_N\frac{S_n}{\sqrt{n}}\| \geq t\right\} \leq \exp\left\{\frac{-t^2\log 3}{24\tau^2(\varepsilon)}\right\} + \frac{4C'}{\sqrt{n}}\left\{E\|X_{1,n}\|^3 + \|m_n\|^3\right\}$$
$$+ nP\left\{\|X_1\| > \sqrt{n}\right\} .$$

Therefore (4.53) holds if for $t = \varepsilon\sqrt{2\,LLn_r}/4$ we have

$$(4.66)\qquad \sum_r n_r P\{\|X_1\| > n_r\} < \infty$$

and

$$(4.67)\qquad \sum_r \frac{E\|X_{1,n_r}\|^3}{\sqrt{n_r}} < \infty$$

as all other terms on the right side of (4.65) sum over the subsequence $n_r = 2^r$, and also

$$\exp\left\{\frac{-\varepsilon^2(2\,LLn_r)\log 3}{16\cdot 24\cdot\tau^2(\varepsilon)}\right\} = \exp\{-2\,LLn_r\} = \frac{1}{(r\log 2)^2}$$

sums over $r \geq 1$.

To verify (4.66) observe that $P\{\|X_1\| > \sqrt{k}\} \downarrow 0$ as k increases to infinity and hence

$$\infty > E\|X_1\|^2 \geq \sum_k P\{\|X_1\|^2 > k\} = \sum_k P\{\|X_1\| > \sqrt{k}\}$$
$$= \sum_{r=1}^{\infty}\sum_{j=n_{r-1}+1}^{n_r} P\{\|X_1\| > \sqrt{j}\}$$
$$\geq \sum_{r=1}^{\infty}(n_r - n_{r-1})\,P\{\|X_1\| > \sqrt{n_r}\}$$
$$= \frac{1}{2}\sum_{r=1}^{\infty} n_r P\{\|X_1\| > \sqrt{n_r}\} .$$

To proof (4. 67) let $a_n = E\|X_{1,n}\|^3$ and note that

$$a_n \leq \sum_{k=1}^{n} k^{3/2} P\{k-1 \leq \|X_1\|^2 \leq k\} .$$

Hence

$$\begin{aligned}\sum_{n=1}^{\infty} \frac{a_n}{n^{3/2}} &\leq \sum_{n=1}^{\infty} n^{-3/2} \sum_{k=1}^{n} k^{3/2} P\{k-1 \leq \|X_1\|^2 \leq k\} \\ &= \sum_{k=1}^{\infty} k^{3/2} P\{k-1 \leq \|X_1\|^2 \leq k\} \sum_{n=k}^{\infty} n^{-3/2} \\ &= O\Big(\sum_{k=1}^{\infty} kP\{k-1 \leq \|X_1\|^2 \leq k\}\Big) \\ &= O(E\|X_1\|^2) < \infty .\end{aligned}$$

Now $a_n \nearrow$ as n increases so

$$\begin{aligned}\infty > \sum_{n=1}^{\infty} \frac{a_n}{n^{3/2}} \geq \sum_{r=1}^{\infty} \sum_{j=n_r+1}^{n_{r+1}} a_j/j^{3/2} &\geq \sum_{r=1}^{\infty} (n_{r+1} - n_r) \frac{a_{n_r}}{(n_{r+1})^{3/2}} \\ &\geq \frac{1}{8} \sum_{r=1}^{\infty} \frac{a_{n_r}}{\sqrt{n_r}} .\end{aligned}$$

Thus (4. 67) holds and the proof is complete.

Another application of Corollary 3. 1 is given in our next result which establishes the law of the iterated logarithm for C(S) valued random variables under conditions exactly the same as those used to establish the central limit theorem in this setting.

Let S denote a compact metric space with metric d. Let $C(S)$ denote the space of real-valued continuous functions on S, and for $f \in C(S)$ define $\|f\|_\infty = \sup_{t \in S} |f(t)|$. If S is a pseudo-metric space with pseudo-metric ρ, then $N(\rho, S, \varepsilon)$ denotes the minimal number of balls of ρ-radius less than ε which cover S. The ε-entropy of (S, ρ) is

$$H(\rho, S, \varepsilon) = \log N(\rho, S, \varepsilon)$$

where $\log x$ denotes the natural logarithm of x.

If S is a metric space under d and ρ is a pseudo metric on S we say ρ <u>is continuous with respect to d</u> if for every $\varepsilon > 0$ there exists $\delta > 0$ such that $d(s,t) < \delta$ implies $\rho(s,t) < \varepsilon$. If S is compact under d (with topology τ_d) then it is easy to see that ρ is continuous with respect to d iff τ_d is stronger than τ_ρ.

<u>Theorem 4.2.</u> Let X be a $C(S)$ valued random variable such that

$$E(X(s)) = 0 \text{ and } E(X^2(s)) < \infty \qquad (s \in S) .$$

Suppose there exists a non-negative random variable M such that for given $s, t \in S$ and sample point ω we have

$$|X(s,\omega) - X(t,\omega)| \leq M(\omega)\, \rho(s,t)$$

with $E(M^2) < \infty$ and ρ a pseudo-metric on S such that ρ is continuous with respect to d. If

(a) $\int_0 H^{1/2}(S, \rho, u)\, du < \infty$,

(b) $X_1, X_2, \ldots$ are independent identically distributed such that $\mathcal{L}(X_k) = \mathcal{L}(X)$, and if

(c) K is the unit ball of $H_{\mathcal{L}(X)}$, then

$$\text{(4.68)} \qquad P\left\{\lim_n d\left(\frac{S_n}{\sqrt{2n\, LLn}}, K\right) = 0\right\} = 1$$

and

$$\text{(4.69)} \qquad P\left\{C\left(\left\{\frac{S_n}{\sqrt{2n\, LLn}} : n \geq 1\right\}\right) = K\right\} = 1 \quad .$$

Here $d(x, K) = \inf_{y \in K} \|x - y\|_\infty$ and, of course, the clustering is relative to the sup-norm $\|\cdot\|_\infty$.

<u>Examples.</u> Let $S = [0, 1]$ and assume $\{X(s) : 0 \leq s \leq 1\}$ is a stochastic process with continuous sample paths. Further, assume

$$\text{(4.70)} \qquad E(X(s)) = 0 \quad \text{and} \quad E(X^2(s)) < \infty \qquad (s \in S) \quad ,$$

and that

$$\text{(4.71)} \qquad |X(t, \omega) - X(s, \omega)| \leq M(\omega)|\log|s - t||^{-\alpha}$$

where $\alpha > 1/2$ and $E(M^2) < \infty$. Then $\{X(s) : 0 \leq s \leq 1\}$ satisfies the conditions of Theorem 4.2 and hence X satisfies the LIL.

To see that the conditions of Theorem 4.2 hold we introduce the natural metric ρ determined by the right hand side of (4.71). Then we show that

$$\int_0 H^{1/2}(S, \rho, u)\, du < \infty$$

with respect to this metric.

Let

$$(4.72) \qquad \phi(t) = \begin{cases} \dfrac{1}{|\log t|^{\alpha}} & 0 \le t \le \dfrac{1}{e^3} \\ \dfrac{1}{3^{\alpha}} & t \ge \dfrac{1}{e^3} \\ 0 & t = 0 \end{cases} .$$

Then $\phi(t)$ is continuous, increasing, and concave downward on $[0, \infty)$. We define the metric ρ on $[0, 1]$ by

$$(4.73) \qquad \rho(s, t) = \phi(|t - s|) \qquad (s, t \in [0, 1]) .$$

To check that ρ defined in (4.73) actually is a metric we need only verify the triangle inequality for ρ. Now the triangle inequality for ρ is equivalent to showing that

$$(4.74) \qquad \phi(t_1) \le \phi(t_2) + \phi(t_3)$$

whenever t_1, t_2, t_3 are in $(0, 1]$ and $t_1 \le t_2 + t_3$. Since (4.74) is easy to check we omit the details.

Thus ρ is a metric which is continuous with respect to the usual metric $d(s,t) = |s - t|$ on $[0,1]$ and since $\alpha > 1/2$ we have

(4.75) $$\int_0 H^{1/2}([0,1], \rho, u)\, du < \infty \quad .$$

To verify (4.75) notice that

(4.76) $$N([0,1], \rho, u) = O(\frac{1}{2h})$$

where $|\log h|^{-\alpha} = u$. Hence $h = \exp\{-(\frac{1}{u})^{1/\alpha}\}$ so $H([0,1], \rho, u) = O((\frac{1}{u})^{1/\alpha})$, and (4.75) holds since $\int_0 (\frac{1}{u})^{1/2\alpha}\, du < \infty$ when $\frac{1}{2\alpha} < 1$.

A situation where (4.71) easily follows can be seen by integrating a stochastic process $\{Y(s) : 0 \le s \le 1\}$. That is, assume $E(Y(s)) = 0$ and $E(Y^2(s)) < \infty$ for $s \in [0,1]$, $\{Y(s) : 0 \le s \le 1\}$ has continuous sample paths, and $E[\sup_{0 \le s \le 1} |Y(s)|]^2 < \infty$. Then

$$X(t,\omega) = \int_0^t Y(s,\omega)\, ds \qquad (0 \le t \le 1)$$

clearly satisfies (4.71) as well as the other conditions of Theorem 4.2.

One can also show that Brownian motion satisfies the condition of Theorem 4.2. That is, let $\{X(t) : 0 \le t \le T\}$ be standard. Brownian motion and take γ to be a positive constant such that $0 < \gamma < 1/2$. Then for any $h > 0$ we have

(4.77) $$P\{\omega : |X(t,\omega) - X(s,\omega)| > h|s - t|^{\gamma} \text{ for some } s,t \in [0,T]\}$$
$$\le Ch^{\frac{-4}{1-2\gamma}}$$

where $C = 2\left\{\frac{(1-2^{-\gamma})^2(1-2\gamma)\,e}{16}\right\}^{\frac{-2}{1-2\gamma}}$.

Letting

$$M(\omega) = \sup_{\substack{s,t \in [0,T] \\ s \neq t}} \frac{|X(t,\omega) - X(s,\omega)|}{|s-t|^{\gamma}} \tag{4.78}$$

we have from (4.77) that

$$P\{\omega : M(\omega) > h\} \le \frac{C}{h^4} \tag{4.79}$$

as $\frac{4}{1-2\gamma} \ge 4$. Therefore $E(M^2) < \infty$, and $\{X(t) : 0 \le t \le T\}$ clearly satisfies the other conditions of Theorem 4.2 with $\rho(s,t) = |s-t|^{\gamma}$.

Further examples are included in [4] and [8] as well as in the references indicated there.

The proof of Theorem 4.2 is given in [14] and proceeds by showing (1.6) holds with $B = C(S)$. There are applications to $C(S)$ valued random variables with subgaussian increments in [14], and in [13] there are results for $C[0,1]$ valued random variables. The results in [13] were proved first and the basic approach used in the proof of Theorem 4.2 is in the same spirit as that of [13], but the exact details for the proof are quite different.

As a final application of Corollary 3.1 we mention some results for $D[0,1]$-valued random variables.

As usual $D[0,1]$ denotes the space of real-valued functions on $[0,1]$ which are right continuous on $[0,1)$ and have left-hand limits on $(0,1]$. The cylinder sets of $D[0,1]$ induced by the maps $x \to x(t)$ induce a sigma algebra which we denote by $\mathcal{D}$.

For each $x \in D[0,1]$ we define the norm

$$\|x\|_\infty = \sup_{0 \le t \le 1} |x(t)| .$$

Theorem 4. 3. Let $X_1, X_2, \ldots$ be independent identically distributed $(D[0,1], \mathcal{B})$ valued random variables such that each $\{X_k(t) : 0 \le t \le 1\}$ is a martingale. Further, assume there exists a $\delta > 0$ such that

$$(4.80) \qquad E(X_k(t)) = 0 \text{ and } E|X_k(t)|^{2+\delta} < \infty \qquad (0 \le t \le 1) ,$$

and the covariance function

$$R(s,t) = E(X_k(s)\, X_k(t))$$

is continuous on $[0,1] \times [0,1]$. Let K_R denote the unit ball of the reproducing kernel Hilbert space H_R determined by the covariance function R. Then

$$(4.81) \qquad P\left(\lim_n d\left(\frac{S_n}{\sqrt{2n\, LLn}}, K_R \right) = 0 \right) = 1$$

and

$$(4.82) \qquad P\left(C\left(\left\{ \frac{S_n}{\sqrt{2n\, LLn}} : n \ge 1 \right\} \right) = K_R \right) = 1$$

where the cluster set and distances are computed in the sup-norm $\|\cdot\|_\infty$.

If the processes $\{X_k(t) : 0 \le t \le 1\}$ are independent increment processes, then (4. 81) and (4. 82) hold with only a second moment condition in (4. 80) rather than the $(2+\delta)^{th}$ moment.

The proof of Theorem 4. 3 appears in [12] and depends on an application of Corollary 3. 1 to a related sequence of random variables with values in $C[0, 1]$.

5. <u>The functional law of the iterated logarithm.</u> The primary emphasis of the previous results in these lectures involved the LIL for sequences of i.i.d. Banach space valued random variables. Here we will examine the functional law of the iterated logarithm (FLIL) of Strassen for such sequences. In [17] we obtained a functional law for Hilbert space valued random variables, but our result here is more general, and demonstrates that it is the LIL for sequences which is more fundamental in the Banach space setting. This is analogous to the relationship between the central limit theorem and the invariance principle for Banach space valued random variables which was pointed out in [10].

Again, let B denote a real separable Banach space with norm $\|\cdot\|$, and assume $X_1, X_2, \ldots$ are i.i.d. B-valued random variables such that $E(X_k) = 0$ and $E\|X_k\|^2 < \infty$. We say the sequence $\{X_k\}$ satisfies the LIL with limit set K if (1.4) and (1.5) hold.

Of course, Corollary 3.1 asserts that the limit set K is always the unit ball of the Hilbert space $H_{\mathcal{L}(X_1)}$ constructed in Lemma 2.1. To describe the functional law of the iterated logarithm based on the sequence $\{X_k\}$ we need some additional terminology. However, we point out that the unit ball K of $H_{\mathcal{L}(X_1)}$ will be involved in our description of the limit set for the functional law of the iterated logarithm as well.

Let $C_B[0, 1]$ denote the functions $f : [0, 1] \to B$ which are continuous, and define $\|f\|_{\infty, B} = \sup_{0 \le t \le 1} \|f(t)\|$. Then $C_B[0, 1]$ is a real separable Banach space in the norm $\|\cdot\|_{\infty, B}$. Let $\{\alpha_k\} \subseteq B^*$ be such that $\{S\alpha_k : k \ge 1\}$ is a C.O.N.S. in $H_{\mathcal{L}(X_1)} \subseteq B$ where the mapping $S : B^* \to B$ is as in Lemma 2.1. The limit set involved in the FLIL for the i.i.d. sequence $\{X_k\}$ is the set

$$(5.1)\qquad \mathcal{K} = \left\{ \begin{array}{l} f \in C_B[0,1] : f(t) \in H_{\mathcal{L}(X_1)} \quad (0 \le t \le 1), \\ \qquad f(t) = \sum_k \int_0^t \frac{d}{ds} \alpha_k(f(s))\, ds\, S\alpha_k, \\ \text{and } \sum_k \int_0^1 [\frac{d}{ds} \alpha_k(f(s))]^2\, ds \le 1 \end{array} \right\} .$$

Now we make precise what we mean when we say the sequence $\{X_k\}$ satisfies the FLIL. For each sample point ω we define the polygonal functions

$$(5.2)\qquad Z_n(t,\omega) = \begin{cases} \dfrac{S_k(\omega)}{\sqrt{2n\, LLn}} & (t = k/n,\ k = 0, 1, \ldots, n) \\ \text{linear elsewhere} & (0 \le t \le 1) \end{cases} .$$

Then for each sample point ω the sequence $\{Z_n(\cdot, \omega)\}$ is a subset of $C_B[0,1]$ and we say $\{X_k\}$ satisfies the FLIL with limit set $\mathcal{K}$ if

$$(5.3)\qquad P\{\omega : \lim_n d(Z_n(\cdot, \omega), \mathcal{K}) = 0\} = 1 ,$$

and

$$(5.4)\qquad P\{\omega : C(\{Z_n(\cdot, \omega) : n \ge 1\}) = \mathcal{K}\} = 1$$

where

$$d(f, A) = \inf_{g \in A} \|f - g\|_{\infty, B}$$

and

$$C(\{f_n\}) = \text{all limit points of } \{f_n\} \text{ in } C_B[0,1] \quad .$$

If $\Psi : C_B[0,1] \to B$ is defined by $\Psi(f) = f(1)$, then it follows rather easily from the definition of $\mathcal{K}$ that $\Psi(\mathcal{K}) = K$. Hence if the i.i.d. sequence $\{X_k\}$ obeys the functional LIL with limit set $\mathcal{K}$, then it also satisfies the LIL with limit set K. Theorem 5.1 provides a converse for this fact provided $\{X_k\}$ is an i.i.d. sequence of B-valued random variables such that $E(X_k) = 0$ and $E\|X_k\|^2 < \infty$.

Theorem 5.1. Let $\{X_k\}$ be a sequence of i.i.d. B-valued random variables such that $E(X_k) = 0$ and $E\|X_k\|^2 < \infty$. If $\{X_k\}$ satisfies the law of the iterated logarithm with limit set K, then $\{X_k\}$ satisfies the functional law of the iterated logarithm with limit set $\mathcal{K}$ as described in (5.1).

Proof. Fix $\varepsilon > 0$. For $N \geq 1$ and $x \in B$ let

$$\Pi_N(x) = \sum_{k=1}^{N} \alpha_k(x)\, S\alpha_k \quad \text{and} \quad Q_N(x) = x - \Pi_N x \tag{5.5}$$

where $\{\alpha_k\} \subseteq B^*$ and $\{S\alpha_k : k \geq 1\}$ is a C.O.N.S. in $H_{\mathcal{L}(X_1)} \subseteq B$. Then by the argument used in the proof of Theorem 3.1 we have an N_0 such that $N \geq N_0$ implies

$$Q_N K \subseteq \{x \in B : \|x\| < \varepsilon/2\} \quad . \tag{5.6}$$

Hence, since (1.4) holds and Q_N is continuous from B into B we have

$$(5.7) \qquad P\left\{\omega : \overline{\lim_k} \left\|Q_N \frac{S_k(\omega)}{\sqrt{2k\, LLk}}\right\| \le \varepsilon/2\right\} = 1 \quad .$$

Combining (5.2) and (5.7) we easily see that

$$(5.8) \qquad P\{\omega : \overline{\lim_n} \|Q_N Z_n(\cdot\,,\omega)\|_{\infty, B} \le \varepsilon/2\} = 1 \quad .$$

For the moment assume $E\|\Pi_N X_k\|^3 < \infty$ and observe that the random variables $\{\Pi_N X_k : k \ge 1\}$ take values in the finite-dimensional Banach space $\Pi_N B$. Since all norms on a finite-dimensional Banach space are equivalent we can put a Euclidean norm on $\Pi_N B$. Then the functional law of the iterated logarithm of [17] applies to the sequence $\{\Pi_N X_k : k \ge 1\}$. Further, the limit set is $\Pi_N \mathcal{K}$ so we have

$$(5.9) \qquad P\{\omega : \lim_k d(\Pi_N Z_k(\cdot\,,\omega), \Pi_N \mathcal{K}) = 0\} = 1$$

and

$$(5.10) \qquad P\{\omega : C(\{\Pi_N Z_k(\cdot\,,\omega) : k \ge 1\}) = \Pi_N \mathcal{K}\} = 1$$

where

$$d(x, A) = \inf_{y \in A} \|x - y\|_{\infty, \Pi_N B}$$

and the cluster set is computed with respect to the norm $\|\cdot\|_{\infty, \Pi_N B}$.

Combining (5.8), (5.9), and (5.10) completes the proof provided $E\|\Pi_N X_k\|^3 < \infty$.

Since the condition $E\|X_k\|^2 < \infty$ need not imply $E\|\Pi_N X_k\|^3 < \infty$ we must truncate as in the proof of Theorem 4.1. That is, we carry out the proof of the FLIL for the sequence $\{\Pi_N X_k\}$ with N fixed as in [17, Theorem 3.2]. At various points in the proof we replace a non-Gaussian probability by a Gaussian probability plus an error estimate, but now we must truncate and use an argument of C. Heyde which was employed in Theorem 4.1. The details are lengthy, but straightforward once one is armed with Heyde's technique and the basic outline of the proof provided by [17, Theorem 3.2]. Hence they are omitted.

6. <u>An application to operator valued random variables</u>. Here we combine the functional law of the iterated logarithm of Theorem 5.1 and our results on the LIL to obtain an application to operator valued random variables and the functional behavior of solutions of certain random evolutions. The terminology random evolution is used in the sense indicated by T. Kurtz in [19]*. The results presented here, in fact, were motivated by those in [19], but we hasten to point out that their domain of applicability is considerably less general. Of course, the mode of development of the random evolution used in [19] parallels the law of large numbers (LLN) for operator valued random variables, whereas that employed here is that of the LIL. Since the classical LIL is much more delicate for Banach space valued random variables than the classical LLN, perhaps this loss of generality is to be expected. Nevertheless, it would be of interest to improve these results to a setting approaching the scope of [19].

Let H denote a real separable Hilbert space, and let L(H) denote the Banach space of bounded linear operators from H to H with the uniform operator norm giving the topology. That is, if $A \in L(H)$, then the norm of A is

$$\|A\| = \sup_{\|x\|_H \le 1} \|Ax\|_H , \tag{6.1}$$

where $\|\cdot\|_H$ denotes the norm on H.

---

* The paper by R. Hersh and R. J. Griego entitled <u>Random evolutions, Markov chains, and systems of partial differential equations</u> which appeared in the Proc. Nat. Acad. Sci. U.S.A., 62 (1969), pp. 305-308 as well as recent work by Hersh, Griego, G. Papanicolaou, M. Pinsky, and others also deals with random evolutions.

Let B denote a closed separable subalgebra of L(H) consisting of self adjoint commuting operators. Then, it is known that there exists a fixed bounded resolution of the identity $E(\lambda)$ for H on [0, 1] such that $A \in B$ implies

(6.2) $$A = \int_0^1 f(\lambda)\, dE(\lambda)$$

for some bounded Borel function f on [0, 1] and $\|A\| \le \|f\|_\infty \equiv \sup_{0 \le t \le 1} |f(t)|$ (equality holds if f is continuous). For details see [1, p. 82], [21, p. 355-360]. We describe this situation by saying B has the spectral resolution $\{E(\lambda) : 0 \le \lambda \le 1\}$.

Now assume A is a B-valued random variable. Then for each sample point ω we have

(6.3) $$A(\omega) = \int_0^1 f(\lambda, \omega)\, dE(\lambda)$$

where $f(\cdot, \omega)$ is a bounded Borel function on [0, 1]. In our next theorem we will examine B-valued random variables A provided A is the spectral integral of certain types of stochastic processes $\{f(\lambda, \cdot) : 0 \le \lambda \le 1\}$.

One of the types of processes we deal with has what we call subgaussian increments. That is, a stochastic process $\{f(\lambda) : 0 \le \lambda \le 1\}$ has subgaussian increments if there exists a constant $A > 0$ such that for t real

(6.4) $$E\{\exp\{t[f(u) - f(v)]\}\} \le \exp\{At^2\tau^2(u, v)\} \qquad (u, v \in [0, 1]) .$$

where $\tau^2(u, v) = E\{f(u) - f(v)\}^2$ $(u, v \in [0, 1])$.

Theorem 6.1. Let B denote a closed separable subspace of L(H) consisting of self-adjoint commuting operators having spectral representation $\{E(\lambda) : 0 \leq \lambda \leq 1\}$. Let $\{f(\lambda) : 0 \leq \lambda \leq 1\}$ denote a stochastic process such that $E(f(\lambda)) = 0$ $(0 \leq \lambda \leq 1)$ and $R(u, v) = E(f(u) f(v))$ is continuous on the square $[0, 1] \times [0, 1]$. Assume that one of the following conditions holds for $\{f(\lambda) : 0 \leq \lambda \leq 1\}$:

(6.5) $\{f(\lambda) : 0 \leq \lambda \leq 1\}$ has continuous sample paths on $[0, 1]$ such that for almost every sample point ω

$$|f(u, \omega) - f(v, \omega)| \leq M(\omega) \, \big|\log |u - v|\big|^{-\rho} \qquad (u, v \in [0, 1])$$

where $\rho > 1/2$ and $E(M^2) < \infty$.

(6.6) $\{f(\lambda) : 0 \leq \lambda \leq 1\}$ is an independent increment process with sample paths in $D[0, 1]$.

(6.7) $\{f(\lambda) : 0 \leq \lambda \leq 1\}$ is a martingale with sample paths in $D[0, 1]$, and for some $\delta > 0$, $E|f(u)|^{2+\delta} < \infty$, $0 \leq u \leq 1$.

(6.8) $\{f(\lambda) : 0 \leq \lambda \leq 1\}$ has sample paths in $C[0, 1]$ and has subgaussian increments such that for fixed $\Lambda > 0$

$$\tau(u, v) \leq \Lambda \big|\log |u - v|\big|^{-\rho} \qquad (u, v \in [0, 1])$$

where $\rho > 1/2$.

6.9) $\{f(\lambda) : 0 \le \lambda \le 1\}$ is a Gaussian process with sample paths in $C[0, 1]$.

f $f_1, f_2, \ldots$ are independent identically distributed copies of $\{f(\lambda) : 0 \le \lambda \le 1\}$ n the probability space $(\Omega, \mathcal{F}, P)$ and we define

6.10)
$$A_k(\omega) = \int_0^1 f_k(\lambda, \omega)\, dE(\lambda) \qquad (k \ge 1, \omega \in \Omega) \quad ,$$

ien $A_1, A_2, \ldots$ are i.i.d. B-valued random variables (operators) such that

5.11)
$$E(A_k) = 0 \quad \text{and} \quad E\|A_k\|^2 < \infty \quad .$$

urthermore, if K is the unit ball of $H_{\mathcal{L}(A_1)} \subseteq B$, then

5.12)
$$P\left\{\omega : \lim_n d\left(\frac{A_1(\omega) + \ldots + A_n(\omega)}{\sqrt{2n\, LLn}}, K\right) = 0\right\} = 1$$

nd

5.13)
$$P\left\{\omega : C\left(\left\{\frac{A_1(\omega) + \ldots + A_n(\omega)}{\sqrt{2n\, LLn}} : n \ge 1\right\}\right) = K\right\} = 1$$

here the distance to K and the clustering throughout K is computed in terms of ie uniform operator norm

Proof. That $A_1, A_2, \ldots$ are i.i.d. on $(\Omega, \mathcal{F}, P)$ follows from their definition 1 (6.10), and that $f_1, f_2, \ldots$ are independent copies of the process $\{f(\lambda) : 0 \le \lambda \le 1\}$. ince the sample paths of each f_k are in $C[0, 1]$ or $D[0, 1]$ whenever one of 5.5)-(6.9) hold we have

$$(6.14) \qquad \|A_k(\omega)\| = \| \int_0^1 f_k(\lambda,\omega)\, dE(\lambda)\| \le \|f_k(\cdot,\omega)\|_\infty$$

and hence in each case we can prove $E\|A_k\|^2 < \infty$ since $E\|f_k\|_\infty^2 < \infty$. Further, the Bochner integral

$$E(A_k) = 0 \quad .$$

To see this apply a continuous linear functional T to $E(A_k)$ getting $T(E(A_k)) = E(T(A_k))$. Hence if $T(A) = (Ax, y)_H$ for $x, y \in H$ then

$$T(A_k(\omega)) = \int_0^1 f(\lambda,\omega)\, d(E(\lambda)(x), y)_H$$

where $(E(\lambda)(x), y)_H$ is a finite variation signed measure on $[0, 1]$. Hence $E(T(A_k)) = \int_0^1 E(f(\lambda,\omega))\, d(E(\lambda)(x), y) = 0$. Since the linear functionals of the form $T(A) = (Ax, y)$ separates points in $L(H)$ we have $E(A_k) = 0$.

Thus by Lemma 2.1 the Hilbert space $H_{\mathcal{L}(A_1)}$ is defined, and by Corollary 3.1 the unit ball K of $H_{\mathcal{L}(A_1)}$ is the limit set which must be used in (6.12) and (6.13). To get (6.12) and (6.13) to hold we need only show that

$$(6.15) \qquad P\left\{\omega : \left\{\frac{A_1(\omega) + \ldots + A_n(\omega)}{\sqrt{2n\, LLn}} : n \ge 1\right\} \text{ is conditionally compact in } B\right\} =$$

Now (6.15) holds since under the assumptions on the stochastic processes $\{f_k\}$ we have from Theorem 4.2, Theorem 4.3, [14, Theorem 2], and [18] that

(6.16)
$$P\left\{\omega : \text{every subsequence of } \left\{\frac{f_1(\cdot,\omega) + \ldots + f_n(\cdot,\omega)}{\sqrt{2n\,LLn}} : n \geq 1\right\} \text{ has a convergent subsequence in the sup-norm on } [0,1]\right\} = 1$$

Now (6.14) and (6.16) combine to show that

$$P\left\{\omega : \text{every subsequence of } \left\{\frac{A_1(\omega) + \ldots + A_n(\omega)}{\sqrt{2n\,LLn}} : n \geq 1\right\} \text{ has a convergent subsequence in the uniform operator norm}\right\} = 1 ,$$

and hence (6.15) holds. This completes the proof.

Corollary 6.1. Let B be as in Theorem 6.1 and assume $A_1, A_2, \ldots$ are i.i.d. B-valued random variables satisfying the conditions of Theorem 6.1. Let $\mathcal{K}$ denote the limit set constructed in $C_B[0,1]$ as in (5.1) from the Hilbert space $H_{\mathcal{L}(A_1)} \subseteq B$. Then, the polygonal partial sum processes

(6.17)
$$Z_n(t,\omega) = \begin{cases} \dfrac{A_1(\omega) + \ldots + A_k(\omega)}{\sqrt{2n\,LLn}} & (t = k/n,\ k = 0, \ldots, n) \\ \text{linear elsewhere} & (0 \leq t \leq 1) \end{cases}$$

converge to $\mathcal{K}$ and cluster throughout $\mathcal{K}$ with probability one. That is, (5.3) and (5.4) hold where for $f \in C_B[0,1]$

(6.18)
$$\|f\|_{\infty,B} = \sup_{0 \leq t \leq 1} \sup_{\|x\|_H \leq 1} \|f(t)(x)\|_H ,$$

and, of course, H is the Hilbert space which the operators in B act on, and $\|\cdot\|_H$ is the norm on H.

Proof. Combine Theorem 5.1 and Theorem 6.1.

Our next result applies to the functional behavior of a process whose environment varies randomly. As mentioned at the beginning of the section, this type of result is motivated by the results in [19] on random evolutions and the random Trotter product formula.

Let H denote a Hilbert space as before and let $\delta > 0$. Let $t_k = k\delta$ for $k = 0, 1, 2, \ldots,$ and assume for each sample point ω in some probability space $A_1(\omega), A_2(\omega), \ldots$ are bounded linear operators on H. Let $X_\delta(t, \omega)$ denote the solution of the differential equation

$$\text{(6.19)} \qquad \frac{d\Phi(t)}{dt} = \begin{cases} A_1(\omega)\,\Phi(t) & 0 \le t \le t_1 \\ \\ A_k(\omega)\,\Phi(t) & t_{k-1} \le t \le t_k,\ k = 1, 2, \ldots \end{cases}$$

$$\Phi(0) = x_0 \in H \quad .$$

Then, for $t_{k-1} \le t \le t_k$ we have

$$\text{(6.20)} \qquad X_\delta(t, \omega) = e^{A_k(\omega)(t - t_{k-1})} \cdot e^{\delta A_{k-1}(\omega)} \cdots e^{\delta A_1(\omega)}(x_0) \quad .$$

One natural interpretation of $X_\delta(t, \omega)$ is that it is the state of a process such that at fixed times $t_1, t_2, \ldots$ the random environments, namely the operators $A_1(\omega), A_2(\omega), \ldots,$ change. As in [19], we are interested in the asympotic behavior of $X_\delta(t, \omega)$ as the rate of change of the environment increases.

Theorem 6.2. Let $A_1, A_2, \ldots$ denote i.i.d. B-valued random variables satisfying the conditions of Theorem 6.1, and assume $\mathcal{K}$ is the limit set constructed in $C_B[0,1]$ from the Hilbert space $H_{\mathcal{L}(A_1)} \subseteq B$ as in (5.1). Let $\delta_n = \frac{1}{\sqrt{2n\, LLn}}$ in (6.20) and let

$$e^{\mathcal{K}} = \{\exp\{f\} : f \in \mathcal{K}\} \quad .$$

Then, $e^{\mathcal{K}} \subseteq C_B[0,1]$ and

$$P\{\omega : \lim_n d(X_{\delta_n}(n\delta_n(\cdot), \omega), e^{\mathcal{K}}) = 0\} = 1 \tag{6.21}$$

and

$$P\{\omega : C(\{X_{\delta_n}(n\delta_n(\cdot), \omega) : n \geq 1\}) = e^{\mathcal{K}}\} = 1 \tag{6.22}$$

where the convergence to $e^{\mathcal{K}}$ and the clustering throughout $e^{\mathcal{K}}$ is computed in the norm $\|\cdot\|_{\infty, B}$ given in (6.18).

Proof. If $f \in C_B[0,1]$ then $\exp\{f\}$ also is in $C_B[0,1]$ so $e^{\mathcal{K}} \subseteq C_B[0,1]$. Further, since the operators are assumed to commute we have (6.21) and (6.22) immediately from (6.20), Corollary 6.1, and that

$$\|\exp\{f\} - \exp\{g\}\|_{\infty, B} \leq \|f - g\|_{\infty, B} \exp\{\max(\|f\|_{\infty, B}, \|g\|_{\infty, B})\} \quad .$$

Remark: If we assume the random operators $A_1, A_2, \ldots$ commute, and take values in the Hilbert space of Hilbert-Schmidt operators on H, then the results of this section hold with the uniform operator norm replaced by the Hilbert-Schmidt norm of the operator. Furthermore, we can apply the results of [17] directly to the operators $A_1, A_2, \ldots$. That is, the operators themselves can be viewed as Hilbert space valued random variables (namely, with values in the Hilbert space of Hilbert-Schmidt operators), and hence Theorem 4. 1 gives the desired result.

7. Some recent developments. In the months between the writing of these lectures and their appearance in print a great deal has happened.

First of all, G. Pisier has shown that if B is a type 2 Banach space, i.e. B satisfies property (g) of Lemma 4.1, and $\{X_k : k \geq 1\}$ satisfies the conditions of Theorem 4.1, then (4.9) and (4.10) hold. This result improves Theorem 4.1 as can be seen from Lemma 4.1 (g). Furthermore, combining Pisier's results with some recent ideas of Joel Zinn regarding type 2 maps one can obtain the results for C(S) valued random variables of Theorem 4.2.

In view of the results in Corollary 3.1 and Section four we say a B-valued random variable X satisfies the LIL if for $X_1, X_2, \ldots$ independent copies of X we have a limit set K in B such that (1.1) and (1.2) hold. In case $E(X) = 0$ and $E\|X\|^2 < \infty$ we see from Lemma 2.1 that the limit set K must be compact.

When B is a finite dimensional Banach space then a result of V. Strassen and an easy application of Corollary 3.1 implies X satisfies the central limit theorem (CLT) and the LIL if and only if $E(X) = 0$ and $E\|X\|^2 < \infty$. Hence the LIL and the CLT for X taking values in finite dimensional spaces are equivalent. If B is infinite dimensional the relationship between these two theorems is still somewhat unclear.

However, in a recent note I produced an example of a random variable X which obeys the LIL and yet fails to satisfy the CLT. Previous examples, such as those of R. M. Dudley mentioned in Section one, where the CLT failed also had the property that the LIL failed. In still another recent paper N. Jain produced an example of a random variable X such that $E\|X\|^2 = \infty$, X satisfies the CLT, and X fails the LIL. Without going into the details of these examples it should be emphasized that they are not pathological, and though they answer some questions regarding the relationship of the CLT and LIL they also raise many others.

In closing some recent results of J. Crawford should also be mentioned. Crawford assumes X is a B-valued random variable such that $E(X) = 0$ and $E\|X\|^2 < \infty$. He then shows that if X' is an independent copy of X and if the symmetric random variable $X - X'$ obeys the LIL, then X also satisfies the LIL. Furthermore, he has shown that if X is symmetric, and X^Y equals X for $\|X\| \leq Y$ and 0 otherwise, then X satisfying the LIL implies that X^Y also obeys the LIL.

As one can easily see there are many differences between the CLT and the LIL in the infinite dimensional setting. Some progress has been made in understanding these theorems and their differences in this general setting, but a definitive state of affairs appears to be a long way from being achieved.

Bibliography

[1] N. I. Akhiezer and I. M. Glazman, Theory of linear operators in Hilbert space, Vol. II, Frederick Ungar Publishing Co., New York (1953).

[2] T. W. Anderson, The integral of a symmetric unimodal function over a symmetric convex set and some probability inequalities, Proceedings Amer. Math. Soc., 6 (1955), 170-176.

[3] J. Crawford, personal communication.

[4] J. L. Devary, Regularity properties of second order processes, thesis submitted to the University of Minnesota (1975).

[5] R. M. Dudley and V. Strassen, The central limit theorem and ε-entropy, Lecture Notes in Mathematics, 89, 223-233, Berlin, Heidelberg, New York, Springer (1969).

[6] X. Fernique, Integrabilite des vecteurs Gaussiens, C. R. Acad. Sci. Paris, 270 (1970), 1698-1699.

[7] R. Fortet and E. Mourier, Les fonctions aléatoires comme élements aléatoires dans les espaces de Banach, Studia Math., 15 (1955), 62-79.

[8] A. M. Garsia and E. Rodemich, Monotonicity of certain functionals under rearrangement, Ann. Inst. Fourier, Grenoble, 24 (1974), 67-116.

[9] L. Gross, Lectures in modern analysis and applications II, Lectures Notes in mathematics, 140, Berlin, Heidelberg, New York, Springer (1970).

[10] J. Kuelbs, The invariance principle for Banach space valued random variables, Journal of Multivariate Analysis, 3 (1973), 161-172.

[11] J. Kuelbs, An inequality for the distribution of a sum of certain Banach space valued random variables, Studia Mathematica, 52 (1974), 69-87.

[12] J. Kuelbs, A strong convergence theorem for Banach space valued random variables, submitted for publication.

[13] J. Kuelbs, The law of the iterated logarithm in C[0, 1], to appear in Z. Wahrscheinlichkeitstheorie und Verw. Gebiete.

[14] J. Kuelbs, The law of the iterated logarithm in C(S), submitted for publicatio

[15] J. Kuelbs, Some results for probability measures on linear topological vector spaces with an application to Strassen's log log law, Journal of Functional Analysis, 14 (1973), 28-43.

[16] J. Kuelbs, Strassen's law of the iterated logarithm, Ann. Inst. Fourier, Grenoble, 24 (1974), 169-177.

[17] J. Kuelbs and T. Kurtz, Berry-Essen estimates in Hilbert space and an application to the law of the iterated logarithm, Annals of Probability 2 (1974), 387-407.

[18] J. Kuelbs and R. LePage, The law of the iterated logarithm for Brownian motion in a Banach space, Trans. Amer. Math. Soc., 185 (1973), 253-264.

[19] T. Kurtz, A random Trotter product formula, Proceedings of Amer. Math. Soc., 35 (1972), 147-154.

[20] T. L. Lai, Reproducing kernel Hilbert spaces and the law of the iterated logarithm for Gaussian processes, preprint.

[21] F. Riesz and B. Sz-Nagy, Functional Analysis, Frederick Ungar Publishing Co., New York (1955).

[22] V. Strassen, An invariance principle for the law of the iterated logarithm, Z. Wahrscheinlichkeitstheorie und Verw. Gebiete, 3 (1964), 211-226.

[23] W. A. Woyczynski, Strong laws of large numbers in certain linear spaces, Ann. Inst. Fourier, Grenoble, 24 (1974), 205-223.

J. Kuelbs
Department of Mathematics
University of Wisconsin
Madison, Wisconsin 53706